W0112612

MAPPING THE TRIANGLE

International Conference on Nuclear Structure

Grand Teton National Park, Wyoming   22–25 May 2002

EDITORS
Ani Aprahamian
University of Notre Dame, Notre Dame, Indiana
Jolie A. Cizewski
Rutgers University, Piscataway, New Jersey
Stuart Pittel
University of Delaware, Newark, Delaware
N. Victor Zamfir
Yale University, New Haven, Connecticut

SPONSORING ORGANIZATIONS
National Science Foundation
Physics Department of Yale University
Institute of Physics

Melville, New York, 2002
AIP CONFERENCE PROCEEDINGS ■ VOLUME 638

Editors:

Ani Aprahamian
Department of Physics
University of Notre Dame
Notre Dame, IN 46556
USA
E-mail: Aprahamian.1@nd.edu

Jolie A. Cizewski
Department of Physics and Astronomy
Rutgers University
136 Frelinghuysen Road
Piscataway, NJ 08854-8019
USA
E-mail: Cizewski@physics.rutgers.edu

Stuart Pittel
Bartol Research Institute
University of Delaware
Newark, DE 19716
USA
E-mail: Pittel@bartol.udel.edu

N. Victor Zamfir
Wright Nuclear Structure Laboratory
Physics Department
Yale University
P.O. Box 20814
New Haven, CT 06520-8124
USA
E-mail: Victor.Zamfir@yale.edu

L.C. Catalog Card No. 2002113019
ISBN 0-7354-0093-8
ISSN 0094-243X
Printed in the United States of America

MAPPING THE TRIANGLE

Other Related Titles from AIP Conference Proceedings

610 Nuclear Physics in the 21st Century: International Nuclear Physics conference, INPC 2001
Edited by Eric Norman, Lee Schroeder, and Gordon Wozninak, April 2002, 0-7354-0056-3

594 Hadrons and Nuclei: First International Symposium
Edited by Il-Tong Cheon, Taekeun Choi, Seung-Woo Hong, and Su Houng Lee, November 2001, 0-7354-0037-7

588 Physics with an Electron Polarized Light-Ion Collider: Second Workshop, EPIC 2000
Edited by Richard G. Milner, October 2001, 0-7354-0028-8

561 Tours Symposium on Nuclear Physics IV: Tours 2000
Edited by M. Arnould, M. Lewitowicz, Yu. Ts. Oganessian, H. Akimune, M. Ohta, H. Utsunomiya, T. Wada, and T. Yamagata, April 2001, 1-56396-996-3

549 Intersections of Particle and Nuclear Physics: 7th Conference, CIPANP2000
Edited by Zohreh Parsa and William J. Marciano, December 2000, 1-56396-978-5

529 Capture Gamma-Ray Spectroscopy and Related Topics: 10th International Symposium
Edited by Stephen Wender, July 2000, 1-56396-952-1

518 Proton Emitting Nuclei: PROCON'99—First International Symposium
Edited by Jon C. Batchelder, May 2000, 1-56396-937-8

512 Nuclear Physics at Storage Rings: Fourth International Conference: STORI99
Edited by Hans-Otto Meyer and Peter Schwandt, June 2000, 1-56396-928-9

495 Experimental Nuclear Physics in Europe: ENPE 99, Facing the Next Millennium
Edited by Berta Rubio, Manuel Lozano, and William Gelletly, November 1999, 1-56396-907-6

481 Nuclear Structure 98
Edited by C. Baktash, September 1999, 1-56396-858-4

455 ENAM 98: Exotic Nuclei and Atomic Masses
Edited by B. M. Sherrill, D. J. Morrissey, and C. N. Davids, December 1998, 1-56396-804-5

To learn more about these titles, or the AIP Conference Proceedings Series, please visit
the webpage **http://proceedings.aip.org**

CONTENTS

*Italicized name indicates author who presented the paper.

*Italicized name indicates author who presented the paper.

*Italicized name indicates author who presented the paper.

*Italicized name indicates author who presented the paper.

Preface

This volume contains the Proceedings of the International Conference on Nuclear Structure *Mapping the Triangle* held from the 22nd through the 25th of May 2002 at the Jackson Lake Lodge in the Grand Teton National Park, Wyoming. The conference, highlighting recent achievements in the very exciting and rapidly evolving topic of nuclear structure physics, honored the 60th birthday of Rick Casten, one of the leading contributors to this field.

The format of the conference included several keynote talks as well as additional invited presentations. There was also a Poster Session where other participants in the conference were able to share their recent results. All speaker and poster contributions are included in these Proceedings, with page guidelines defined by the type of presentation. In addition, session chairs were invited to submit contributions. The session contributions are ordered to reflect the conference program, and are then followed by the poster contributions. Throughout the volume are photos taken during the conference and the banquet, which we feel appropriately reflect the spirit and joy of the participants. In the beginning of the Proceedings can be found several pages of photos of Rick Casten's life, interspersed with verse prepared by Lee Riedinger, the chairman of the Organizing Committee. This was part of a presentation made by Lee at the conference banquet, where many other conference participants also shared comments on Rick.

While Lee's verse gives a nice overview of Rick's life and loves, it does not do adequate justice to the enormous breadth and importance of his contributions to nuclear structure physics during a career spanning more than three decades. Nor can we here, considering that Rick's CV contains almost 400 publications. Nevertheless we felt it important to say a few words summarizing a few of the areas that he has helped to shape.

Ever since the introduction of the Interacting Boson Model by Arima and Iachello in 1974, Rick has been one of the major contributors to its experimental tests and theoretical development. He and his collaborators were the first to establish the empirical existence of the O(6) dynamical symmetry. They also made many seminal contributions to understanding the phenomenology of the model, including its other two dynamical symmetry limits, SU(3) and U(5) . With Dave Warner, he introduced the consistent Q-formalism, which is now used in virtually all phenomenological applications. Likewise of profound importance have been his studies related to the valence residual proton-neutron interaction and its role in the development of collectivity and phase/shape transitions in nuclei. His work, as highlighted by the pioneering development of the $N_p N_n$ scheme and the tripartite classification of nuclear structure, has led to a substantial revision of our understanding of how, where and why collectivity arises in nuclei. These new methods serve to simplify and correlate vast quantities of nuclear data and to unify our understanding of the evolution of nuclear structure. Most recently, studies of transitional nuclei by Rick and his collaborators have served to motivate the development of X(5) and E(5) as critical point symmetries in nuclei.

Throughout his career, Rick has made many important experimental contributions to nuclear structure physics. From his initial studies of Coulomb excitation, to one- and two-particle transfer, to (n,γ) measurements, to beta decay, to his current efforts with Coulomb excitation of unstable beams, he has remained focused on the nuclear spectroscopic properties that need to be measured to push the frontiers of the field.

Rick's research, as impressive as it has been, does not tell the full story of his enormous impact on the field. In the mid-90s, Rick left Brookhaven National Laboratory to assume a position as Professor of Physics at Yale and as Director of the Wright Nuclear Structure Laboratory. In less than a decade, Rick has returned Yale to a position as one of the leading academic nuclear physics laboratories in the country.

Likewise of great importance has been Rick's leadership role in the US Radioactive Beam Program. Ever since its inception, Rick has served as Chairman of the IsoSpin Lab Steering Committee. The fact that the Rare Isotope Accelerator is now the US's highest priority new construction project in nuclear physics is testimony to the job he has done in leading this effort. In his presentation at this conference, Brad Sherrill reviewed the history behind this ten-year effort, making clear that Rick was the glue that held it together.

The Editors of these Proceedings were part of the Organizing Committee of the conference. We would like to express our appreciation to our fellow organizers – Hans Börner, Franco Iachello, Witek Nazarewicz, Lee Riedinger and Dave Warner – for making it such a joy to put it on. Our special thanks also go to Karen DeFelice, Jackie Mooney, Mary Anne Shultz and Kay Thacker, for their incredible behind the scenes efforts in support of the conference and to Mark Caprio for his innumerable photos taken during the Conference.

We would like to acknowledge the National Science Foundation, the Physics Department of Yale University, the Institute of Physics, and Airgas East for their generous support of the conference, so critical to its success.

Finally, we would like to thank the contributors to these Proceedings for providing their manuscripts in a timely manner and in accord with the page and style guidelines we imposed. We believe that the contributions are uniformly excellent as were the talks on which they were based. This reflects both the vitality of the field of nuclear structure physics and the enormous regard and respect that the community has for our honoree and friend, Rick Casten.

Ani Aprahamian
Jolie Cizewski
Stuart Pittel
Victor Zamfir

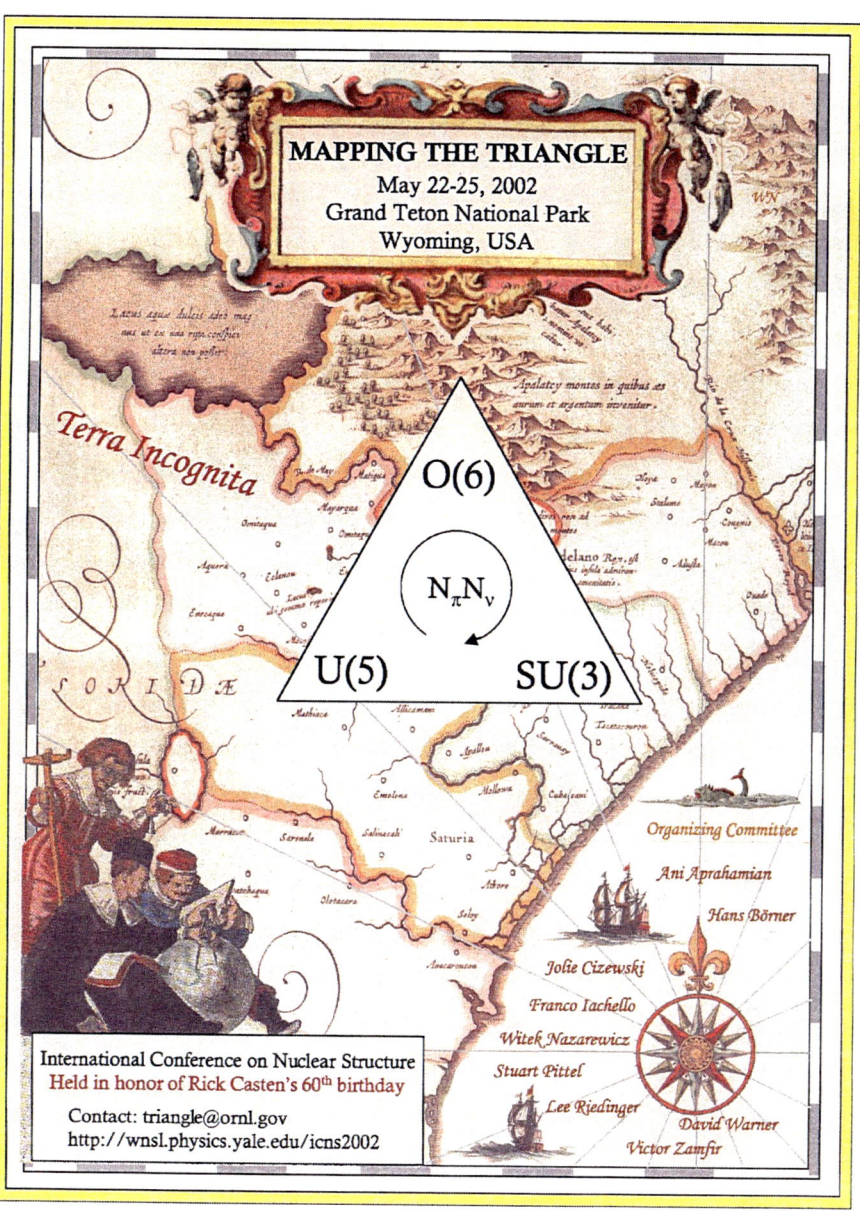

MAPPING THE TRIANGLE
May 22-25, 2002
Grand Teton National Park
Wyoming, USA

Happy 60th to Rick

The time has come and I have the stage
To tell about Rick's advancing age.

We're here to roast him and what's even worse,
I will try to do this in my questionable verse.

As a baby you were as cute as could be;
I'd say a ladies man before you were three.

His first girlfriend was young and cute;
She probably thought him smart and astute.

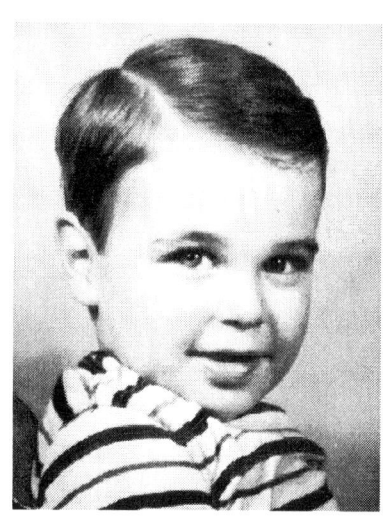

He explained to her the world and its seven seas
Before getting around to the birds and the bees.

Rick grew up as a New York City liver,
Thinking the earth ended at the Hudson River.

The sixth floor of a 12-story building was his beat
At one seventy seven East Seventy Seventh Street.

He went by subway to a Quaker school,
Where he learned the non-violent Golden Rule.

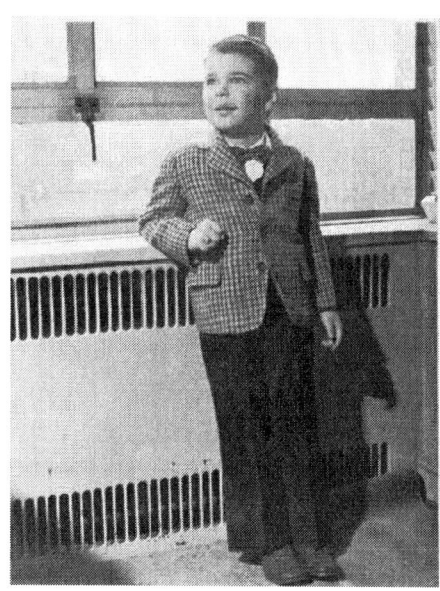

Rick grew to be a dashing young man,
Great student and a full career plan.

To avoid causing his mother anguish and loss,
Rick went off to college at Catholic Holy Cross.

At Yale his nuclear physics career was laid
By Bromley, later Bush's top science aide.

Rick married Jo Ann in June of sixty-four;
Cute, smart, tall – who could ask for more.

Jo Ann ordered his life, kept Rick on his toes;
Dressed him each morning in matching clothes.

Next Rick gave postdocing a try
For two years at the NBI.

Then they left Denmark and gave the west a stab,
Postdocing a few more years at the Los Alamos lab.

No grass or tall buildings, few green trees;
Nobody speaking much Easternese.

There Rick caught the strangest virus that has ever been;
It resulted in western clothes and hair on his chin.

Even stranger than Rick becoming a western hick,
Jo Ann turned into a groovy hippy chick.

I never saw them during their western stay;
Seeing this, I think it was better this way.

Rick left the west to fulfill a dream
To live and work near his favorite team.

Science was not why to BNL he came,
But rather a chance to follow each Yankee game.

Rick surprised his friends in various spaces
By working at a reactor of all places.

Working at a reactor gave Rick a wonderful chance
To justify going often to Grenoble France.

The mountains, skiing, French cuisine;
He loved spending time with Hans' expert team.

But, a physics group in this picture I don't understand;
I'd say it is a large heavy metal rock and roll band.

At reactors Rick searched and focused his nose on
The new physics of an interacting boson.

Franco taught him how to do it his way
And Rick's middle name became IBA.

The beloved collective model fought for survival
when Rick's group had their annual IBA revival.

Rick found bosons everywhere he looked;
Franco's ideas had him completely hooked.

But, we must break the glee that here he posed on;
These are not large orange interacting bosons.

Rick became an advisor, wise and astute,
For students who were female and cute.

Many jealous friends could never say
Why Jo Ann would let him operate that way.

The first of these tenures was Jolie C.'s;
Rick taught her about bosons, birds, and bees.

Rick was not satisfied with any proposer
Until Jolie found a unique mailman composer.

Jolie's Ziggy loves this spouse conference life;
During sessions he spends much time with my wife.

The second to come along was Ani A.;
From Clark, Daeg Brenner sent her away.

She wanted an advisor young and exciting;
At age forty Rick seemed rather inviting.

His ideas were great, approaches bold;
But in the end she considered him old.

The years have passed, Ani's career is stellar;
For Rick I'm afraid we must now call him elder.

Sixty does not sound old as an elder should be,
But it is compared to a young person like me.

Noemie was even so imprudent
To ask if I were once Rick's student.

To you that may seem like a crazy lie,
But I will love Noemie til the day I die.

Rick started a pre-Gordon Conference ritual;
The workshop at Yale has become habitual.

The physics is good, much fellowship and libration,
Good preparation for Gordon Conference migration.

Spend three days in New Haven – everyone should,
Then the dorms at Colby Sawyer actually look good

Rick and Victor are like steady dates,
Long-time inseparable physics mates.

They have long been a productive pair,
Victor's solid roots and Rick's stylish flare.

But looking at this photo I've made many tries,
But can't identify these two young thin guys.

Life is more than physics, as strange as that may sound;
The Castens have made close friendships all around.

We all admire their lives, travel, and flare,
Though we lack the energy of this globe-trotting pair.

While we worry about fixing the roof and getting ahead,
They jet off for a long weekend at fancy Club Med.

Still I think I can speak for the rest;
The Castens as friends are simply the best.

The toughest job I ever did try
Was to teach Rick to wear a tie.

He can do it and even look good,
Far better than a physicist should.

I have not tried to delve and pry
About why he resists wearing a tie.

My guess is that he doesn't want to try
To be just another good-looking physics guy.

Tennis, golf, baseball – Rick's talent is more than I can stand;
He even has appeared once in a rock and roll band.

He strummed his guitar without the use of a pick,
In hopes he'd be pursued by a cute groupie chick.

But the chicks did not want to put him down,
Since he played guitar more like Bozo the Clown.

Rick looks like a spy, thinking man, playful clown,
Even a ski racing maniac going down.

A guy who could carry the diplomatic tag,
Or one who walks in the rain in a plastic bag.

Jo Ann has been there in fun and in strife;
His friend, buddy, partner for life.

They travel, they hike, no mountains too steep;
They even sell maps, when others would sleep.

They have a Long Island house too nice to sell,
Especially after their contractor put them through hell.

If you want to burn your ears and hear a sailor curse,
Ask Jo Ann how that remodeling affected her purse.

Rick is a great man, we love him to death,
His mind, his spirit, his incredible breadth.

In a life so rich one thing is sad,
Rick lost too early a talented dad.

It would have been fairer for this to be done
For his dad to see the growth of his amazing son.

Rick's ideas are so fertile, productivity so great;
His vast energy is a really remarkable trait.

He's an amazing man from beginning to end;
I feel lucky to call him a very close friend.

Sixty for some is an advanced age;
For Rick it's just turning a new page.

As I close this verse and with a fond goodbye,
And a Happy Birthday, you wild and crazy guy.

In the end I return to one of my great whys,
Who the heck are these two young, very thin guys?

Lee Riedinger
May 24, 2002

Critical Point Symmetries In Nuclei And Other Quantum Systems

F. Iachello

Center for Theoretical Physics, Sloane Physics Laboratory,
Yale University, New Haven, CT 06520-8120

Abstract.
   A new concept, recently introduced, critical point symmetry, is briefly discussed and applied to the study of shape phase transitions in nuclei. Experimental examples are presented. Implications to other fields, especially molecular physics are mentioned.

INTRODUCTION

In the last 25 years, an important role in the description of physical systems has been played by the concept of dynamic symmetry, implicitly contained in the work of Pauli, Fock, Bargmann in the 1930's, of Gell'Mann, Ne'eman, Barut and Bohm in the 1960's and explicitly spelled out in 1978 [1] and used extensively in systems ranging from nuclei [2] to molecules [3], from hadrons [4] to polymers [5], etc.. This is a situation in which all properties of the system can be written explicitly in terms of quantum numbers, thus providing benchmarcks for the analysis of experimental data. Dynamic symmetries have led to major discoveries in nuclear physics, such as the occurrence of $SO(6)$ symmetry [6], [7]. A crucial role in the experimental finding of dynamic symmetries in nuclei has been played by Rick Casten. It is a pleasure to dedicate to him, on the occasion of his 60th birthday, the introduction of a new concept, that of critical point symmetry [8]. Critical point symmetries extend the concept of dynamic symmetry to the most difficult situation that one can encounter in physics, namely that of systems at the critical point of a phase transition.

SHAPE PHASE TRANSITIONS

Many physical systems (molecules, nuclei, atomic clusters,...) are characterized in their equilibrium configuration by a shape. These shapes are in many cases rigid. However, there are several situations in which the system is rather floppy and undergoes a phase transition between two different shapes. A challenging problem is how to describe properties of the system in the phase transition region and in particular at the phase transition point. Shape phase transitions are a special case of so-called "quantum" phase transitions, that is phase transitions that occur as a function of some parameter, g, called control parameter , that appears in the quantum Hamiltonian. They are different but

CP638, *Mapping the Triangle: International Conference on Nuclear Structure*
edited by A. Aprahamian, J. A. Cizewski, S. Pittel, and N. V. Zamfir
© 2002 American Institute of Physics 0-7354-0093-8/02/$19.00

closely related to thermodynamic phase transitions, where the control parameter is the temperature, T. Quantum phase transitions occur in all fields of physics. An example of a quantum phase transition in condensed matter physics is provided by the Ising model in a transverse magnetic field. The Hamiltonian for this system is

$$H = -Jg \sum_i \hat{\sigma}_i^x - J \sum_{\langle ij \rangle} \hat{\sigma}_i^z \hat{\sigma}_j^z. \tag{1}$$

Here J is a scale and g the control parameter. This model has two phases: when $g < g_c$ the system is a ferromagnet, while when $g > g_c$ the system is a paramagnet. A straightforward method to study quantum phase transitions is to write down a Hamiltonian that is a combination of Hamiltonians describing the two phases

$$H = \alpha_1 H_1 + \alpha_2 H_2 = (Scale) \left[H_1 + g H_2 \right] \tag{2}$$

and to study numerically the solutions of H as a function of g. The state of the system is characterized by an order parameter. For the Ising model, the order parameter is the magnetization, M. Phase transitions occur at some critical value $g = g_c$ where the order parameter changes abruptly. In finite systems, the order parameter changes continuously, but it has been found in numerical simulations that even for $N \sim 10$ (where N is the number of particles), the change is sudden enough to be detected experimentally.

PHASE DIAGRAM OF NUCLEI

The model Hamiltonian for shape phase transitions is obtained by quantizing (here bosonizing) the shape degrees of freedom (for low-lying excitations of nuclei, where quadrupole shapes play the dominat role, the quantization is obtained by introducing s and d bosons.) The corresponding Hamiltonian is that of the Interacting Boson Model [9]

$$H = \varepsilon \hat{n}_d + \kappa \hat{Q}^\chi \cdot \hat{Q}^\chi \tag{3}$$

with

$$
\begin{aligned}
\hat{n}_d &= \left(d^\dagger \cdot d \right) \\
\hat{Q}^\chi &= \left(d^\dagger \times s + s^\dagger \times d \right)^{(2)} + \chi \left(d^\dagger \times d \right)^{(2)}.
\end{aligned}
\tag{4}
$$

This Hamiltonian has three exactly solvable limits, usually denoted by the first algebra in the chain of subalgebras originating from $U(6)$

$$
\begin{aligned}
U(6) &\supset U(5) \supset SO(5) \supset SO(3) \supset SO(2), \\
U(6) &\supset SU(3) \supset SO(3) \supset SO(2), \\
U(6) &\supset SO(6) \supset SO(5) \supset SO(3) \supset SO(2).
\end{aligned}
\tag{5}
$$

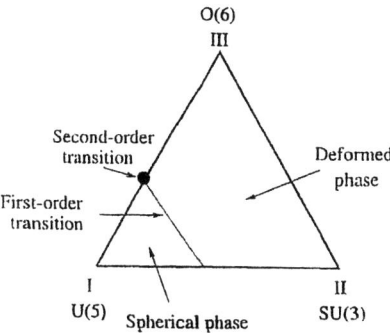

FIGURE 1. Phase diagram of nuclei in the Interacting Boson Model-1. Adapted from [14].

When studying phase transitions an important question is the order of the phase transition. For algebraic Hamiltonians such as that of the Interacting Boson Model, a study of the nature of phase transitions can be done by introducing the classical limit of the boson Hamiltonian through coherent (or intrinsic) states [10],[11],[12],

$$| N; \alpha_\mu \rangle = \left(s^\dagger + \sum_\mu \alpha_\mu d_\mu^\dagger \right)^N | 0 \rangle. \tag{6}$$

Here the shape variables α_μ appear as c-numbers. Instead of the variables α_μ, one can introduce the Bohr variables $(\beta, , \gamma, \theta_1, \theta_2, \theta_3)$. One can then evaluate the energy functional

$$E(\beta, \gamma) = \frac{\langle N; \beta, \gamma \mid H \mid N; \beta, \gamma \rangle}{\langle N; \beta, \gamma \mid N; \beta, \gamma \rangle}. \tag{7}$$

This energy functional depends only on the intrinsic variables β, γ. The order of the phase transition (Erhenfest classification) can then be determined using an algorithm due to Gilmore [13]:

(a) Evaluate the energy functional and the energy per particle, $\varepsilon = E/N$

(b) Find the minimum value ε_{min} which occurs for some value β_e, γ_e

(c) Study ε_{min} and its derivatives with respect to the control parameter, g. If ε_{min} is discontinuous the transition is called zeroth order, if $\frac{\partial \varepsilon_{min}}{\partial g}$ is discontinuous it is called first order, if $\frac{\partial^2 \varepsilon_{min}}{\partial g^2}$ is discontinuous is called second order, ...

A study of the shapes corresponding to the three dynamic symmetries, reveals that: (I) $U(5)$ corresponds to spherical shape, (II) $SU(3)$ corresponds to axially deformed shape, (III) $SO(6)$ corresponds to γ−unstable shape. Gilmore's algorithm applied to the Interacting Boson Model then gives: A second order transition when going from $U(5)$ to $SO(6)$ (spherical to γ-unstable deformed); a first order transition when going from $U(5)$ to $SU(3)$ (spherical to axially deformed); no phase transition between γ-unstable deformed and axially deformed ($SO(6)$ to $SU(3)$). The situation can be summarized in the phase diagram [14] shown in Fig.1.

3

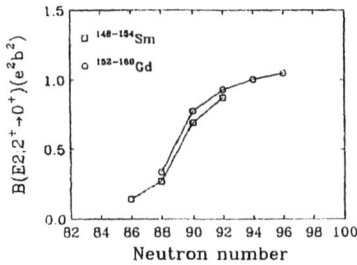

FIGURE 2. The $B(E2; 2^+ \to 0^+)$ versus neutron number for the Sm and Gd isotopes.

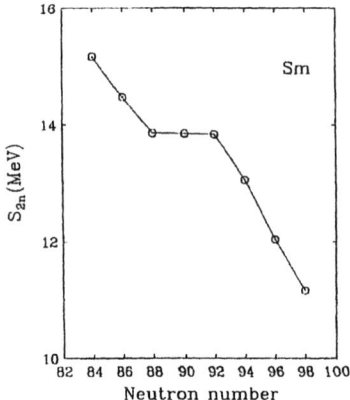

FIGURE 3. The two-neutron separation energies versus neutron number for the Sm isotopes.

In the last 20 years, evidence has been found for the occurrence of these phase transitions in nuclei, based on the analysis of binding energies, excitation energies, electromagnetic transition rates,... Direct evidence is shown in Fig. 2. Here the $B(E2; 2^+ \to 0^+)$ (proportional to the square of the order parameter $\langle n_d \rangle^2 \sim \beta^2$) is plotted versus the control parameter, or rather a function of it, the neutron number, proportional to the control parameter, $g \sim N\kappa/\varepsilon$). The sudden increase at neutron number 90 is indication of a phase transition. In Fig. 3 some additional indirect evidence is shown. Here the two-neutron separation energy (a function of the order parameter) is plotted against neutron number. In this plot, the occurrence of the phase transition is even clearer.

CRITICAL POINT SYMMETRIES

New and more accurate experimental data are shedding some light into the spectroscopy of nuclei at or around the phase transition point [15]. An important question [16] is whether or not it is possible to extend the concept of dynamic symmetry, originally

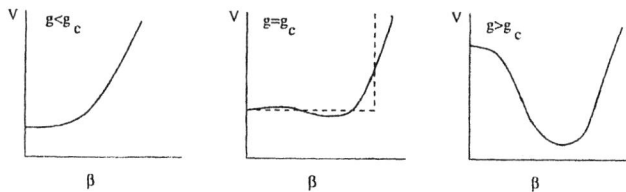

FIGURE 4. Potential as function of the variable β, below, at and above the critical value $g = g_c$.

developed for rigid situations, to floppy transitional situations. I am happy to report on the occasion of this conference that it appears to be possible to formulate the concept of dynamic symmetry (here meant as explicit solvability in terms of quantum numbers) for systems at the critical point of a phase transition [8]. The reason is that, as one can see from the schematic Hamiltonian of Eq.(2), at $g = g_c$ the spectrum of H depends only upon a scale.(Scale invariance). The simplest way to impose scale invariance is to replace the potential that appears in the differential equation $H\psi = E\psi$ that describes the classical limit of the boson Hamiltonian by a square-well potential. As one can see from Fig. 4 and from numerical solutions of the boson Hamiltonian [17] this is a good approximation at $g = g_c$. For quadrupole shapes, a convenient differential equation is the Bohr equation [18], written in terms of variables $(\beta, \gamma, \theta_1, \theta_2, \theta_3)$, with H given by

$$
H = -\frac{\hbar^2}{2B}[\frac{1}{\beta^4}\frac{\partial}{\partial\beta}\beta^4\frac{\partial}{\partial\beta} + \frac{1}{\beta^2\sin 3\gamma}\frac{\partial}{\partial\gamma}\sin 3\gamma\frac{\partial}{\partial\gamma}
$$
$$
-\frac{1}{4\beta^2}\sum_\kappa \frac{Q_\kappa^2}{\sin^2(\gamma - \frac{2}{3}\pi\kappa)}] + V(\beta, \gamma). \tag{8}
$$

Two solutions have been worked out:

(i) The spherical to γ-unstable transition ($U(5) - SO(6)$ transition). As mentioned above this is a second order phase transition in the variable β. At $g = g_c$, the potential can be approximated by a 5-dim square well,

$$
\begin{aligned}
u(\beta) &= 0 \quad for \ \beta \leq \beta_W \\
u(\beta) &= \infty \quad for \ \beta > \beta_W.
\end{aligned} \tag{9}
$$

The solutions of the Bohr Hamiltonian with this potential are:

$$
\Psi(\beta, \gamma, \theta_i) = c_{s,\tau}\beta^{-3/2}J_{\tau+3/2}(k_{s,\tau}\beta)\Phi_{\tau, v_\Delta, L, M}(\gamma, \theta_i) \tag{10}
$$

with

$$
k_{s,\tau} = \frac{x_{s,\tau}}{\beta_W} \tag{11}
$$

where $x_{s,\tau}$ is the s-th zero of $J_{\tau+3/2}(z)$ and the functions Φ are eigenfunctions of the quadratic invariant of $SO(5)$ with quantum number τ. The corresponding eigenvalues are given by the simple expression

5

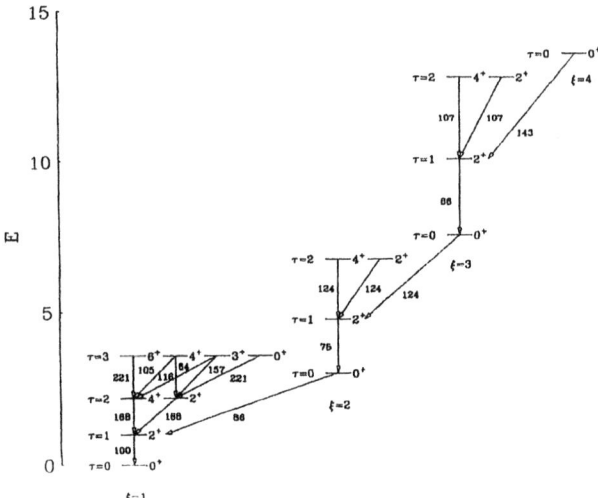

FIGURE 5.   Spectrum of $E(5)$ symmetry.

$$E_{s,\tau} = \frac{\hbar^2}{2B}k_{s,\tau}^2 = A\left(x_{s,\tau}\right)^2. \tag{12}$$

This solution has been called $E(5)$. One can see directly from the last equation that the eigenvalues are given in terms only of a scale, A. Its spectrum is shown in Fig. 5. In addition to energies, other properties can be calculated explicitly. For example, $E2$ electromagnetic transition rates can be calculated using

$$
\begin{aligned}
T^{(E2)} &= \alpha_{2\mu} \\
\alpha_{2\mu} &= \beta\left[D^{(2)}_{\mu,0}\cos\gamma + \frac{1}{\sqrt{2}}\left(D^{(2)}_{\mu,+2} + D^{(2)}_{\mu,-2}\right)\sin\gamma\right]
\end{aligned} \tag{13}
$$

and properties of the Bessel functions. An improved transition operator, including terms proportional to β^2, can also be used [19]. Caprio [20] has also analyzed solutions of a finite square-well. An experimental example of $E(5)$ symmetry has been found by Casten and Zamfir [21] in ^{134}Ba and is shown in Fig.6.

(ii) The spherical to axially deformed transition ($U(5)$ to $SU(3)$). This phase transition being of first order has coexistence of two different shapes (spherical and axially deformed). In addition it involves simultaneously both variables β and γ. It is thus much more difficult to treat analytically. An approximate solution can be obtained with potential

$$
\begin{aligned}
u(\beta,\gamma) &= 0 \quad for\ \beta \leq \beta_W; \gamma \leq \gamma_W \\
u(\beta,\gamma) &= \infty \quad for\ \beta > \beta_W; \gamma > \gamma_W.
\end{aligned} \tag{14}
$$

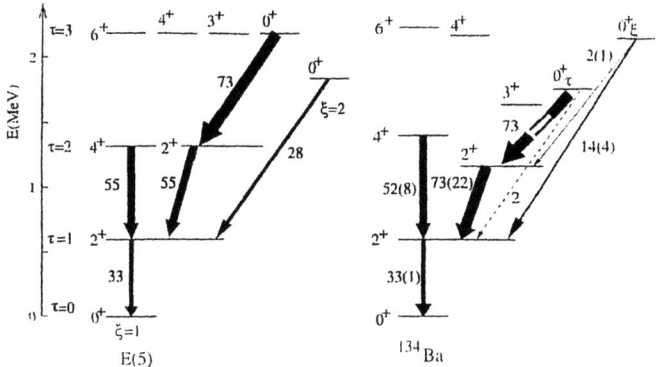

FIGURE 6. Comparison between $E(5)$ and the experimental spectrum of ^{134}Ba. Adapted from [21].

This solution will be henceforth called $X(5)$. [In the preliminary account [22] a harmonic oscillator in γ was used, but for sake of uniformity of description and of connecting $X(5)$ to $E(5)$ it has been found to be more appropriate to use a square-well both in β and γ [23].] Wave functions can be written as

$$\Psi(\beta,\gamma,\theta_i) = c_{s,L}\beta^{-3/2}J_\nu(k_{s,L}\beta)J_{\nu'}(k_{s',K}\gamma)D^L_{MK}(\theta_i) \tag{15}$$

and energy eigenvalues as

$$E(s,L,s',K,M) = A\left(x_{s,L}\right)^2 + B\left(x_{s',K}\right)^2 - 0.89AK^2 \tag{16}$$

where $x_{s,L}$ is the s-th zero of $J_\nu(z)$ with

$$\nu = \left(\frac{L(L+1)}{3} + \frac{9}{4}\right)^{1/2} \tag{17}$$

in general an irrational number and $x_{s',K}$ is the s'-th zero of $J_{\nu'}(z)$ with

$$\nu' = \frac{K}{2}. \tag{18}$$

The β part of the spectrum has properties that depend only on the scale A. Particularly interesting are energy ratios which are parameter free

$$E_{4_1}/E_{2_1} = 2.91$$
$$E_{0_2}/E_{2_1} = 5.67 \tag{19}$$

In a landmark paper, Casten and Zamfir [24] have found that the spectrum and electro-magnetic transition rates of ^{152}Sm are well described by the $X(5)$ solution. Recently, Krücken et al [25] have found other examples in ^{150}Nd and Bizzeti and Bizzeti-Sona

7

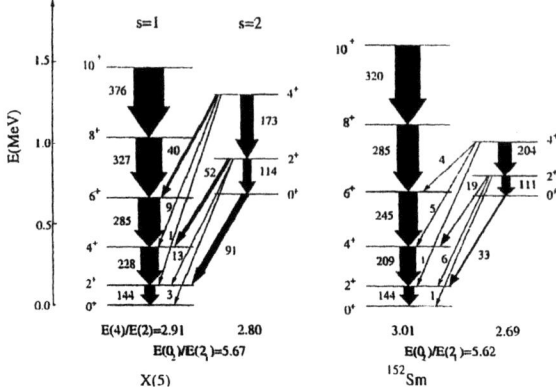

FIGURE 7. Comparison between $X(5)$ and the β part of the experimental spectrum of ^{152}Sm. Adapted from [24].

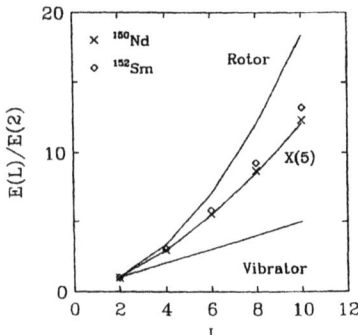

FIGURE 8. Energies of the ground band as a function of the angular momentum L in ^{152}Sm and ^{150}Nd. Adapted from [24].

[26] in ^{104}Mo. Fig.7 shows a comparison between the experimental spectrum of ^{152}Sm and $X(5)$. For the two ratios above, the experimental values are 3.01 and 5.62 respectively in good accord with the theoretical quantities. The agreement between experiment and $X(5)$ is ever more dramatic if the energy levels in the ground band are plotted against the angular momentum L, as shown in Fig.8.

OTHER QUANTUM SYSTEMS

Scale invariance is expected to play a role in all systems at the critical point of a phase transition. Another system where this has been found is molecules. The pertinent model for this system is the Vibron Model [27]. Shape phase transitions occur in molecules when the shape of the molecule changes (from example from linear to non-linear or from planar to a-planar). Linear-bent phase transitions have been recently studied in triatomic

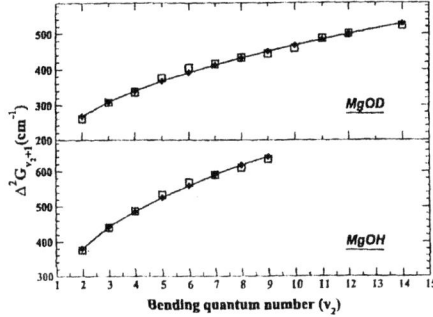

FIGURE 9. Differences in energies of the vibrational levels of $MgOD$ and $MgOH$ as function of the vibrational quantum number v_2. From [28].

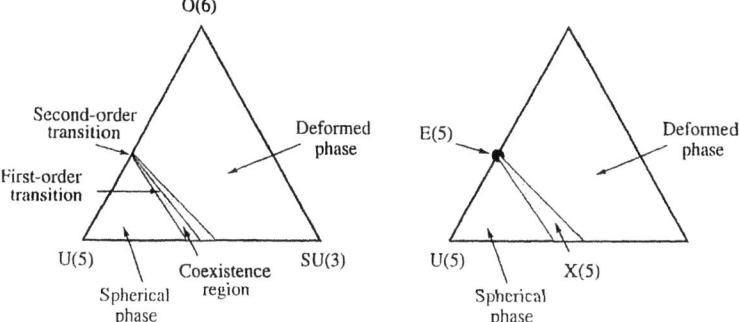

FIGURE 10. A summary of the new solutions for critical points. On the left, the "old" solutions, $U(5), SU(3), O(6)$. On the right the "new" solutions, $E(5)$ and $X(5)$.

molecules. These are second order phase transitions in $n = 2$ dimensions. Spectra at the critical point are described by $E(2)$. Very recently, Perez-Bernal and Vaccaro [28] have found that several atoms in the family of molecules XOH with X a heavy atom can be described by $E(2)$. Two examples are shown in Fig.9.

CONCLUSIONS

A new concept, called critical point symmetry, has been introduced. This concept provides benchmarks for the experimental analysis of spectra of physical systems. Two benchmarks have been introduced in nuclear physics, $E(5)$ and $X(5)$, as summarized in Fig. 10.

Applications to other systems are feasible and are being pursued (molecules, atomic clusters, macromolecules, polymers,...). It is yet another example of how concepts developed in nuclear physics can be used in other fields.

ACKNOWLEDGEMENTS

This work was supported in part under D.O.E. Grant No. DE-FG-02-91ER40608.

REFERENCES

1. F. Iachello, in *Group Theoretical Methods in Physics,* Lecture Notes in Physics, **Vol. 94**, A. Böhm, ed., Lange Springer, Berlin, (1979), p.420.
2. F. Iachello and A. Arima, *Phys. Lett.* **53B**, 309 (1974).
3. F. Iachello, *Chem. Phys. Lett.* **78**, 581 (1981).
4. F. Iachello, *Nucl. Phys.* **A497**, 23c (1989).
5. F. Iachello and P. Truini, *Ann. Phys. (N.Y.)* **276**, 120 (1999).
6. F. Iachello, in *Lecture Notes in Physics,* University of Groningen, The Netherlands (1976); A. Arima and F. Iachello, *Phys. Rev. Lett.* **40**, 385 (1978).
7. J.A. Cizewski, R.F. Casten, G.J. Smith, M.L. Stelts, W.R. Kane, H.G. Börner and W.F. Davidson, *Phys. Rev. Lett.* **40**, 167 (1978).
8. F. Iachello, *Phys. Rev. Lett.* **85**, 3580 (2000).
9. F. Iachello and A. Arima, *The Interacting Boson Model*, Cambridge University Press, Cambridge (1987).
10. A.E.L. Dieperink, O.Scholten and F. Iachello, *Phys. Rev. Lett.* **44**, 1747 (1980).
11. J. Ginocchio and M. Kirson, *Phys. Rev. Lett.* **44**, 1740 (1980).
12. A. Bohr and B.R. Mottelson, *Phys. Scr.* **22**, 468 (1980).
13. R. Gilmore, *J. Math. Phys.* **20**, 891 (1978).
14. D.H. Feng, R. Gilmore and S.R. Deans, *Phys. Rev.* **C23**, 1254 (1981).
15. H.J. Harter, P. von Brentano, A. Gelberg and R.F. Casten, *Phys. Rev.* **C32**, 631 (1985).
16. R.F. Casten, private communication.
17. N.V. Zamfir, private communication.
18. A. Bohr, *Mat. Fys. Medd. K. Dan. Vidensk Selsk.* **26**, No. 14 (1952).
19. J.M. Arias, *Phys. Rev. C* **63**, 034308 (2001).
20. M.A. Caprio, *Phys. Rev. C* **65**, 031304 (2001).
21. R.F. Casten and N.V. Zamfir, *Phys. Rev. Lett.* **85**, 3584 (2000).
22. F. Iachello, *Phys. Rev. Lett.* **87**, 052502 (2001).
23. F. Iachello, to be published.
24. R.F. Casten and N.V. Zamfir, *Phys. Rev. Lett.* **87**, 052503 (2001).
25. R. Krücken et al, *Phys. Rev. Lett.*, in press.
26. P.G. Bizzeti and A. Bizzeti-Sona, submitted.
27. F. Iachello and R.D. Levine, *Algebraic Theory of Molecules*, Oxford University Press, Oxford (1995).
28. F. Perez-Bernal and P. Vaccaro, in preparation.

Mixed Symmetry in the Symmetry Triangle

N. Pietralla[*], C.Fransen[*†], A.F. Lisetskiy[*], P. von Brentano[*] and V. Werner[*]

[*]Institut für Kernphysik, Universität zu Köln, 50937 Köln, Germany
[†]Dept. of Physics and Astronomy, University of Kentucky, Lexington, KY 40506, U.S.A.

Abstract. Recent investigations of off-yrast low-spin states of ^{94}Mo yielded evidence for one quadrupole phonon and two-phonon states with mixed proton-neutron symmetry. The mixed-symmetry assignments are based on the observation of strong $M1$ transitions and the measurement of their absolute $M1$ matrix elements $\approx 1\,\mu_N$. Corresponding structures were identified very recently in the $N = 52$ neighboring even-even isotones, ^{96}Ru and ^{92}Zr, too. Progress on this topic is reviewed.

Collective phenomena of the nuclear quantum system and the evolution of nuclear structure are two major themes of the outstanding research work of Rick Casten, who contributed to the field like nobody else. It is our pleasure and a great honor for us to contribute the present paper to the conference on the occasion of Rick's sixtieth birthday.

One goal of modern nuclear structure physics is to uncover the principles that govern why and how collectivity emerges from the underlying dynamics of nucleons and how it evolves as a function of the nucleon numbers. Rick Casten's $N_p N_n$-scheme [1] highlights the role of valence nucleons and underlines the impact of the proton-neutron degree of freedom on the formation and evolution of nuclear collectivity.

While microscopic calculations for collective phenomena in terms of the nuclear shell model are nowadays starting to become accessible even for heavy nuclei [2], the interacting boson approximation [3] offers a simple tool to interpolate between various benchmarks of nuclear collectivity and is, therefore, very well suited for investigating the evolution of structure.

The discovery of nuclei with a soft triaxiality [O(6) nuclei], like ^{196}Pt [4] or nuclei in the xenon/barium region [5], supported strongly the relevance of the interacting boson model formulated initially [3] even without the distinction between proton bosons and neutron bosons (IBM-1). The $N_p N_n$-scheme for heavy nuclei suggests, however, the importance of the proton-neutron degree of freedom.

The IBM-2 takes both into account, proton bosons and neutron bosons [6]. With respect to the IBM-1 it enlarges the model space to states with non-symmetric coupling of proton and neutron bosons. The latter states are considered to be of mixed-symmetry character, which can be quantified by the F-spin quantum number. The F-spin dynamical symmetry limit [7] of the IBM-2 contains the IBM-1 Hilbert space as the sub-space of states with maximum F-spin quantum number, $F = F_{max} = N/2$, where $N = N_\pi + N_\nu$ is the sum of the proton boson number, N_π, and the neutron boson number, N_ν. Mixed symmetry states (MSSs) are characterized by the F-spin quantum numbers $F < F_{max}$. We will restrict ourselves to MSSs with $F = F_{max} - 1$. MSSs form a whole class of col-

CP638, *Mapping the Triangle: International Conference on Nuclear Structure*
edited by A. Aprahamian, J. A. Cizewski, S. Pittel, and N. V. Zamfir
© 2002 American Institute of Physics 0-7354-0093-8/02/$19.00

lective states with similar wave functions [8]. They are connected among themselves by strong electric quadrupole ($E2$) transitions (in the absence of further selection rules) and can decay by strong magnetic dipole ($M1$) transitions to symmetric states with $F = F_{max}$. This feature enables us to uniquely identify MSSs because $M1$ transitions between symmetric states are forbidden.

The last statement follows from the structure of the $M1$ transition operator

$$
\begin{aligned}
T(M1) &= \sqrt{\frac{3}{4\pi}} \left(g_\pi L_\pi + g_\nu L_\nu \right) \\
&= \sqrt{\frac{3}{4\pi}} \left[\frac{g_\pi N_\pi + g_\nu N_\nu}{N} L^{tot} + (g_\pi - g_\nu) \frac{N_\pi N_\nu}{N} \left(\frac{L_\pi}{N_\pi} - \frac{L_\nu}{N_\nu} \right) \right]
\end{aligned}
\tag{1}
$$

and from the fact [7] that the matrix element of every one-body operator $\hat{O}_\rho = b_\rho^+ \tilde{b}_\rho$ for $\rho \in \{\pi, \nu\}$ between any two symmetric states is proportional to the number of bosons, b, with isospin label ρ:

$$
\langle F_{max}, \alpha \| \hat{O}_\rho \| F_{max}, \beta \rangle = c_{\alpha\beta} N_\rho .
\tag{2}
$$

The angular momentum operators $L_\rho = \sqrt{10}[d_\rho^+ \times \tilde{d}_\rho]^{(1)}$ are one-body operators and, therefore, the last term in Eq. (1) has vanishing matrix elements between symmetric states. On the other hand, $L^{tot} = L_\pi + L_\nu$ represents the total angular momentum operator in the IBM-2 which cannot induce transitions between different states, either, because it is diagonal. Thus, the search for MSSs focuses on the observation of strong $M1$ transitions with large matrix elements of the order of $1\,\mu_N$ since the relevant difference of the boson g-factors, $g_\pi - g_\nu$, is roughly of that size.

In the early 1980s Richter and coworkers discovered a strong $M1$ excitation mode in heavy deformed even-even nuclei in electron scattering experiments performed in Darmstadt [9, 10]. This $M1$ mode is called scissors mode due to its geometrical picture in rotors. The scissors mode was subsequently investigated in great detail using the photon scattering technique, mostly in experiments by Kneißl and collaborators in Stuttgart [11]. The scissors mode is known to be usually fragmented over several $J^\pi = 1^+$ states at energies around 3 MeV [12] and its total $M1$ strength to the ground state correlates to the collective $B(E2; 2_1^+ \to 0_1^+)$ value [13]. Increasing sensitivity in the photon scattering experiments in the 1990s made the discovery of the scissors mode even in odd-mass nuclei [14] and in O(6) nuclei [15, 16] possible.

Vibrational behavior represents another benchmark of nuclear structure and the phonon concept is a simple and useful scheme for the understanding of vibrational excitations. The Q-phonon scheme [17] is an approximate scheme in the IBM and suggests to approximate the wave functions of many yrast and near-yrast states, even outside of analytically solvable limits [18, 19], by simple operators acting on the ground state. The proton and neutron $E2$ operators, $Q_\rho = s_\rho^+ \tilde{d}_\rho + d_\rho^+ s_\rho + \chi_\rho [d_\rho^+ \times \tilde{d}_\rho]^{(2)}$, represent the building blocks of the Q-phonon scheme in the sd-IBM-2 [20]. In the vibrational F-spin limit they correspond to the formation of the fundamental one-phonon states

$$
|2_1^+\rangle = Q_s |0_1^+\rangle \tag{3}
$$
$$
|2_{ms}^+\rangle = Q_m |0_1^+\rangle \tag{4}
$$

where $Q_s = Q_\pi + Q_\nu$ and $Q_m = Q_\pi/N_\pi - Q_\nu/N_\nu$. The normalization constants are suppressed in Eqs. (3,4) for brevity. The fundamental character of the 2^+_{ms} state is particularly obvious in the Q-phonon scheme.

Experimental hints at 2^+_{ms} states from the observation of small $E2/M1$ multipole mixing ratios of $2^+ \rightarrow 2^+_1$ transitions have first been discussed by Hamilton *et al.* [21] for vibrational $N = 84$ isotones. However, small mixing ratios do not unambiguously identify large $M1$ matrix elements (the $E2$ strength can be very small, instead) and are, thus, not a reliable signature for MSSs. Unambiguous identifications of 2^+_{ms} states were possible in a handful of nuclides from lifetime measurements using Coulomb excitation [22] or Doppler shift attenuation methods, *e.g.*, [23, 24, 25]. A step forward in the investigation of 2^+_{ms} states was done by the demonstration that this fundamental state can be well investigated also in high-resolution photon scattering as it was first done for ^{136}Ba at Stuttgart University [26].

The subsequent combination of photon scattering data on the $N = 52$ nuclide ^{94}Mo with very clean and high-statistics off-beam γ-ray spectroscopy following β-decay performed at the University of Cologne led to an unprecedented richness of information on the lowest 1^+_{ms} and 2^+_{ms} states [27]. In particular, a strong γ transition between these MSSs could be observed for the first time representing first direct evidence that these states belong to a class of states with similar wave functions.

Further significant progress resulted from the addition of in-beam γ-ray spectroscopy on ^{94}Mo studied in the ^{91}Zr(α, n) reaction at the University of Cologne. Besides a considerable enlargement of the known level scheme of ^{94}Mo the combination of in-beam and off-beam $\gamma\gamma$-coincidence studies resulted in the observation of many new branching ratios and of unambiguous assignments of multipole mixing ratios. Further lifetime information could be deduced from the analysis of Doppler shifts. This information enabled us to discover further MSSs with spin and parity quantum numbers 3^+_{ms} [28] and $2^+_{2,ms}$ [29].

The observed MSSs of ^{94}Mo arrange into a multiphonon structure and lead to the first measurement of the anharmonicity, $\varepsilon = \langle E_x(J) \rangle / [E_x(2^+_1) + E_x(2^+_{ms})] - 1$, and the energy splitting, $\Delta_J = \langle E_x(J) - E_x(J_0) \rangle$, of the two-phonon multiplet with mixed symmetry. The energy splitting of the largest fragments of the MS two-phonon multiplet is small [30]. It amounts to only 10% of the excitation energy. The energy anharmonicity is even smaller making the MS two-phonon multiplet a beautiful example of inhomogeneous phonon coupling besides that of the better known quadrupole-octupole coupling.

Very recent investigations of ^{94}Mo at the University of Kentucky using inelastic neutron scattering [31, 32] confirm in general the previous findings, in many cases with considerably smaller uncertainties for the level lifetimes. The new data initiated detailed theoretical studies on the formation of mixed-symmetry collectivity using microscopic models such as the nuclear shell model [33] and the quasiparticle phonon model [34] and led to new developments in the framework of the IBM with respect to the formulation of new $M1$ sum rules [35] and the coupling of the octupole degree of freedom to the proton-neutron degree of freedom [36].

Other recent γ-ray spectroscopy work on ^{92}Zr and ^{96}Ru [37, 38] led to the observation of corresponding structures in these $N = 52$ isotones to ^{94}Mo, too, and start the formation of a broader data base on MSSs in the $A \approx 100$ mass region. Figure 1 displays some

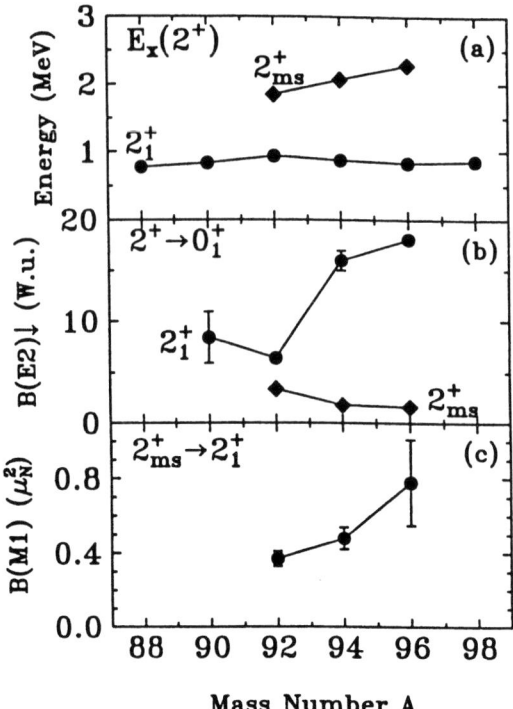

FIGURE 1. Some properties of the fundamental one-phonon 2^+ states in $N = 52$ isotones as a function of the mass number. Part (a) shows the excitation energies, part (b) displays the $B(E2; 2^+ \rightarrow 0_1^+)$ value in Weisskopf units [1 W.u. = 0.06 $A^{4/3} e^2 \text{fm}^4$], and part (c) gives the $M1$ transition strength between these 2^+ states in μ_N^2.

known properties of the one-phonon quadrupole excitations in $N = 52$ isotones. We hope that the present and future data on MSSs in this mass region will help to further reveal the impact of the proton-neutron degree of freedom on the formation and evolution of nuclear collectivity.

ACKNOWLEDGMENTS

We are grateful to Rick Casten for many years of productive collaboration. We further thank all the coauthors of our articles summarized in this short review for their contributions. This paper has been supported by the DFG under contract Nos. Br 799/10-1 and Pi 393/1-2.

REFERENCES

1. R.F. Casten, Nucl. Phys. **A443**, 1 (1985).
2. N. Shimizu, T. Otsuka, T. Mizusaki, M. Honma, Phys. Rev. Lett. **86**, 1171 (2001).
3. A. Arima and F. Iachello, Phys. Rev. Lett. **35**, 1069 (1975).
4. R.F. Casten and J.A. Cizewski, Nucl. Phys. **A309**, 477 (1978).
5. R.F. Casten and P. von Brentano, Phys. Lett. **B152**, 22 (1985).
6. A. Arima, T. Otsuka, F. Iachello, and I. Talmi, Phys. Lett. **B66**, 205 (1977).
7. P. Van Isacker, K. Heyde, J. Jolie, and A. Sevrin, Ann. Phys. (NY) **171**, 253 (1986).
8. F. Iachello, Phys. Rev. Lett. **53**, 1427 (1984).
9. D. Bohle, A. Richter, W. Steffen, A.E.L. Dieperinck, N. LoIudice, F. Palumbo, and O. Scholten, Phys. Lett. **B137**, 27 (1984).
10. A. Richter, Prog. Part. Nucl. Phys. **34**, 261 (1995).
11. U. Kneissl, H.H. Pitz, and A. Zilges, Prog. Part. Nucl. Phys. **37**, 349 (1996).
12. N. Pietralla, P. von Brentano, R.-D. Herzberg, U. Kneissl, N. LoIudice, H. Maser, H.H. Pitz, and A. Zilges, Phys. Rev. C **58**, 184 (1998).
13. C.Rangacharyulu, A.Richter, H.J.Wörtche, W.Ziegler, R.F.Casten, Phys. Rev. C **43**, R949 (1991).
14. I. Bauske *et al.*, Phys. Rev. Lett. **71**, 975 (1993).
15. P. von Brentano *et al.*, Phys. Rev. Lett. **76**, 2029 (1996).
16. H.Maser, N.Pietralla, P.von Brentano, R.-D.Herzberg, U.Kneissl, J.Margraf, H.H.Pitz, and A.Zilges, Phys. Rev. C **54**, R2129 (1996).
17. T. Otsuka and K.-H. Kim, Phys. Rev. C **50**, R1768 (1994).
18. N.Pietralla, P.von Brentano, R.F.Casten, T.Otsuka, and N.V.Zamfir, Phys. Rev. Lett. **73**, 2962 (1994).
19. N.Pietralla, T.Mizusaki, P.von Brentano, R.V.Jolos, T.Otsuka, V.Werner, Phys. Rev. C **57**, 150 (1998).
20. K.-H. Kim, T. Otsuka, P. von Brentano, A. Gelberg, P. Van Isacker, R.F. Casten, in *Capture γ-Ray Spectroscopy and Rel.Topics*, etd. G. Molnàr (Springer, Budapest, 1996), Vol. **I**, p. 195.
21. W.D. Hamilton, A. Irbäck, and J.P. Elliott, Phys. Rev. Lett. **53**, 2469 (1984).
22. W.J. Vermeer, C.S. Lim, R.H. Spear, Phys. Rev. C **38**, 2982 (1988).
23. K.P. Lieb, H.G. Börner, M.S. Dewey, J. Jolie, S.J. Robinson, S. Ulbig, and Ch. Winter, Phys. Lett. **B215**, 50 (1988).
24. B. Fazekas, T. Belgya, G. Molnár, A. Veres, R.A. Gatenby, S.W. Yates, T. Otsuka, Nucl. Phys. **A548**, 249 (1992).
25. I. Wiedenhöver, A. Gelberg, T. Otsuka, N. Pietralla, J. Gableske, A. Dewald, and P. von Brentano, Phys. Rev. C **56**, R2354 (1997).
26. N. Pietralla *et al.*, Phys. Rev. C **58**, 796 (1998).
27. N. Pietralla, C. Fransen *et al.*, Phys. Rev. Lett. **83**, 1303 (1999).
28. N. Pietralla, C. Fransen, P. von Brentano, A. Dewald, A. Fitzler, C. Frießner, and J. Gableske, Phys. Rev. Lett. **84**, 3775 (2000).
29. C. Fransen, N. Pietralla, P. von Brentano, A. Dewald, J. Gableske, A. Gade, A. Lisetskiy, and V. Werner, Phys. Lett. **B 508**, 219 (2001).
30. N. Pietralla, in *Nuclear Structure Physics*, edts. R.F. Casten, J. Jolie, U. Kneissl, K.P. Lieb (World Scientific, Singapore, 2001), p.243.
31. C. Fransen *et al.*, in *Nuclear Structure Physics*, edts. R.F. Casten, J. Jolie, U. Kneissl, K.P. Lieb (World Scientific, Singapore, 2001), p.351.
32. C. Fransen, Z. Ammar, D. Bandyopadhyay, N. Boukharouba, P. von Brentano, A. Dewald, J. Gableske, J. Jolie, U. Kneissl, S.R. Lesher, A.F. Lisetskiy, M.T. McEllistrem, M. Merrick, N. Pietralla, H.H. Pitz, N. Warr, V. Werner, and S.W. Yates, in preparation.
33. A.F.Lisetskiy, N.Pietralla, C.Fransen, R.V.Jolos, P.von Brentano, Nucl. Phys. **A 677**, 100 (2000).
34. N. LoIudice and Ch. Stoyanov, Phys. Rev. C **62**, 047302 (2000).
35. N.A.Smirnova, N.Pietralla, A.Leviatan, J.N.Ginocchio, C.Fransen, Phys. Rev. C **65**, 024319 (2002).
36. Nadya A. Smirnova, Norbert Pietralla, Takahiro Mizusaki, and Piet Van Isacker, Nucl. Phys. **A 678**, 235 (2000).
37. H. Klein, A.F. Lisetskiy, N. Pietralla, C. Fransen, A. Gade, and P. von Brentano, Phys. Rev. C **65**, 044315 (2002).
38. N. Pietralla, C.J. Barton III., R. Krücken, C.W. Beausang, M.A. Caprio, R.F. Casten, J.R. Cooper, A.A. Hecht, H. Newman, J.R. Novak, and N.V. Zamfir, Phys. Rev. C **64**, 031301(R) (2001).

Experiments on critical point nuclei

M. A. Caprio

Wright Nuclear Structure Laboratory, Yale University, New Haven, Connecticut 06520

Abstract. A new generation of experiments studying nuclei in spherical-deformed transition regions has been encouraged by the introduction of innovative theoretical approaches to the treatment of these nuclei, including the critical point models $E(5)$ and $X(5)$. Experiments and results in the A=100 and A=150 regions are discussed.

INTRODUCTION

Intriguing phenomenological observations regarding nuclei in the spherical-deformed shape transitional regions, as well as new theoretical constructs for the understanding of these nuclei, have provided the impetus for a series of experiments designed to carefully measure key structural observables.

The evolution of observables across the transitional regions exhibits behaviors characteristic of first and second order thermodynamic phase transitions. A new family of models [1, 2, 3] recently proposed by Iachello, based upon the dynamical symmetries exhibited by square-well potentials (Fig. 1), provide a simple description of critical point nuclei. The critical point models — $E(5)$ for the spherical to deformed-γ-soft transition and $X(5)$ for the spherical to axially-symmetric rotor transition — give analytic solutions and yield essentially parameter-free predictions for behavior at the critical point.

Recent experiments performed in the A=100 (Ru, Pd) and A=150 (Nd, Sm, Dy) regions, populating the low-spin states of interest through β decay, neutron capture, and Coulomb excitation, have provided a wealth of new spectroscopic and lifetime information. The important structural signatures in the transitional regions, beyond the basic yrast level properties, involve γ-ray transitions between low-spin, non-yrast levels. Reliable information on γ-ray branching ratios, multipolarities, and absolute matrix elements (or level lifetimes) is crucial. The γ-rays of interest, however, often originate from weakly populated levels, are of low transition energy (and thus $E_\gamma^{2\lambda+1}$ suppressed), or have small matrix elements. They are consequently "weak" transitions from an experimental viewpoint, requiring sensitive spectroscopy for their study.

THE $A = 100$ PD-RU REGION

Nuclei described by the $E(5)$ Hamiltonian [1] should exhibit a few basic structural signatures [Fig. 2(a)]: The $E(5)$ model predicts an $R_{4/2} \equiv E(4_1^+)/E(2_1^+)$ ratio of 2.20. The scheme of excited levels follows a multiplet structure (0^+, 2^+, 4^+-2^+, 6^+-4^+-3^+-0^+, ...), universal to all γ-soft problems [4]. Excited families of levels exhibiting the

CP638, *Mapping the Triangle: International Conference on Nuclear Structure*
edited by A. Aprahamian, J. A. Cizewski, S. Pittel, and N. V. Zamfir
© 2002 American Institute of Physics 0-7354-0093-8/02/$19.00

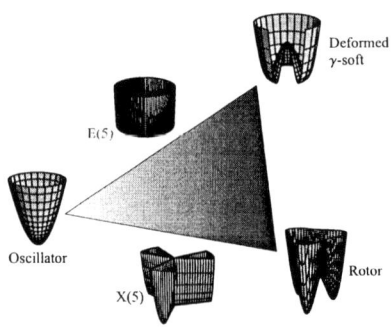

FIGURE 1. $E(5)$ γ-soft and $X(5)$ γ-localized square-well potentials for transitional nuclei.

same multiplet structure are predicted, starting with $E(0_2^+)=3.03E(2_1^+)$. Levels within a family are connected by $E2$ strengths according to a phonon-like scheme of allowed and forbidden transitions, and weak inter-family transitions are allowed as well. These predictions of the $E(5)$ model, made using an ideal infinite square well potential, are extremely robust when carried over to a realistic finite-depth potential [5], suggesting that it is meaningful to directly compare them with experiment.

A nucleus with $R_{4/2} \approx 2.20$ in a known spherical to γ-soft transition region is immediately of interest as a prospective $E(5)$ nucleus. It must then be seen to what extent the excited levels follow the energy and transition properties expected for the multiplet members and whether or not an excited 0^+ level with its associated family of levels can be found. $E(5)$ candidate nuclei have been noted in the $A = 130$ [6] and $A = 100$ [7, 8] regions.

The nucleus ^{102}Pd has recently been studied through γ-ray spectroscopy following β decay at the Yale Moving Tape Collector (MTC) [8]. This nucleus has an $R_{4/2}$ value of 2.29 and levels at the correct energies to be the 4^+-2^+ members of the "2-phonon" multiplet, the 6^+-4^+-3^+ members of the "3-phonon" multiplet, and the 0^+ level heading the first excited family. However, several of the transitions predicted by $E(5)$ to be

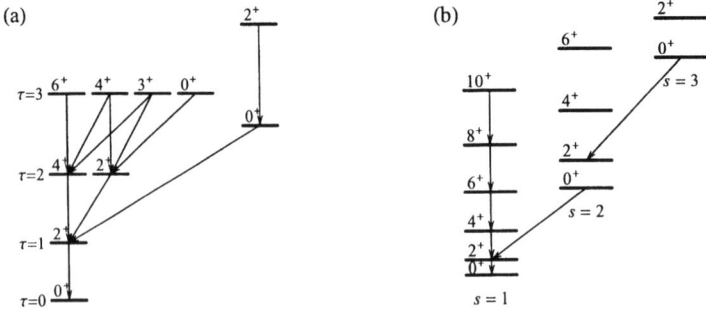

FIGURE 2. Level schemes showing the lowest-lying levels in the solutions for the (a) $E(5)$ and (b) $X(5)$ Hamiltonians. More detailed level schemes are provided in Refs. [1, 2].

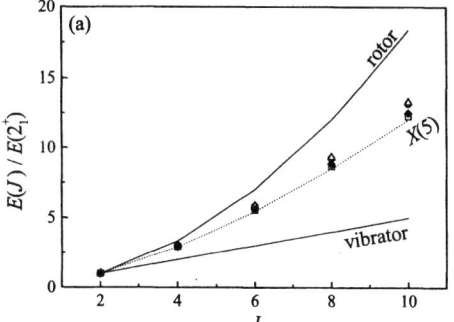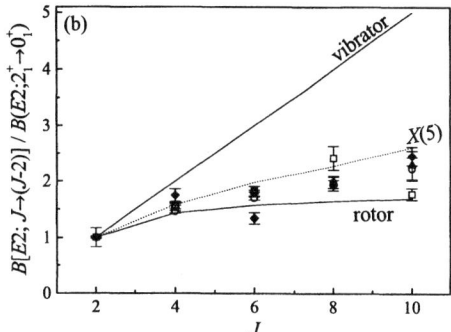

FIGURE 3. Yrast (a) energies and (b) $B(E2)$ values, normalized to those for the first 2^+ state, for the $N=90$ isotones ^{150}Nd (squares), ^{152}Sm (circles), ^{154}Gd (triangles), and ^{156}Dy (diamonds), along with model predictions.

enhanced transitions had been missing from the known decay scheme. The 3^+ level of the 3-phonon multiplet had only been known to decay by a transition to the one-phonon 2^+ [forbidden in $E(5)$], but the new work showed that the transitions to the 2-phonon 2^+ and 4^+ levels [allowed in $E(5)$] are present and have much larger $B(E2)$ values (assuming pure $E2$ multipolarity, which, however, is unknown). The 3-phonon 4^+ level, which had already been known to decay preferentially to the 2-phonon 2^+ level, was confirmed to have a substantial decay to the 2-phonon 4^+ level as well. This nucleus also provides an example of an important consideration relating to the interpretation of the excited states, especially low-lying 0^+ states. The lowest excited 0^+ level is a 15 ns isomer, decaying primarily by $E0$ decay to the ground state, with properties which are inexplicable within the $E(5)$ model space. However, an inspection of the systematics across the $Z=40$ shell closure indicates this level to be a collective intruder level originating in the preceeding shell.

Measurements on ^{100}Ru and ^{102}Ru have been carried out in neutron capture reactions at the Institut Laue-Langevin, using the gamma-ray induced Doppler shift (GRID) method [9] to obtain lifetime information, primarily on levels in the family built upon the first excited 0^+ state. Experiments at the Yale Moving Tape Collector are also planned.

THE $N = 90$ REGION

The $X(5)$ model [2] predicts a band structure resembling that of a rotor [Fig. 2(b)], but with substantially different energy ratios and $B(E2)$ strengths along the bands (Fig. 3). The model also predicts comparatively strong interband transitions, with branching properties differing from those of a true rotor. The model gives an $R_{4/2}$ energy ratio of 2.90. The $N=90$ transition region is a likely place for the identification of $X(5)$ nuclei, as several of the $N=90$ nuclei have yrast band energies and $B(E2)$ strengths reasonably close to those predicted by $X(5)$, and the first candidates for $X(5)$ nuclei, ^{152}Sm and ^{150}Nd [10, 11], were identified in this region.

Much of the recent interest in transitional phenomena originated with the careful remeasurement of weaker transitions in the otherwise-well-studied nucleus ^{152}Sm, in a pair of spectroscopy studies of ^{152}Sm populated in ^{152}Eu source decay [12, 13], leading to a variety of unexpected results and inspiring the subsequent theoretical work. Follow-up experiments, including recoil distance method (RDM), GRID, and Compton polarimetry measurements, have been performed as well [14, 15]. Since ^{152}Sm has been studied in such detail, it provides a benchmark "case study" for model comparisons.

Excellent agreement has been found between the properties of the neighboring isotone, ^{150}Nd, and the $X(5)$ predictions. This is partly evident from the yrast properties in Fig. 3. Much of the relevant data was provided by an experiment in which low-spin excited states of ^{150}Nd were populated in-beam through Coulomb excitation and their lifetimes were obtained by the RDM method at the Yale NYPD/SPEEDY setup [11]. This experiment yielded crucial γ-ray spectroscopy information as well, including branching ratio information for several of the low-lying 2^+ and 4^+ states.

The N=90 nucleus ^{156}Dy has been studied through γ-ray spectroscopy following β decay with a compact array of Clover and LEPS detectors at the Yale Moving Tape Collector [16, 17]. Collective nuclei such as ^{156}Dy populated at Q_β values of a few MeV produce several hundred identifiable transitions in their decay, many of them yielding overlapping or unresolved peaks in singles spectra, and so the previous generation of experiments, which relied heavily on singles data, faced a tremendous challenge in properly identifying γ-ray transitions and measuring their intensities. The high-statistics coincidence data obtained in the new experiment provided extensive information on the placements and intensities of transitions. Several outstanding issues, such as wildly conflicting claims in the literature for the intensities of the $2_2^+ \rightarrow 0_2^+$ and $4_2^+ \rightarrow 2_2^+$ transitions, were resolved, and it was found that much of the higher-lying literature level scheme (above ∼1200 keV) was in error. General agreement with the $X(5)$ predictions is good, with some discrepancies in quantitative detail. The $X(5)$ model predicts a characteristic branching pattern for the decay of members of the first-excited band to the yrast band, in which the spin-*descending* branches are highly suppressed. From the new data, as depicted in Fig. 4, it is seen that ^{156}Dy exhibits an extreme form of this suppression.

Interpretation of ^{156}Dy is hindered by a lack of lifetime information — at low-spins, lifetimes are known only for the yrast band members and the 2_3^+ level. Also, the literature $B(E2)$ value for the yrast $6^+ \rightarrow 4^+$ transition is surprizingly low (Fig. 3), below even the rotor prediction, suggesting a need for experimental verification. A combination RDM and Doppler shift attenuation method (DSAM) lifetime experiment in (HI, xn) has recently been carried out by a Köln/Legnaro/Yale collaboration at GASP and is currently being analyzed.

Experiments continuing the study of the N=90 region are underway, including at the Yale MTC, where a fast electronic scintillation timing (FEST) [18] lifetime system is now being used. It will be interesting to investigate candidate $X(5)$ nuclei in other regions as well [19], especially as new regions become accessible at radioactive beam facilities.

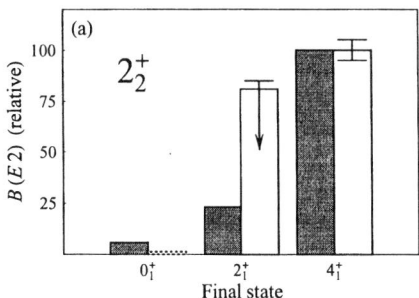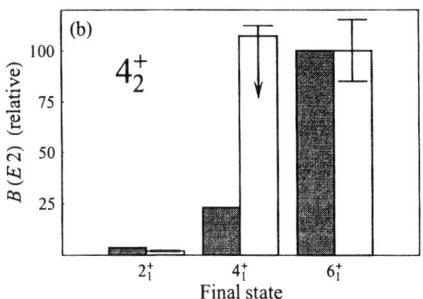

FIGURE 4. Comparison of relative $B(E2)$ branching strengths predicted in $X(5)$ (shaded bars) with the values observed in ^{156}Dy (unshaded bars). Transitions strengths are shown for branches from the (a) 2_2^+ and (b) 4_2^+ levels. The downward arrows on the measured $B(E2)$ values for the spin-unchanging transitions indicate the possibility of $M1$ contamination. The $2_2^+ \to 0_1^+$ transition in ^{156}Dy is extremely weak (in spite of its being the most energetically preferred branch), and only a limit on its intensity could be obtained.

ACKNOWLEDGMENTS

Valuable discussions with F. Iachello, N. V. Zamfir, R. F. Casten, and D. S. Brenner are gratefully acknowledged. I would also like to thank my collaborators at Yale and the Institut Laue-Langevin for the results mentioned here. This work was supported by the US DOE under grant DE-FG02-91ER-40609.

REFERENCES

1. Iachello, F., *Phys. Rev. Lett.*, **85**, 3580 (2000).
2. Iachello, F., *Phys. Rev. Lett.*, **87**, 052502 (2001).
3. Iachello, F. (in these proceedings).
4. Wilets, L., and Jean, M., *Phys. Rev.*, **102**, 788 (1956).
5. Caprio, M. A., *Phys. Rev. C*, **65**, 031304(R) (2002).
6. Casten, R. F., and Zamfir, N. V., *Phys. Rev. Lett.*, **85**, 3584 (2000).
7. Frank, A., Alonso, C. E., and Arias, J. M., *Phys. Rev. C*, **65**, 014301 (2001).
8. Zamfir, N. V., et al., *Phys. Rev. C*, **65**, 044325 (2002).
9. Börner, H. G., and Jolie, J., *J. Phys. G*, **19**, 217 (1993).
10. Casten, R. F., and Zamfir, N. V., *Phys. Rev. Lett.*, **87**, 052503 (2001).
11. Krücken, R., et al., *Phys. Rev. Lett.* (in press).
12. Casten, R. F., et al., *Phys. Rev. C*, **57**, R1553 (1998).
13. Zamfir, N. V., et al., *Phys. Rev. C*, **60**, 054312 (1999).
14. Klug, T., et al., *Phys. Lett. B*, **495**, 55 (2000).
15. Zamfir, N. V., et al., *Phys. Rev. C* (in press).
16. Caprio, M. A., et al., *Rom. J. Phys.*, **46**, 41 (2001).
17. Caprio, M. A., et al. (in preparation).
18. Mach, H., Gill, R. L., and Moszyński, M., *Nucl. Instrum. Methods*, **A280**, 49 (1989).
19. Brenner, D. S. (in these proceedings).

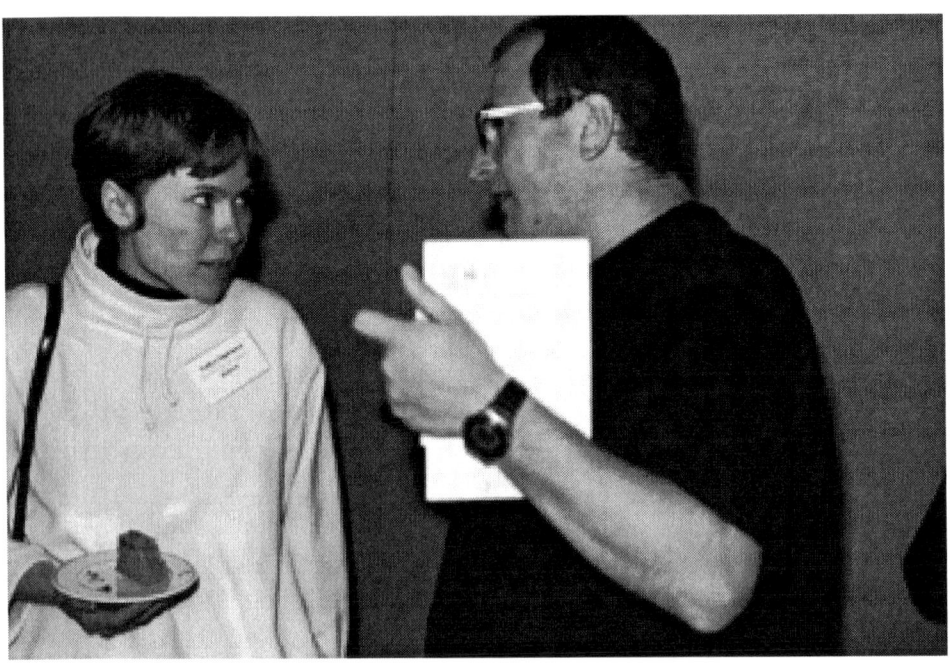

The Geometry of the IBM with Configuration Mixing

A. Frank[*†], O. Castaños[**], P. Van Isacker[‡] and E. Padilla[**]

[*]*Instituto de Ciencias Nucleares, UNAM, Apdo. Postal 70-543, 04510 México,D.F.*
[†]*Centro de Ciencias Físicas, UNAM, Apdo. Postal 139-B, 62251 Cuernavaca, Morelos, México.*
[**]*Instituto de Ciencias Nucleares, UNAM, Apdo. Postal 70-543, 04510 México, D.F.*
[‡]*Grand Accélerateur National d'Ions Lourds, Boite Postale 55027, F-14076 Caen Cedex 5, France*

Abstract.
We introduce a matrix formulation in a coherent state basis which can be applied to configuration mixing IBM Hamiltonians. The potential energy matrices give rise to eigenpotentials which may display the phenomenon of shape coexistence . Some examples are discussed.

INTRODUCTION

The application of the IBM and its extensions in nuclear structure has had great success and constitutes a paradigm for algebraic methods in physics [1, 2]. On the other hand current experiments are able to probe regions of the nuclear table where diverse kinds of complex phenomena can occur, such as that of shape coexistence and shape transitions in nuclei. There is now strong evidence for these phenomena in many regions of the nuclear chart[3]. They appear to be due to intruder configurations, interpreted as particle-hole excitations across major shells in the shell model[4, 5]. In the framework of the Interacting Boson Model, a configuration mixing Hamiltonian was first proposed in reference[6], where a simple Hamiltonian of the form

$$H = \hat{H}^N + \hat{H}^{N+2} + \hat{H}_{\text{mix}} \tag{1}$$

was introduced. In (1), $\hat{H}^N$ corresponds to the normal configuration with N bosons, while $\hat{N}^{N+2}$ corresponds to the intruder Hamiltonian associated with $2p - 2h$ excitations and is thus described by $N + 2$ bosons. The mixing Hamiltonian adopts the form given below, in equation (6), in the simplest approximation. The $4p4h$, $6p6h$, ..., excitations can be introduced through $N + 4$, $N + 6$..., boson configurations. More sophisticated treatments can be carried out, *e.g.*, in the $n - p$ version of othe IBM, but we shall restrict our considerations here to equation (1).

From a different perspective, although the algebraic analysis of the IBM constitutes a powerful and elegant methodology, it does not directly provide a geometric interpretation of the nuclear shape.

This can be done, however, through the introduction of coherent states[7, 8]. In particular, in ref.[8] a general study of shapes and stability of the IBM-1 Hamiltonian has been carried out, which gives rise, among other results, to a realization of Casten's

CP638, *Mapping the Triangle: International Conference on Nuclear Structure*
edited by A. Aprahamian, J. A. Cizewski, S. Pittel, and N. V. Zamfir

triangle in terms of two essential parameters, *i.e.*, any IBM-1 Hamiltonian corresponds to a concrete point in this space[8].

In a recent development, Iachello[9] and Casten and Zamfir[10] have explored the subject of coexistence, critical point symmetries and shape transitions from a fresh and intriguing point of view[11]. A different way to study these phenomena is in the context of Hamiltonian (1), *i.e.*, in the general situation incuding intruder configurations and their mixing. The question arises as to what is the appropriate geometric interpretation for such generalized boson Hamiltonians. In this paper we introduce a new matrix formulation, which gives rise to concepts such as potential energy matrices and eigenpotentials. We first illustrate our ideas by means of some schematic cases and then proceed to a physical example in the cadmium isotopes.

THE METHOD

The most general potential energy surface associated with the IBM-1 Hamiltonian is given by[7, 8]

$$\varepsilon_1(N, \beta, \gamma) = \frac{N \varepsilon \beta^2}{(1 + \beta^2)} + \frac{N(N-1)}{(1 + \beta^2)^2} \left(a_1 \beta^4 + a_2 \beta^3 \cos 3\gamma + a_3 \beta^2 + \frac{u_0}{2} \right) \quad (2)$$

where $\varepsilon \equiv \varepsilon_d - \varepsilon_s$ and the constant $N\varepsilon_s$ was substracted. The parameters a_1, a_2 and a_3 are given in terms of the strengths of the two boson interactions.

As mentioned above, the intruder configuration corresponds to $2p2h$ excitations described by two additional bosons. The energy surface can be obtained as in (2) but with a different set of parameters and with N replaced by $N + 2$. Finally, the mixing operator takes the form

$$H_{\text{mix}} = w_0 \, (s^{\dagger 2} + s^2) + \frac{w_2}{\sqrt{5}} \, (d^\dagger \cdot d^\dagger + \tilde{d} \cdot \tilde{d}). \quad (3)$$

We now define the potential energy matrix for this case:

$$E(\beta, \gamma, \Delta) = \begin{pmatrix} \varepsilon_1(N, \beta, \gamma) & w(\beta) \\ w(\beta) & \Delta + \varepsilon_2(N+2, \beta, \gamma) \end{pmatrix}, \quad (4)$$

where $\varepsilon_i(N, \beta, \gamma)$, $i = 1, 2$ denote the energy surfaces of the normal and intruder configurations, respectively. The parameter Δ essentially corresponds to the energy expended in rising two particles from the lower shell.

The non-diagonal matrix elements are given by

$$w(\beta) = \langle N+2, \beta, \gamma | H_{\text{mix}} | N, \beta, \gamma \rangle,$$

$$= \frac{\sqrt{(N+2)(N+1)}}{1 + \beta^2} \left\{ w_0 + \frac{w_2}{\sqrt{5}} \beta^2 \right\}. \quad (5)$$

The analysis of this potential energy matrix is equivalent to a two-level system, which can be written as follows

$$E(\beta,\gamma) = \varepsilon_1(N,\beta,\gamma) + \begin{pmatrix} 0 & w(\beta) \\ w(\beta) & 2g(\beta,\gamma) \end{pmatrix}, \tag{6}$$

where

$$g(\beta,\gamma) \equiv \frac{\varepsilon_2(N+2,\beta,\gamma) - \varepsilon_1(N,\beta,\gamma)}{2} + \frac{\Delta}{2}. \tag{7}$$

The solution of the eigenvalue problem leads to the eigen-potentials

$$E_\pm(\beta,\gamma) = \varepsilon_1(N,\beta,\gamma) + g(\beta,\gamma) \pm \sqrt{g^2 + w^2}, \tag{8}$$

and the corresponding eigenfunctions

$$X_+ = \frac{1}{\sqrt{2R}} \begin{pmatrix} \sqrt{R-1} \\ \sqrt{R+1} \end{pmatrix}, \tag{9}$$

$$X_- = \frac{1}{\sqrt{2R}} \begin{pmatrix} -\sqrt{R+1} \\ \sqrt{R-1} \end{pmatrix}, \tag{10}$$

with $R = \sqrt{1 + (w/g)^2}$.

MIXING BETWEEN DYNAMICAL SYMMETRIES

In this section, we describe the normal and intruder configurations in terms of the exact dynamical symmetry limits of the IBM and determine how the corresponding energy surfaces are affected when they are mixed by $2p2h$ excitations of the sort described in the previous section.

The energy surface can be rewritten in the form

$$\varepsilon(\beta,\gamma) = \frac{1}{(1+\beta^2)^2} \left(\beta^4 + r_1\beta^2(\beta^2+2) - r_2\beta^3 \cos 3\gamma \right), \tag{11}$$

where r_1 and r_2 are the independent parameters, which are functions of the IBM parameters [8].

In Figs. 1a-1c we show the eigenpotentials corresponding to schematic calculations where $\hat{H}^N$ and $\hat{H}^{N+2}$ are taken to correspond to each of the dynamical symmetries of the IBM. We note that shape coexistence can occur in all cases.

CADMIUM ISOTOPES

It has been established that the even cadmium nuclei with $A = 110$, 112, and 114 present strong mixing between their normal and $2p2h$ intruder configurations. In this Section we follow the work of Lehmann and Jolie [12], where they consider the normal and intruder

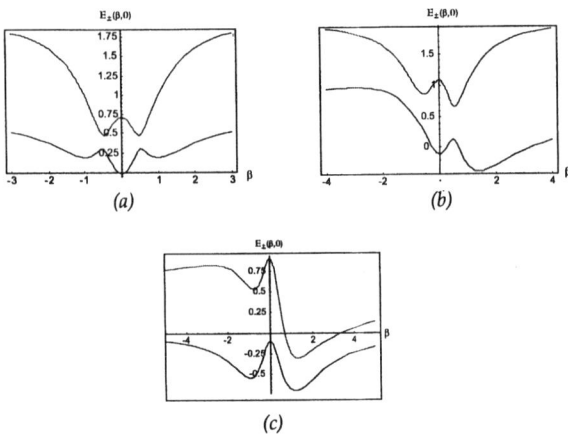

FIGURE 1. Eigenpotentials for (a) $U(5) - O(6)$, (b) $O(6) - SU(3)$ and (c) $U(5) - SU(3)$

configurations as associated to the U(5) and O(6) dynamical symmetries, respectively. Then the corresponding energy surfaces are given by the expressions

$$\varepsilon_1(N,\beta,\gamma) \;=\; \frac{N\varepsilon\beta^2}{(1+\beta^2)}, \tag{12}$$

$$\varepsilon_2(N+2,\beta,\gamma) \;=\; \frac{(N+2)\bar{\varepsilon}\beta^2}{(1+\beta^2)} + \frac{(N+2)(N+1)}{(1+\beta^2)^2}\bar{a}_3\beta^2, \tag{13}$$

The parameters for the calculation were taken without modification from the fits of ref. [12]. In Figs. 2a, 2b and 2c we show the results for ^{110}Cd, ^{112}Cd and ^{114}Cd, respectively. We see from our geometric analysis the way in which coexistence develops gradually and can be clearly observed for ^{114}Cd.

CONCLUSIONS

More detailed calculations are required to gauge our potential energy matrix formalism. However, we have shown here that the present framework can generalize in a systematic fashion the standard coherent-state analysis to more complex situations.

A careful geometrical analysis of the kind carried out in [8]would be quite interesting and should give rise toa corresponding generalization of Casten's triangle.

We believe that this approach can be applied to other algebraic systems, such as the vibronic models of molecules [2].

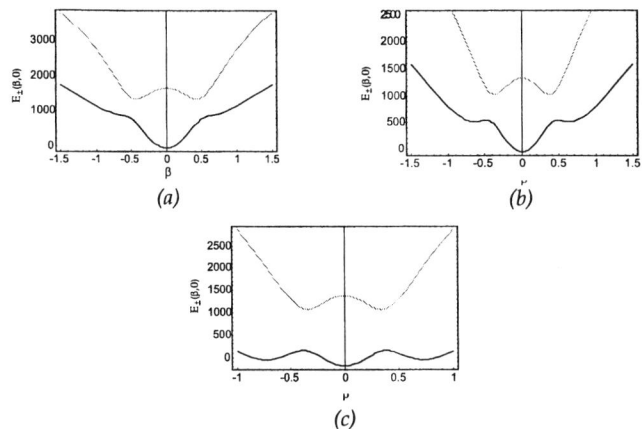

FIGURE 2. Eigenpotentials for (a) ^{110}Cd, (b) ^{112}Cd and (c) ^{114}Cd.

ACKNOWLEDGMENTS

It is a pleasure to dedicate this work to Rick Casten, whose fundamental contributions to nuclear physics have always inspired us. Rick's triangle may yet become a higher dimensional geometrical object, consistent with the many dimensions of his extraordinary personality.

This work was supported in part by ICN-UNAM under projects Conacyt 32635-E and 32397-E DGAPA-in106400.

REFERENCES

1. Iachello F. and Arima A., "The Interacting Boson Model". Cambridge University Press, Cambridge (1987).
2. Frank A. and Van Isacker P., "Algebraic Methods in Molecular and Nuclear Structure Physics", Wiley Interscience (1994).
3. Wood, J.L., Heyde, K., Nazariewicz W., Huyse, M., and Van Duppen, P., *Phys. Rep.* **215**, (1992) 101, and refs. therein.
4. Heyde, K., Van Isacker, P., Waroquier, M., Wood, J.L., and Meyer, R.A., *Phys. Rep.* **102** (1983) 291 and refs. therein.
5. Deleze, M., Drissi, S., Jolie, J., Kern, J., and Vederlet, J.P., *Nucl. Phys.* **A554** (1993) 1.
6. Duval, P.D., and Barrett, B.R., *Nucl. Phys.* **A376** (1982) 213.
7. Ginocchio, J.N., and Kirson, M.W., *Phys. Rev. Lett.* **44**, 1744 (1980) ; Dieperink, A., Scholten, O., and Iachello, F., *Phys. Rev. Lett.* **44**, 1747 (1980).
8. López Moreno E., and Castaños O., *Phys. Rev.* **C54**, 2374 (1996).
9. Iachello F., *Phys. Rev. Lett.* **85**, 3580 (2000); *Phys. Rev. Lett.* **87**, 052502 (2001).
10. Casten, R.F., and Zamfir, V., *Phys. Rev. Lett.* **85**, 3584 (2000); *Phys. Rev. Lett.* **87**, 052503 (2001).
11. Iachello, F., These proceedings.
12. Lehmann, H., Jolie, J., *Nucl. Phys.* **A578** (1995) 623.

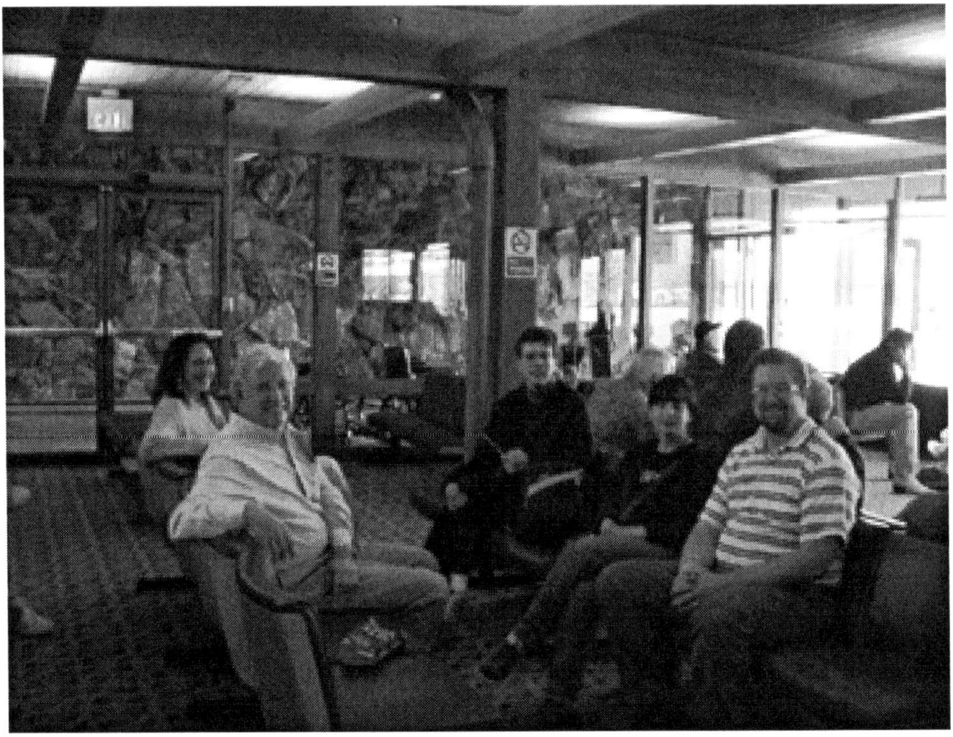

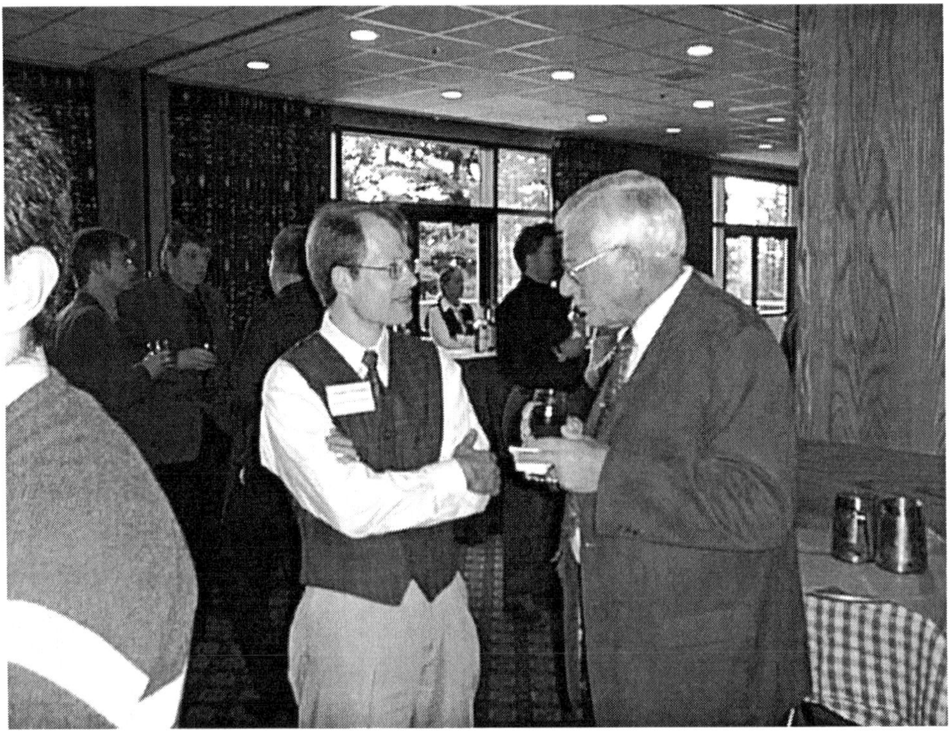

Shell-Model Description of Weakly Bound Nuclei

Nicolas Michel*, Witold Nazarewicz† and Marek Płoszajczak*

*GANIL, CEA/DSM-CNRS/IN2P3, BP 5027, F-14076 Caen Cedex 05, France
†Department of Physics and Astronomy, The University of Tennessee, Knoxville, Tennessee 37996
Physics Division, Oak Ridge National Laboratory, P.O. Box 2008, Oak Ridge, Tennessee 37831
Institute of Theoretical Physics, Warsaw University, ul. Hoża 69, PL-00681, Warsaw, Poland

Abstract. This work presents the first continuum shell-model study of weakly bound neutron-rich nuclei with *several* valence nucleons. For the single-particle basis, we take the complex-energy Berggren ensemble representing the bound single-particle states, narrow resonances (Gamow states), and the non-resonant continuum. In our calculations, we consider the shell-model Hamiltonian consisting of a one-body finite Woods-Saxon potential and a residual two-body interaction. The proposed Gamow shell model, which is a straightforward extension of the traditional shell model based on a discrete basis, is shown to be an excellent tool for the microscopic description of weakly bound systems.

INTRODUCTION

It is a great pleasure to dedicate this paper to Rick Casten on the occasion of this conference in his honor. The subject of our work, i.e., the shell-model treatment of exotic nuclei, is of particular relevance to Rick's anniversary. Firstly, the nuclear shell model has been a guiding principle behind most of his scientific work [1]. Secondly, the physics of exotic nuclei far from the beta stability line has been one of Rick's main scientific interests and activities [2].

In many respects, weakly bound nuclei are much more difficult to treat theoretically than well-bound systems [3]. The major theoretical difficulty and challenge is the treatment of the particle continuum. For weakly bound nuclei, the Fermi energy lies very close to zero, and the decay channels must be taken into account explicitly. As a result, many cherished approaches of nuclear theory such as the conventional shell model and the pairing theory must be modified. But there is also a splendid opportunity: the explicit coupling between bound states and continuum, and the presence of low-lying scattering states invite strong interplay and cross-fertilization between nuclear structure and reaction theory. Many methods developed by reaction theory can now be applied to structure aspects of loosely bound systems. And, of course, nuclear structure effects can clearly manifest themselves in reactions involving exotic nuclei.

The treatment of continuum states is an old problem which, in the context of excited states near or above the decay threshold, has been a playground of the continuum shell model (CSM) [4]. In the CSM, including the recently developed Shell Model Embedded in the Continuum (SMEC) [5], the scattering states and bound states are treated on an equal footing. So far, most applications of the CSM, including SMEC, have been used to describe limiting situations in which there is coupling to *one-nucleon decay*

CP638, *Mapping the Triangle: International Conference on Nuclear Structure*
edited by A. Aprahamian, J. A. Cizewski, S. Pittel, and N. V. Zamfir
© 2002 American Institute of Physics 0-7354-0093-8/02/$19.00

channels only. However, by allowing only one particle to be present in the continuum, it is impossible to apply the CSM to 'Borromean systems' for which A- and $(A$-$2)$-nucleon systems are particle-stable but the intermediate $(A$-$1)$-system is not. Various approaches, including the hyperspherical harmonic method or the coupled-channel approach, have been developed to study structure and reaction aspects of three-body weakly bound nuclei [6]. However, most of these models utilize the particle-core coupling which does not allow for the exact treatment of core excitations and the antisymmetrization between the core nucleons and the valence particles.

The reason for limiting oneself to only one particle in the continuum in the CSM has been two-fold. First, the number of scattering states needed to properly describe the underlying dynamics can easily go beyond the limit of what present computers can handle. Second, treating the continuum-continuum coupling, which is always present when two or more particles are scattered to unbound levels, is difficult. There have been only a few attempts to treat the multi-particle case and, unfortunately, the proposed numerical schemes, due to their complexity, have never been adopted in microscopic calculations involving multiconfiguration mixing. Consequently, an entirely different approach is called for. Recently, we formulated and tested the multiconfigurational shell model in the complete Berggren basis [7], the so-called Gamow Shell Model (GSM). (For application to two-particle resonant states, see also Ref. [8].) In this paper, GSM is applied to systems containing several valence neutrons.

BERGGREN BASIS AND COMPLETENESS RELATIONS

The Gamow states (sometimes called Siegert or resonant states) [9] are generalized eigenstates of the time-independent Schrödinger equation with complex energy eigenvalues $E = E_0 - i\Gamma/2$, where Γ stands for the decay width (which is zero for bound states). These states correspond to the poles of the S-matrix in the complex energy plane lying on or below the positive real axis; they are regular at the origin and satisfy purely outgoing asymptotics. Figure 1 shows the distribution of Gamow states in the complex momentum plane.

In the following, we consider the Gamow states of a one-body spherical finite potential. The single-particle (s.p.) basis of Gamow states must be completed by means of a set of non-resonant continuum states. This completeness relation [10, 11], reads:

$$\sum_n |\phi_{nj}\rangle\langle\tilde{\phi}_{nj}| + \frac{1}{\pi}\int_{L_+} |\phi_j(k)\rangle\langle\phi_j(k^*)|\,dk = 1, \qquad (1)$$

where ϕ_{nj} are the Gamow states carrying the s.p. angular momentum j, n stands for all the remaining quantum numbers labeling Gamow states, $\phi_j(k)$ are the modified scattering Gamow states, and the contour L_+ in the complex k-plane has to be chosen in such a way that all the poles in the discrete sum in Eq. (1) are contained in the domain between L_+ and the real energy axis (cf. Fig. 1). If $u_{nj}(r)$ stands for the radial part of ϕ_{nj}, then $\tilde{u}_{nj}(r) = u_{nj}(r)^*$ and $\tilde{\phi}_{nj} = \phi_{nj}(u \to \tilde{u})$. If the contour L_+ is chosen reasonably close to the real energy axis, the first term in (1) represents the contribution from bound states and narrow resonances, while the integral part accounts for the non-resonant

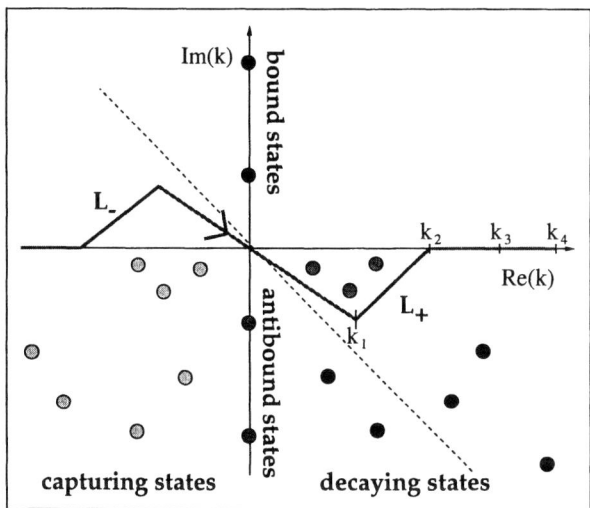

FIGURE 1. Schematic representation of the Gamow basis in the complex momentum plane. The Gamow states (i.e., poles of the S-matrix) are indicated by dots: bound states ($k = i\gamma$), antibound states ($k = -i\gamma$), decaying resonances ($k = \kappa - i\gamma$), and capturing resonances ($k = -\kappa - i\gamma$). The contour L represents the non-resonant continuum (see Eq. 1).

continuum. A number of completeness relations similar to (1) were studied by Lind [11]. In particular, if L_+ coincides with the real axis, one obtains the well-known Newton completeness relation involving bound and scattering states:

$$\sum_{n=\text{bound}} |\phi_{nj}\rangle\langle\tilde{\phi}_{nj}| + \frac{1}{\pi}\int_R |\phi_j(k)\rangle\langle\phi_j(k^*)|\,dk = 1. \tag{2}$$

There have been several applications of resonant states to problems involving continuum [12], but in most cases the so-called *pole expansion*, neglecting the contour integral in Eq. (1), was used. The importance of the contour contribution was investigated in Refs. [13, 14] in the context of the continuum RPA with separable particle-hole interactions where it was concluded that the non-resonant part must be accounted for if one aims at a quantitative description. This can be achieved by discretizing the integral in Eq. (1) [15]. In this work, we use the quadrature based on the four-point interpolation. The number of discretization points in the non-resonant scattering continuum is denoted by N_{cont} in the following.

In our study, Gamow states are determined using the generalized shooting method for bound states which requires an exterior complex scaling [12]. The antisymmetric two-particle wave functions $|\phi_{i_1}^{(1)} \phi_{i_2}^{(2)}\rangle_J$ are obtained in the usual way by coupling the s.p. wave functions of the considered bound, resonance, and scattering Gamow states. The radial integrals entering the Hamiltonian matrix elements were regularized separately by an appropriate choice of the angle of the exterior complex scaling. The resulting

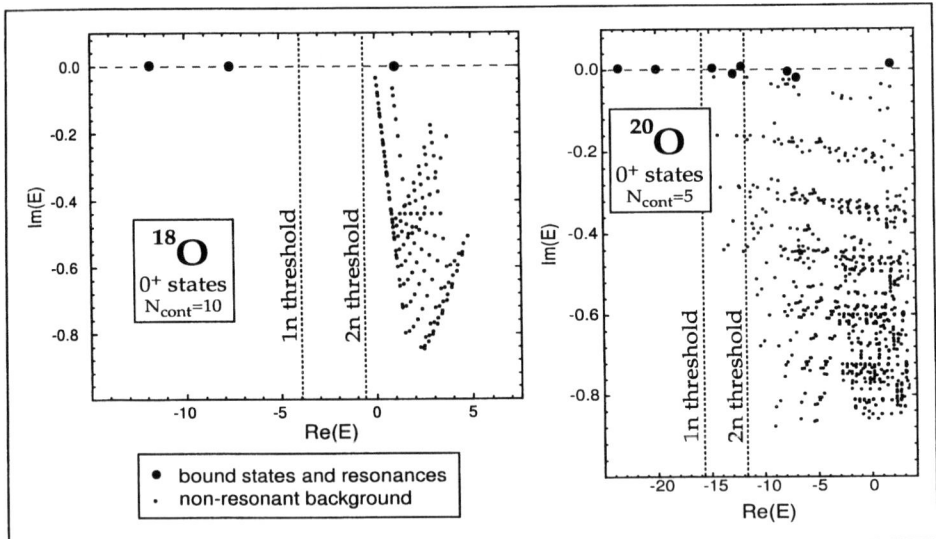

FIGURE 2. Distribution of GSM states in the complex energy plane calculated for ^{18}O (left; (two-particle case) and ^{20}O (right; four particle case). The bound and resonant states lie close to the real energy axis; they are marked by large dots. The remaining states represent the non-resonant continuum. One- and two-neutron thresholds are indicated.

(complex symmetrix) Hamiltonian matrix can be diagonalized using standard methods [7].

SELECTION OF RESONANT STATES

The crucial problem pertaining to the interpretation of the CSM results is the selection of states associated with resonant excitations of the system. Bound states can be clearly identified, because the imaginary part of their energy must be zero. No equally simple criterion exists for resonance states. Fortunately, the coupling between scattering states and resonant states is usually weak; hence, one can determine the resonances using the following two-step procedure. In the first step, the shell-model Hamiltonian is diagonalized in both (i) the full space including the contour, and (ii) the subspace of Gamow states (pole expansion). In the second step, one identifies the eigenstates of (i) which have the largest overlap with those of the second diagonalization. As a representative example, Fig. 2 shows the distribution of shell-model energies in the complex energy plane calculated for ^{18}O and ^{20}O. In both cases the bound states and resonances lie at, or very close to, the real energy axis. The remaining states represent the non-resonant continuum; their distribution varies with N_{cont}. On the other hand, bound and resonant states are stable with respect to N_{cont}. All states below the one-neutron threshold (marked as 1n) are bound. The states above the two-neutron (2n) threshold are unstable to one- and two-neutron emission. As seen in Fig. 2, for ^{18}O, the eigenstates forming the non-resonant background align along regular trajectories in the complex energy plane [8] that reflect the choice of the contour L_+ in the momentum space. In the four-particle case, ^{20}O, this regularity is practically lost.

EXAMPLES OF CALCULATIONS

In the following exploratory GSM calculations, we shall consider two cases: (i) $^{18-22}$O with the inert ^{16}O core and active neutrons in the sd shell, and (ii) $^{6-9}$He with the inert ^{4}He core and active neutrons in the p shell. Our aim is not to give the precise description of these nuclei (for this, one would need a realistic Hamiltonian and a larger configuration space), but rather to illustrate the method, its basic ingredients, and underlying features. The details of these calculations will be published elsewhere [16].

The "Oxygen" case

The s.p. basis was generated by a Woods-Saxon (WS) potential with the radius R_0=3.05 fm, the surface diffuseness d=0.65 fm, the potential depth U_0=−55.8 MeV, and the strength of the spin-orbit term U_{so}=6.06 MeV. With this choice of parameters, the single particle $0d_{5/2}$ and $1s_{1/2}$ states are bound with s.p. energies −4.14 MeV and −3.27 MeV, respectively, and $0d_{3/2}$ is a resonance with the s.p. energy 0.9–i0.048 MeV. Energies of these s.p. states are close to the s.p. states of ^{17}O. The completeness relation requires taking the $s_{1/2}, d_{5/2}$, and $d_{3/2}$ non-resonant continuums. For the $1s_{1/2}$ and $0d_{5/2}$ bound states, their non-resonant continuums can be chosen along the real momentum axis. Since, to the first order, the inclusion of these continuums should only result in the renormalization of the effective interaction, they are ignored for the purpose of the present exercise whose main focus is the neutron emission. On the contrary, $0d_{3/2}$ is a resonance state, so the associated contour has to be complex to produce the correct energy width. The number of discretization points was N_{cont}=5. For the residual interaction, we assumed the δ-force with the strength V_0=-350 MeV fm^3.

Figure 3 displays binding energies of $^{18-22}$O calculated in GSM. In spite of very simple interaction and rather limited configuration space, the agreement with experimental data is good. Also quite reasonable is the predicted spectrum of ^{18}O (Fig. 3, inset). In particular, the states which are predicted to appear above the 1n threshold acquire a positive width [7].

The "Helium" case

A description of the neutron-rich helium isotopes, including Borromean nuclei ^{6,8}He, is a challenge for the GSM. ^{4}He is a well-bound system with the one-neutron emission threshold at 20.58 MeV. On the contrary, the nucleus ^{5}He, with one neutron in the p shell, is unstable with respect to the neutron emission. Indeed, the $J^{\pi} = 3/2_1^-$ ground

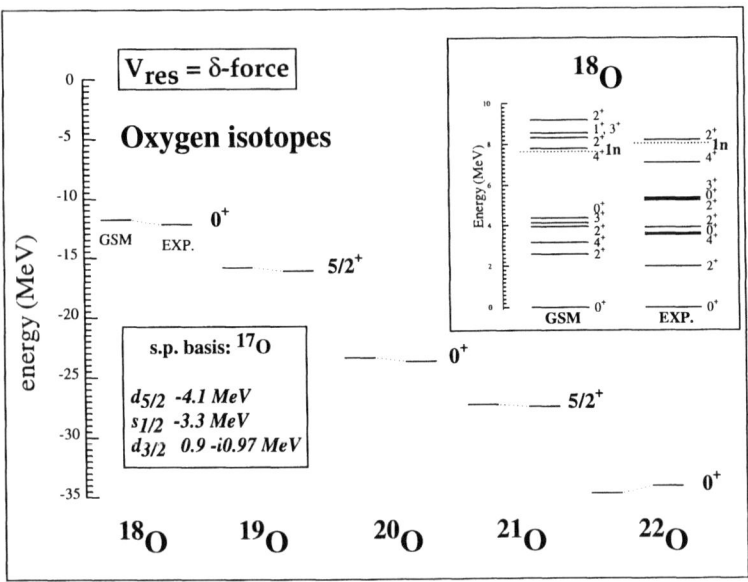

FIGURE 3. Experimental (EXP) and predicted(GSM) binding energies of $^{18-22}$O. The inset shows the calculated two-neutron spectrum of ^{18}O compared to experimental data.

state of ^{5}He lies 890 keV above the neutron emission threshold and its neutron width is large, Γ=600 keV. The first excited state, $1/2_1^-$, is a very broad resonance (Γ=4 MeV) that lies 4.89 MeV above the threshold. ^{6}He, on the contrary, is bound with the two-neutron emission threshold at 0.98 MeV and one-neutron emission threshold at 1.87 MeV. The first excited state 2_1^+ at 1.8 MeV in ^{6}He is neutron unstable with a width Γ=113 keV. In our GSM calculations, the states in ^{5}He are viewed as one-neutron resonances outside of the ^{4}He core. A good fit to $3/2_1^-$ and $1/2_1^-$ states in ^{5}He is obtained by taking the WS potential with R_0 = 2 fm, d=0.65 fm, U_0 = -47 MeV, and U_{so} = 7.5 MeV. With this potential, one finds the single-neutron resonances $p_{3/2}$ and $p_{1/2}$ at E=0.745–i0.32 MeV and E=2.13–i2.94 MeV, respectively. The number of discretization points was N_{cont}=5 for both partial waves. With this choice of N_{cont}=5 the accuracy of calculations (i.e., the size of the false width) was of the order of 100 keV. For the residual interaction, we assumed the surface-delta interaction (SDI) with the strength V_0=-1670 MeV fm^3 and radius R=2 fm.

Our calculations reproduce the most important feature of ^{6}He and ^{8}He: *the ground state is particle-bound, despite the fact that all the basis states lie in the continuum.* In spite of a very crude Hamiltonian, rather limited configuration space, etc., the calculated ground state energies reproduce surprisingly well the experimental data. The neutron separation energy anomaly (i.e., the *increase* of the neutron separation energy when going from ^{6}He to ^{8}He) is reproduced. Also, the energies of excited 2_1^+ states are in fair agreement with the data. As discussed in Refs. [7, 16], the contribution from the non-resonant continuum to the ground state wave functions of Borromean systems ^{6}He

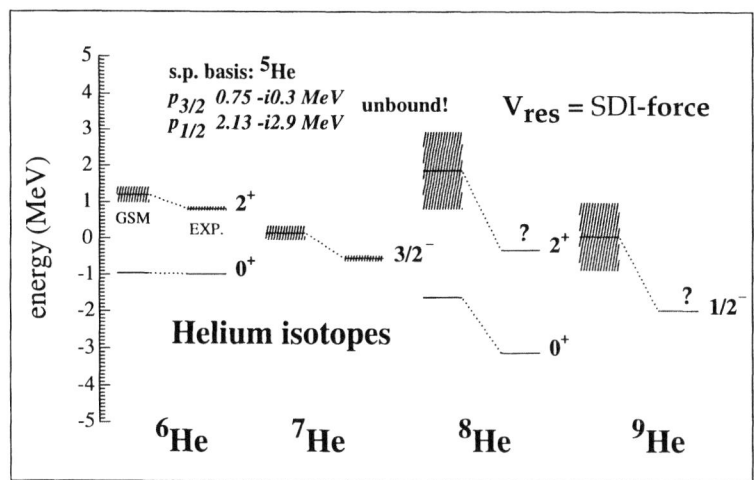

FIGURE 4. Experimental (EXP) and predicted(GSM) binding energies of $^{6-9}$He as well as energies of $J^\pi=2^+$ states in ^{6,8}He. The resonance widths are indicated by shading.

and ^{8}He is large. As seen in Table 1, the contributions from one and two particles in the non-resonant continuum, $L_+^{(1)}$ and $L_+^{(2)}$, dominate the structure of the ground-state wave function of ^{8}He.

TABLE 1. Squared complex amplitudes of different configurations in 0_1^+ and 2_1^+ states of ^{8}He. The sum of squared amplitudes of all Slater determinants including n particles in the non-resonant continuum (n=1,2,3,4) are denoted by $L_+^{(n)}$.

c^2	0_1^+	2_1^+
$0p_{3/2}^4$	0.30–i1.32	0
$0p_{3/2}^3 0p_{1/2}^1$	0	0.81–i0.77
$0p_{3/2}^2 0p_{1/2}^2$	–0.06–i0.16	–0.01–i0.03
$L_+^{(1)}$	1.60+i1.07	0.45 +i0.66
$L_+^{(2)}$	–0.73+i0.63	–0.21+i0.16
$L_+^{(3)}$	–0.13–i0.20	–0.03–i0.24
$L_+^{(4)}$	–0.02–i0.01	~0

CONCLUSIONS

In conclusion, the complex energy Berggren ensemble is applied for the first time in shell-model calculations for many-neutron states near the particle-emission threshold. In addition to the successful inclusion of the continuum-continuum coupling, we succeeded in solving another principal problem of the GSM, i.e., the treatment of the non-resonant part of the continuum. Another problem which has been solved in our work is the selection of resonance states. As a result of the GSM diagonalization, one obtains a multitude of states corresponding to the many-body continuum, some being resonances and some representing the non-resonant background. Our work offers a simple prescription on how to identify the resonance states.

The results of our pilot calculations are very encouraging. It is seen that the contribution from the non-resonant continuum is important, especially for bound and near-threshold states. In particular, pairing correlations due to the continuum-continuum scattering can bind the ground states of ^{6,8}He with a completely unbound basis provided by the s.p. resonances of ^{5}He. In all cases considered, calculations yield neutron resonances above the calculated neutron threshold – a property that is not guaranteed *a priori* by the formalism. Other applications of GSM, including the case of open proton *and* neutron shells, are in progress.

ACKNOWLEDGMENTS

This work was supported in part by the U.S. Department of Energy under Contract Nos. DE-FG02-96ER40963 (University of Tennessee) and DE-AC05-00OR22725 with UT-Battelle, LLC (Oak Ridge National Laboratory).

REFERENCES

1. Casten, R.F. *Nuclear Structure from a Simple Perspective* (Oxford University Press, Oxford 1990).
2. Nazarewicz, W. and Casten, R.F., *Nucl. Phys.* A **682**, 295c (2001); Casten, R.F. and Sherrill, B.M., *Prog. Part. Nucl. Phys.* **45**, S171 (2000).
3. Dobaczewski, J. and Nazarewicz, W., *Phil. Trans. R. Soc. Lond.* A **356**, 2007 (1998).
4. Fano, U., *Phys. Rev.* **124**, 1866 (1961); Mahaux, C. and Weidenmüller, H.A., *Shell-model approach to nuclear reactions* (North-Holland, Amsterdam, 1969).
5. Bennaceur, K. *et al.*, *Nucl. Phys.* A **651**, 289 (1999); ibid. A **671**, 203 (2000); Bennaceur, K. *et al.*, *Phys. Lett.* **B488**, 75 (2000).
6. Danilin, B.V. *et al.*, *Nucl. Phys.* A **632**, 383 (1998); Nielsen, E. *et al.*, *Phys. Rep.* **347**, 373 (2001); Esbensen, H. and Bertsch, G.F., *Phys. Rev.* C **59**, 3240 (1999).
7. Michel, N., Nazarewicz, W., Płoszajczak, M. and Bennaceur, K., *Phys. Rev. Lett.* (2002), in press.
8. Betan, R.I. *et al.*, *Phys. Rev. Lett.* (2002), in press.
9. Gamow, G., *Z. Phys.* **51**, 204 (1928); Siegert, A.F.J., *Phys. Rev.* **56**, 750 (1939).
10. Berggren, T., *Nucl. Phys.* A **109**, 265 (1968); *Nucl. Phys.* A **389**, 261 (1982).
11. Lind, P., *Phys. Rev.* C **47**, 1903 (1993).
12. Vertse, T., Curutchet, P. and Liotta, R.J., *Lecture Notes in Physics* 325 (Springer Verlag, Berlin 1987), p. 179; Dussel, G.G. *et al.*, *Phys. Rev.* C **46**, 558 (1992).
13. Vertse, T., Liotta, R.J. and Maglione, E., *Nucl. Phys.* A **584**, 13 (1995).
14. Lind, P. *et al.*, *Z. Phys.* A **347**, 231 (1994).
15. Liotta, R.J. *et al.*, *Phys. Lett.* B **367**, 1 (1996); Vertse, T. *et al.*, *Phys. Rev.* C **57**, 3089 (1998).
16. Michel, N., Nazarewicz, W., Płoszajczak , M., and Okołowicz, J., to be published.

Electromagnetic Transitions in Isospin Multiplets

Paul D. Cottle

Department of Physics, Florida State University
Tallahassee, FL 32306-4350

Abstract. The comparison of the strengths of corresponding *E2* transitions in $T=1$ isospin multiplets has been an important tool for understanding the isovector dependence of *E2* transitions for many years. The development of radioactive beams and more sensitive detector technology has made this method much more powerful so that more $T=1$ multiplets can be studied in detail and $T=2$ multiplets can be studied for the first time.

In nuclear structure physics, isospin is a powerful symmetry that provides critical insights regarding the complex proton-neutron system. Systematic measurements of the properties of isobaric multiplets – including mirror nuclei - are important for fully exploiting the isospin symmetry to understand the behavior of nuclei far from stability as well as those closer to and at the stability line. The dramatic development of experimental techniques involving beams of radioactive nuclei is pushing the isospin frontier forward, allowing a repertoire of nuclear structure measurements to be performed on nuclei that could not be reached with stable beams and targets. In addition, dramatic advances in the resolving power of γ-ray detectors have provided a means to bring great improvements in precision to the measurement of nuclei that can be studied with stable beams and targets.

Historically, one fruitful venue for studying the isospin dependence of nuclear behavior in mirror nuclei and isobaric multiplets has been the *sd* shell, where detailed measurements of energy spectra and electromagnetic transitions in isospin multiplets have been possible because the line of stable nuclei is at or near the $N=Z$ line. Bernstein, Brown and Madsen [1] demonstrated this by using experimental results for electromagnetic transition rates in $T = \frac{1}{2}$ and $T = 1$ multiplets to extract the isospin dependence of the transition matrix elements for isobaric multiplets and – in the case of $T=1$ multiplets - to test isospin purity. They also used the electromagnetic data on mirror nuclei to point out that a neutron multipole matrix element extracted for a particular transition in one nucleus using a hadronic probe could be checked by measuring the proton multipole matrix element in the corresponding transition of the mirror nucleus using an electromagnetic probe.

Of course, the range of isospin values that Bernstein, Brown and Madsen could address was constrained because experiments were limited to stable beams and targets. Coulomb excitation and hadronic scattering experiments could only be performed on stable nuclei, while measurements of lifetimes of excited states in some very proton-

CP638, *Mapping the Triangle: International Conference on Nuclear Structure*
edited by A. Aprahamian, J. A. Cizewski, S. Pittel, and N. V. Zamfir
© 2002 American Institute of Physics 0-7354-0093-8/02/$19.00

rich nuclei could not be performed. Advances in radioactive beam technology have expanded the number of pairs of mirror nuclei that can be measured.

Our experimental contributions to the study of transitions in isobaric multiplets include measurements of $B(E2;0_{gs}^+ \rightarrow 2_1^+)$ values in ^{38}Ca [2], ^{18}Ne [3], ^{26}Si [4] and ^{32}Ar [5] by means of the scattering of beams of these radioactive isotopes from gold targets (a technique often called "intermediate energy Coulomb excitation" and discussed in [6]) at the National Superconducting Cyclotron Laboratory (NSCL); the excitation of the 2_1^+ states of ^{18}Ne [7] and ^{32}Si [5] via inelastic proton scattering in inverse kinematics, also at the NSCL; and a measurement of the lifetime of the 2_1^+ state of ^{18}Ne using the Doppler Shift Attenuation Method (DSAM) at the Florida State University Superconducting Linear Accelerator Laboratory using the lab's array of large volume "clover" Ge γ-ray detectors [8].

In the convention given by Bernstein, Brown and Madsen [1], the proton multipole transition matrix element, M_p, which for a $0^+ \rightarrow 2^+$ transition can be written as $M_p = [B(E2;0^+ \rightarrow 2^+)]^{1/2}$, is written in the isospin representation as

$$M_p = 1/2[M_0 - M_1] , \quad (1)$$

where M_0 is the isoscalar multipole matrix element and M_1 is the isovector multipole matrix element. The former is constant across an isospin multiplet, while the latter obeys the relation

$$M_1(T_z) = (T_z' / T_z) M_1(T_z). \quad (2)$$

The isoscalar and isovector matrix elements can be determined from the strengths of transitions in the members of an isobaric multiplet by fitting a straight line to the M_p values for the members of the multiplet. For a $T=1$ multiplet, this can include the $0_{gs}^+ \rightarrow 2_1^+$ transitions in the $T_z = \pm 1$ nuclei and the $0_{T=1}^+ \rightarrow 2_{T=1}^+$ transition in the $T_z = 0$ nucleus. Examples of such fits for the $A=18, 22, 26, 30, 34$ and 38 multiplets are shown in Fig. 1. This analysis assumes that the M_p values for a given multiplet all have the same signs (positive or negative). For the present discussion, we assume they are positive (although the results of the fits are – with opposite signs – correct if the signs of the M_p values are negative).

The plots in Fig. 1 in which the point corresponding to the $T_z=0$ nucleus falls off the fitted line highlight a different issue – the isospin purity of the $0_{T=1}^+$ and $2_{T=1}^+$ states in the multiplet. As pointed out in Ref. 1, isospin symmetry dictates that the M_0 value obtained by adding the M_p results for the $T_z = \pm 1$ isotopes is equal to the isoscalar matrix element obtained by doubling the M_p measured for the $T_z=0$ nucleus (see equations 1 and 2). This analysis is shown in Fig. 2. The experimental points for three masses ($A=34, 38$ and 42) have experimental uncertainties that do not overlap. This suggests isospin symmetry breaking to a degree that would be unexpected. The greatest experimental uncertainties for these masses are in the $T_z=0$ nuclei, and Fig. 2 should provide motivation for improving the precision of the measurements of $B(E2;0_{T=1}^+ \rightarrow 2_{T=1}^+)$ in these odd-odd nuclei.

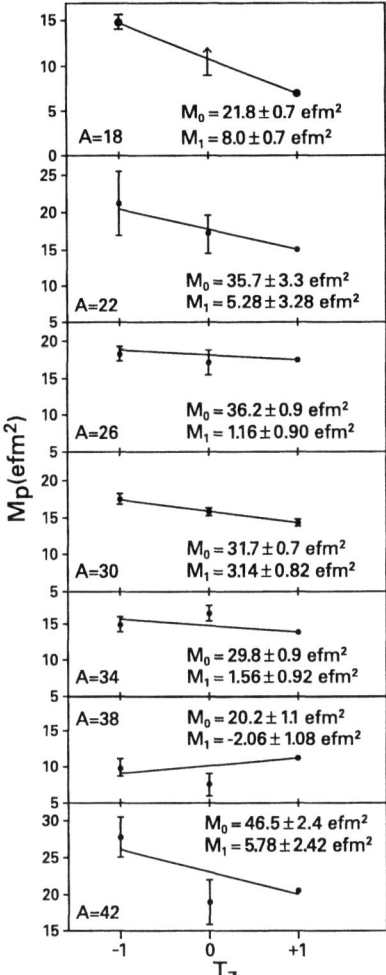

FIGURE 1. Fits to determine M_0 and M_1 from electromagnetic matrix elements for $0^+ \to 2^+$ transitions in $T=1$ isobaric multiplets. Data are taken from Refs. 2, 4, 8 and 9.

Our measurement of $B(E2; 0_{gs}^+ \to 2_1^+)$ in the $T_z = -2$ nucleus ^{32}Ar [5], when taken with previously existing data on the corresponding quantity in ^{32}Si [10] (in a pairing we have dubbed "distant mirror nuclei") allowed us to extract the isospin dependence of M_p in the $A=32$ $T=2$ isobaric multiplet. This was the first time a $T=2$ multiplet was studied in this way. The two experimental results ($|M_p|=16.3(21)\,fm^2$ and $10.6(15)\,fm^2$ for ^{32}Ar and ^{32}Si, respectively) give $|M_0|=26.9(26)\,fm^2$ and $|M_1(T_z=+2)|=5.7(26)\,fm^2$. The M_p values for both nuclei must have the same sign (as must M_0 and $M_1(T_z=+2)$) because otherwise the Endt limit [11] for isovector $E2$ transitions would be violated.

While data on $0_{T=1}^+ \to 2_{T=1}^+$ transitions are available for $T_z=0$ nuclei, it seems unlikely that data on $0_{T=2}^+ \to 2_{T=2}^+$ transitions will be available for $|T_z|<2$ isobars in the foreseeable future. To illustrate, the $0_{T=2}^+$ states in both $|T_z|=1$ $A=32$ nuclei occur above 5 MeV (in the $T_z=0$ member of the multiplet, the $0_{T=2}^+$ state occurs at 12 MeV). Hence, an expansion of the database on $E2$ transitions in $T=2$ isobaric multiplets will consist entirely of measurements of mirror $|T_z|=2$ nuclei. The possibilities for measuring $E2$ transitions in $T=3$ mirror pairs seem even more limited.

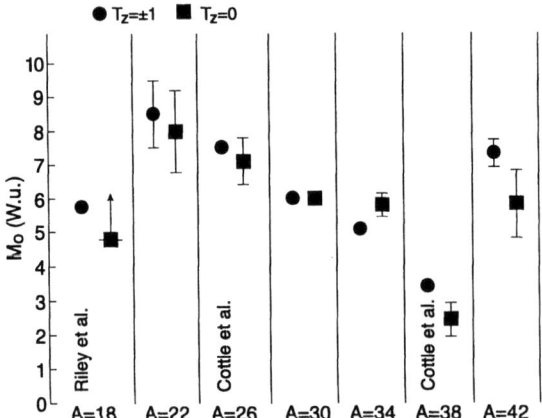

FIGURE 2. A comparison of M_0 values extracted from comparison of M_p values for $0_{gs}^+ \rightarrow 2_1^+$ transitions in $T_z=\pm 1$ nuclei to the M_0 values taken from transitions between $T=1$ states in $T_z=0$ nuclei. Data are taken from Refs. 2, 4, 8 and 9.

Another way to look at the M_p results for ^{32}Ar and ^{32}Si emphasizes the collective aspects of the $2_{T=1}^+$ states in these isobars. Isospin symmetry requires that the neutron multipole matrix element M_n for a transition in one nucleus be equal to M_p for the corresponding transition in the mirror nucleus [1]. The ratio M_n/M_p can provide useful insights regarding the collective behavior reflected in electromagnetic transitions. In even-even nuclei in which both proton and neutron shells are open, M_n/M_p for the $0_{gs}^+ \rightarrow 2_1^+$ transition is generally equal to the hydrodynamical result N/Z [12]. Nuclei with closed shells systematically deviate from the hydrodynamical value. If we take M_n for ^{32}Si to be equal to the ^{32}Ar M_p result described here, then for ^{32}Si $M_n/M_p=1.53(29)$, which is consistent with $N/Z=1.29$.

Equivalent information on ^{32}Si can be obtained by the use of a second (in addition to the electromagnetic measurement) experimental probe which has sensitivity to the neutron transition density, such as the inelastic scattering of protons. The data on the $0_{gs}^+ \rightarrow 2_1^+$ transition obtained with a 42 MeV/nucleon beam of this nucleus at the NSCL was analyzed by means of the distorted wave Born approximation (DWBA) using the computer code ECIS88 [13]. A standard vibrational form factor was used, as were the optical model parameters obtained in a study of the ^{36}S(p,p') reaction at 28 MeV by Hogenbirk *et al.* [14] (the parameter set labeled "OM" in that reference was used). Using this procedure, we obtained the strength parameter $\beta_2=0.28_{-0.03}^{+0.05}$. The measured and calculated angular distributions are shown in Fig. 3. Using the prescription of Ref. 15 (as described in Ref. 5), we obtained $M_n/M_p=1.63_{-0.26}^{+0.27}$, which is consistent with the result from the mirror nucleus analysis.

The results described here offer examples of how developments in the technologies of radioactive beams and γ-ray detection are advancing the isospin frontier, making it possible to measure pairs of mirror nuclei that are more and more distant. A rapid expansion in the database for $T=2$ mirror pairs is likely, and perhaps some $T=3$ pairs can be measured as well.

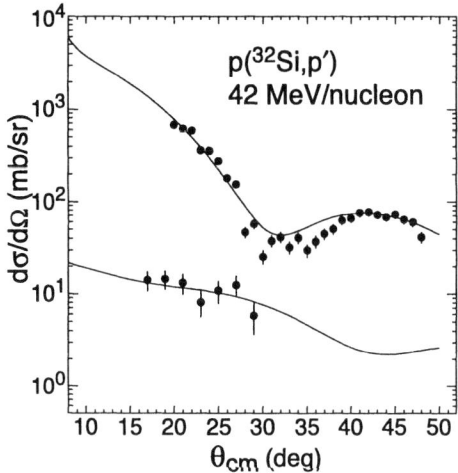

FIGURE 3. Experimental and calculated angular distributions for elastic scattering and inelastic scattering to the 2_1^+ state in the ^{32}Si(p,p') experiment [5].

ACKNOWLEDGMENTS

The work described here was supported by the NSF and the State of Florida.

REFERENCES

1. Bernstein, A.M., Brown, V.R., and Madsen, V.A. Madsen, *Phys. Rev. Lett.* **42**, 425 (1979).
2. Cottle, P.D. *et al.*, *Phys. Rev.* **C60**, 031301(R) (1999).
3. Riley, L.A. *et al.*, *Phys. Rev.* **C62**, 034306 (2000).
4. Cottle, P.D. *et al.*, *Phys. Rev.* **C64**, 057304 (2001).
5. Cottle, P.D. *et al.*, *Phys. Rev. Lett.* **88**, 172502 (2002).
6. Glasmacher, T., *Annu. Rev. Nucl. Part. Sci.* **48**, 1 (1998).
7. Riley, L.A. *et al.*, *Phys. Rev. Lett.* **82**, 4196 (1999).
8. Riley, L.A. *et al.* (to be published).
9. Endt, P.M., *Nucl. Phys.* **A521**, 1 (1990).
10. Ibbotson, R.W. *et al.*, *Phys. Rev. Lett.* **80**, 2081 (1998).
11. Endt, P.M., *At. Data Nucl. Data Tables* **55**, 171 (1993).
12. Bernstein, A.M., Brown, V.R., and Madsen, V.A. Madsen, *Phys. Lett.* **103B**, 255 (1981).
13. Raynal, J., *Phys. Rev.* **C23**, 2571 (1981).
14. Hogenbirk, A. *et al.*, *Nucl. Phys.* **A516**, 205 (1990).
15. Bernstein, A.M., Brown, V.R., and Madsen, V.A. Madsen, *Comments Nucl. Part. Phys.* **11**, 203 (1981).

In-Beam Thin Target Fragmentation of ^{197}Au.

P.H. Regan*, S.D. Langdown*, K. Gladnishki†, S.M. Mullins**,
Zs. Podolyák‡, M. Benatar§, E. Gueorguieva**, P. Kwinana**, J.J. Lawrie**,
G.K. Mabala§, S. Mukherjee**, R.T. Newman**, C.J. Pearson‡,
J.F. Sharpey-Schafer**, F.D. Smit**, S.M. Vincent‡ and A.D. Yamamoto*

*WNSL, Yale University, P.O. Box 208124, 272 Whitney Avenue, New Haven CT 06520-8124, USA
and Dept. of Physics, University of Surrey, Guildford, GU2 7XH, UK
†Dept. of Physics, University of Surrey, Guildford, GU2 7XH, UK and Faculty of Physics,
University of Sofia, 1164 Sofia, Bulgaria
**iThemba LABS, PO Box 722, Somerset West, 7129, South Africa
‡Dept. of Physics, University of Surrey, Guildford, GU2 7XH, UK
§Dept. of Physics, University of Cape Town, Private Bag, Rondebosch, 7701 and iThemba LABS,
PO Box 722, Somerset West, 7129, South Africa

Abstract. This paper reports on a test experiment to investigate the viability of using in-beam gamma-ray spectroscopy with projectile fragmentation reactions to explore the structure of heavy, neutron-rich nuclei. The experiment used a 'thin target' fragmentation reaction with a 30 MeV per nucleon ^{12}C beam on a 30 mg/cm^2 self-supporting gold target. The initial analysis has identified a number of previously known discrete cascades in a variety of in-beam reaction products. The decay of products has also been investigated and an initial comparison of their relative yields with predictions from the EPAX parameterisation begun. The relevance and future viability of this method as a complimentary technique to fragmentation-isomer studies to investigate the near-yrast properties of heavy, neutron-rich nuclei is discussed.

INTRODUCTION

The combination of high-efficiency arrays of hyperpure germanium detectors coupled with projectile fragment separators has led to a number of exciting new spectroscopic results following measurements of decays from isomeric states. Highlights of this type of work include (a) studies using intermediate energy ($\sim$60 MeV/u) beams at GANIL pushing out towards the proton drip line [1] and for neutron-rich nuclei approaching the doubly-magic nucleus $^{78}_{28}$Ni$_{50}$ [2]; and (b) work at GSI providing the first spectroscopy of heavy, neutron-rich nuclei between Erbium (Z=68) and Uranium (Z=92) following the fragmentation of relativistic ($\sim$1 GeV/u) beams of ^{238}U [3] and ^{208}Pb [4].

Although these experimental programs have been very successful in exploring new areas of the nuclear chart, the restriction of requiring an isomeric state that decays within a specific temporal range of tens of nanoseconds up to a few ms makes this technique somewhat restrictive. Coloumb excitation using secondary fragmentation beams has been pioneered by groups at both Riken and MSU [5]. More recently, there have been reports of 'in-beam' gamma-ray spectroscopic studies using both single and double-step fragmentation [6]. These experiments are however limited by the singles event rates in the gamma-ray detectors and current technology precludes investigations of nuclei with

CP638, *Mapping the Triangle: International Conference on Nuclear Structure*
edited by A. Aprahamian, J. A. Cizewski, S. Pittel, and N. V. Zamfir
© 2002 American Institute of Physics 0-7354-0093-8/02/$19.00

A>100 using this method.

This paper reports on the preliminary analysis of a 'thin-target fragmentation' experiment aimed at investigating the feasibility of performing in-beam fragmentation studies, *without* the use of a fragment separator. The ultimate aim of this work is to develop a technique which allows detailed spectroscopy of heavy, neutron-rich nuclei using specific experimental tags such as decays from isomeric states and/or characteristic X rays following internal conversion decay.

EXPERIMENTAL DETAILS, RESULTS AND FORWARD LOOK.

The philosophy behind the feasibility experiment was the use of a non-inverse reaction in which, assuming a simple two-body reaction for the fragmentation process (admittedly a rather gross assumption), the much heavier target-like fragmentation reaction products are emitted almost perpendicular to the beam direction in the laboratory frame. As such they should should stop in the target within a few picoseconds of the reaction and therefore, states populated with apparent lifetimes greater than this should be observed with no Doppler shift and associated loss in resolution due to Doppler broadening effects. (Note that a similar experiment to study decays in light, neutron-rich nuclei has recently been reported by the Berkeley group [7]).

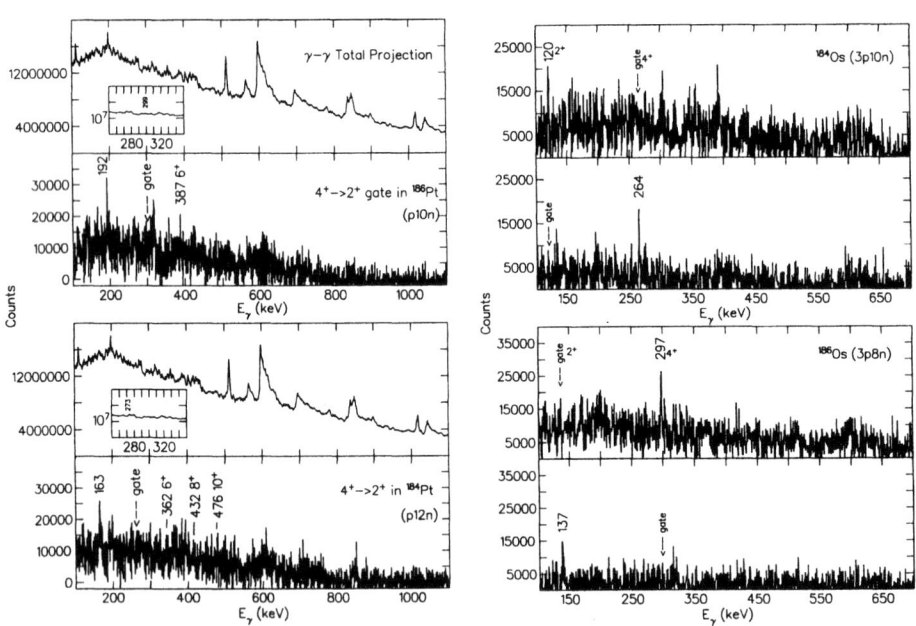

FIGURE 1. In-beam $\gamma - \gamma$ coincidence spectra from the current work.

The experiment was performed at the iThemba LABS, South Africa using a 30 MeV per nucleon ^{12}C beam incident on a self-supporting 30 mg/cm^2 ^{197}Au target. At this energy, the beam loses approximately 1 MeV/*u* through the target. In-beam reaction gamma rays were detected using the AFRODITE gamma-ray spectrometer [8] comprising 7, four-element clover germanium detectors and 8, four-way segmented planar 'LEPS' detectors (which provided enhanced sensitivity for low-energy decays). The experimental master-trigger condition was set such that a valid event required at least three detectors to fire within approximately 200 ns of each other, at least two of which were independent clover modules. Typical beam currents for the one day, in-beam test run were ~50 ppA, corresponding to a typical master-gate event rate of 2 kHz.

Figure 1 shows the gamma-gamma total projection for this work. Note the broadened lines at 596 keV and 691 keV associated with neutron scattering on the ^{74}Ge and ^{72}Ge isotopes in AFRODITE, highlighting the large neutron flux. The main path of the fragmentation 'corridor' flows primarily towards neutron-deficient nuclei and the subsequent cooling of the initial, hot fragmentation products gives rise to significant neutron evaporation. As figure 1 shows however, even with the poor signal to background ratio associated with simple gamma-gamma gating due to the very large number of reaction channels which are open, discrete yrast cascades can be identified. Specifically, figure 1 shows the yrast cascades in ^{184,6}Pt and ^{184,6}Os.

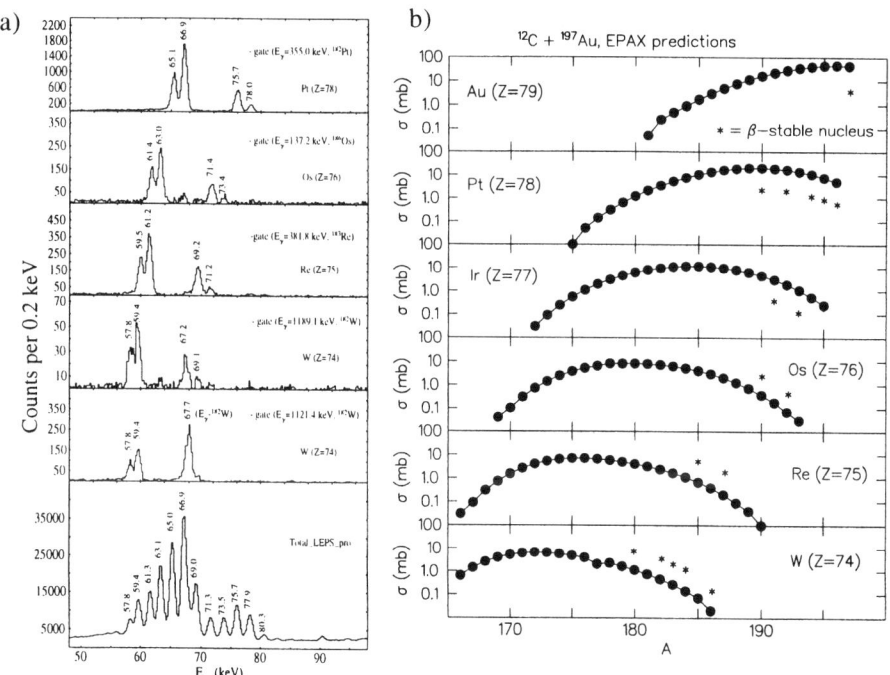

FIGURE 2. (a) Gamma gated LEPS spectra showing characteristic x rays from some of the strongly populated decay products. (b) EPAX calculations [9] for ^{12}C+^{197}Au.

Following the in-beam experiment the ^{197}Au target was left inside the AFRODITE array for four days and the data acquistion master condition set to singles. In this way the activity from the β-decaying target-like reaction products and their daughters could be measured. This data could then be used to provide a clearer picture of which reaction products were formed in-beam. A LEPS-clover coincidence matrix was sorted for the decay data which allow gating on characteristic K_α and K_β X-rays in elements ranging from W (Z=74) to Au (Z=79). Figure 2 highlights the clear separation of the different elements achieved by gating on decay transitions from known elements.

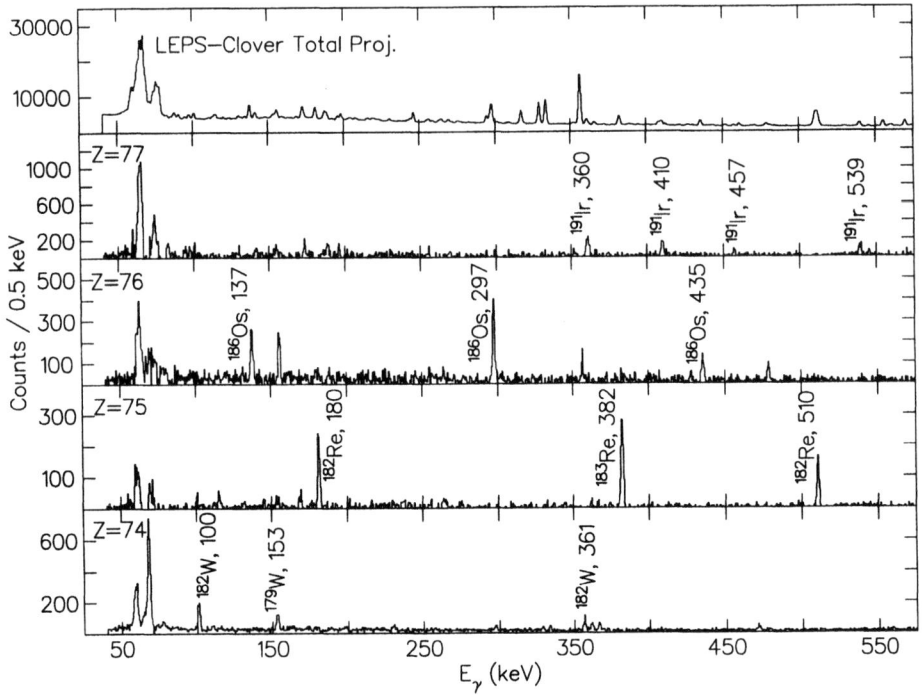

FIGURE 3. Examples of X-ray gated gamma-ray spectra for the decay products.

Elementally pure decay spectra were obtained by gating on characteristic X rays and subtracting a normalised portion of the neighboring element gate due to the usual energy overlap of the $K_{\alpha_1}(Z)$ and $K_{\alpha_2}(Z+1)$ peaks in neighboring elements (see fig. 3). The decay measurements will allow isobaric and elemental [10] yields to be determined.

The initial data suggest that the technique of in-beam, thin-target fragmentation may be a viable mechanism for studying heavy, neutron-rich nuclei. Although only proton-rich nuclei have been identified the current data, the EPAX calculations suggest that neutron-rich nuclei were also populated. Clearly in order to study these nuclei in-beam using this method, a significant increase in experimental sensitivity is required. In nuclei where the yrast cascade is known, this should be achievable by taking higher-fold gamma-ray gates (a gamma-triples analysis of the current data is underway). A major increase in selectivity could also be obtained by gating on transitions depopulating ns→ μs decays between the beam pulses. This would allow correlations above and below the metatable states providing a complementarity between the fragment separator based isomer studies and this type of 'prompt', in-beam work. Other experimental ancillary detectors which may improve sensitivity include the use of a multi-element neutron wall, since the more neutron-rich nuclei (by definition) emitted fewer neutrons than the more populous proton-rich systems.

The use of X-ray gating for in-beam decays to provide elemental selection is also clearly of significant benefit to these types of studies. Ideas being investigated in this area include the development of a LEPS-Clover hybrid detector, which would allow the *simultaneous* detection of both X-rays and gamma-rays. Linked to this is the development of the future generation of gamma-ray tracking arrays such as GRETA and AGATA. Such methods will lead to the possibility of much faster data rates which will be important in obtaining information on the more weakly populated neutron-rich nuclei. To this end significant developments are already taking place on the use of digital electronics for pulse shape analysis, which will soon usurp standard analogue amplifiers (see eg. reference [11]).

On the physics side, advances in this technique will enable a number of questions to be attacked. Currently, isomeric ratio measurements are the main method which has been used to obtain information in the angular momentum distribution of residual nuclei produced in such reactions [12]. However, the selection of discrete, in-beam cascades will provide a much clearer picture of this question over a very wide range of nuclei.

ACKNOWLEDGMENTS

This work is supported by EPSRC(UK) and the National Research Foundation of South Africa. KAG acknowledges support from the Marie Curie scheme (EU). PHR acknowledges support from Yale University via the Department of Energy (grant DE-FG02-91ER-40609) and the Yale Flint and Science Development funds.

REFERENCES

1. Chandler, C. *et al., Phys. Rev.* **C61** 044309, 2000; Grzywacz, R. *et al., Phys. Lett.* **355B** 439, 1995
2. Daugas, J. M. *et al., Phys. Lett.* **476B** 213, 2000; Grzywacz, R. *et al., Phys. Rev. Lett.* **81** 776, 1998
3. Pfützner, M. *et al., Phys. Lett.* **444B** 32, 1998
4. Podolyák, Zs. *et al., Phys. Lett.* **491B** 225, 2000; Caamaño, M. *et al., Nucl. Phys.* **A682** 223c, 2001
5. Glasmacher, T. *et al., Eur. Phys. J.* **A13** 59, 2002; Motobayashi, T. *et al., Phys. Lett.* **346B** 9, 1995
6. Belleguic, M. *et al., Nucl. Phys.* **A682** 136c, 2001; Yoneda, K. *et al., Phys. Lett.* **499B** 233, 2001
7. Lee, I. Y. *et al., Acta Phys. Pol.* **B32** 2499, 2001
8. Newman, R. T. *et al., Balkan Phys. Lett.* Special Issue, 182, 1998
9. Sümmerrer, K. and Blank, B., *Phys. Rev.* **C61** 034607, 2000
10. Erduran, M. N. *et al., to be submitted to Nucl. Inst. Meth. Phys. Res. A*
11. Pearson, C. J. *et al., IEEE Trans. Nuc. Sci. in press*, 2002; Valiente Dobon, J. J. *et al., submitted to Nucl. Inst. Meth. Phys. Res.* **A**, 2002; Vetter, K. *et al., Nucl. Phys.* **A682** 279c, 2001
12. Pfützner, M. *et al., Acta Phys. Pol.* **B32** 2507, 2001; and *Phys. Rev.* **C65** *in press*, 2002

Perspectives for nuclear structure through moment measurements

Andrew E. Stuchbery

Department of Nuclear Physics, Research School of Physical Sciences and Engineering,
Australian National University, Canberra, ACT 0200, Australia, and
NSCL, Michigan State University, East Lansing, Michigan 48824

Abstract. The importance of nuclear moments for understanding nuclear structure is discussed along with some new developments in moment measurement techniques.

The unknown territory in nuclear moment measurements covers more than that part of the nuclear landscape which rare-isotope beams are opening up. To a large degree the moments of short-lived states are uncharted away from the valley of stability - even on the neutron-deficient side which can be accessed with stable beams. This lack of data is significant because the important terms in a wavefunction may not be the largest and differ from operator to operator. Hence studies of an observable like the magnetic moment can bring to light new aspects of nuclear structure in otherwise well studied regions. This is illustrated by recent high-precision measurements of g factors in nuclei near closed shells in which it has been demonstrated that there can be significant collective contributions in apparently single-particle excitations and vice versa [1, 2, 3, 4]. The g factors probe the single-particle configuration, the effectiveness of the shell gaps and the magnitude of the interactions between the protons and neutrons.

Looking back, there have been many times when our moment studies in Australia have related to Rick Casten's work. I want to mention another example where the g factors give an insight not gleaned from other observables, in this case concerning the even osmium isotopes, which Rick first studied experimentally by Coulomb excitation [5], and later revisited, particularly in relation to the Interacting Boson Model [6]. The isotopes ^{188}Os and ^{192}Os were the first cases where $g(2_2^+)/g(2_1^+)$ ratios different from unity were observed in a collective nucleus [7]. These ratios and the mixing ratios for the $2_2^+ \to 2_1^+$ transitions in ^{188,190,192}Os were later explained consistently in terms of F-spin mixing in the Interacting Boson Model [8]. However $g(4_1^+)/g(2_1^+)$ in ^{188}Os appears to differ from unity, which may indicate additional physics such as shape co-existence.

Looking to the future, it is evident that excited-state magnetic moments will be very useful to delineate the structures of exotic nuclei, if they can be measured precisely enough. At this stage I do not see a universally applicable technique for measuring g factors of short-lived states of radioactive ion beams because the perturbation must be increased by more than an order of magnitude to compensate for low statistics. However by selecting the technique case by case, it should be possible to study many nuclei. For example, to measure the g factor of the 2^+ state in ^{34}Si by the transient-field technique

CP638, *Mapping the Triangle: International Conference on Nuclear Structure*
edited by A. Aprahamian, J. A. Cizewski, S. Pittel, and N. V. Zamfir
© 2002 American Institute of Physics 0-7354-0093-8/02/$19.00

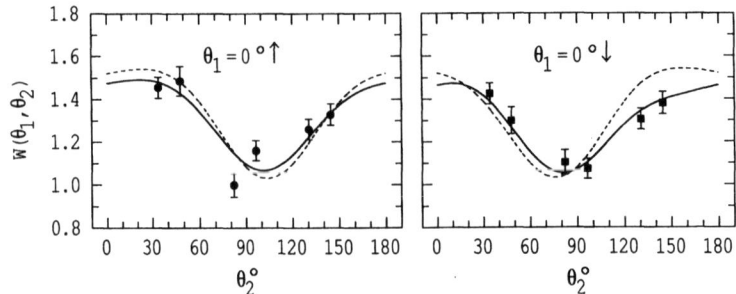

FIGURE 1. Perturbed DCOs for the first-excited state of ^{184}Pt. The dotted lines are for a pure magnetic interaction, while the solid lines include an electric quadrupole interaction. The difference gives a measure of the nuclear quadrupole moment.

would require collecting γ-ray spectra with about 10^5 counts in the $2_1^+ \rightarrow 0_1^+$ photopeak, but if the fields due to H-like or Li-like free ions are used instead it should be possible to measure the g factor with considerably fewer counts.

At ANU we have been developing techniques to measure moments in neutron-deficient nuclei, especially heavy nuclei ($A > 100$). Recently we measured g factors in ^{180,182,184}Pt by means of a perturbed DCO technique [1, 9]. This technique also offers the possibility to measure the quadrupole moments of subnanosecond 2^+ states with a precision of about 15%, as illustrated in Fig. 1. If the electric-field gradient for Pt in Gd is interpolated from literature values for the neighbors, the $Q(2_1^+)$ values in ^{182}Pt and ^{184}Pt are consistent with transition quadrupole moments obtained from the lifetimes. There are very few quadrupole moment measurements on subnanosecond states in unstable nuclei. This technique will complement Coulomb-excitation re-orientation measurements on rare-isotope beams.

REFERENCES

1. A.E. Stuchbery, Nuclear Physics **A682**, 470c (2001).
2. J. Holden, N. Benczer-Koller, G. Jakob, G. Kumbartzki, T.J. Mertzimekis, K.-H. Speidel, A. Macchi-avelli, M. McMahan, L. Phair, P. Maier-Komor, A.E. Stuchbery, W.F. Rogers, A.D. Davies, Physics Letters **B493**,7 (2000).
3. P.F. Mantica, A.E. Stuchbery, D.E. Groh, J.I. Prisciandaro, M.P. Robinson, Physical Review C **63**, 034312 (2001).
4. G. Jakob, N. Benczer-Koller, G. Kumbartzki, J. Holden, T.J. Mertzimekis, K.-H. Speidel, R. Ernst, A.E. Stuchbery, A. Pakou, P. Maier-Komor, A. Macchiavelli, M. McMahan. L. Phair, I. Y. Lee, Physical Review C **65**, 024316 (2002).
5. R.F. Casten, J.S. Greensberg, G.A. Burginyon, D.A. Bromley, Phys. Rev. Lett. **18**, 912 (1967).
6. R.F. Casten, J.A. Cizewski, Nucl. Phys. **A 309**, 477 (1978).
7. A.E. Stuchbery, I. Morrison, L.D. Wood, R.A. Bark, H. Yamada, H.H. Bolotin, Nuclear Physics **A435**, 635 (1985).
8. S. Kuyucak, A.E. Stuchbery, Physics Letters **B 348**, 315 (1995).
9. M.P. Robinson, A.E. Stuchbery, R.A. Bark, A.P. Byrne, G.D. Dracoulis, S.M. Mullins, A.M. Baxter, Physics Letters **B530**, 74 (2002).

Experiments with Radioactive Nuclear Beams for Nuclear Structure

C. J. Barton*, D. Shapira†, R. F. Casten**, M. A. Caprio**, J. R. Cooper**, N. V. Zamfir**, C. W. Beausang**, J.R. Novak**, R. Krücken**, D. S. Brenner‡, R. L. Gill§, T.A. Lewis†, R. Lemmon* and D. D. Warner*

*CLRC Daresbury Laboratory, Daresbury, Warrington WA4 4AD, United Kingdom
†Nuclear Physics Division, Oak Ridge National Laboratory, Oak Ridge, TN 37831
**Wright Nuclear Structure Laboratory, Yale University, New Haven, CT 06520
‡Clark University, Worcester, MA 01610
§Brookhaven National Laboratory, Upton, NY 11973

Abstract.
    Radioactive Nuclear Beams (RNBs) are opening new regions of the nuclear landscape to nuclear structure studies. Early experiments with RNBs rely on reactions with large cross sections, inverse kinematics, and very efficient detector geometries in order to measure observables that are very sensitive to structural features. A Coulomb excitation experiment extracted the $B(E2; 0_1^+ \rightarrow 2_1^+)$ values of the neutron rich RNBs ^{132,134}Te at the HRIBF with the GRAFIK through-well NaI(Tl) detector. In addition, other experiments with RNBs, such as Coulomb excitation of octupole states and reorientation experiments, inelastic scattering, and single-particle transfer, will be discussed.

Important techniques for radioactive nuclear beam experiments

Stressing the nuclear many-body system with extreme ratios of neutron to proton number is yielding a better understanding of the evolution of nuclear structure. The increasing availability of facilities that can accelerate radioactive nuclear beams (RNBs) will greatly enhance these studies. Challenges posed by RNBs involve low beam intensities and high background levels. Measuring observables which are very sensitive and selective to structural features and using inverse kinematics, where the beam nucleus is heavier than the target nucleus, overcome these challenges.

A few observables, such as the 2_1^+, the ratio $4_1^+/2_1^+$, and the $B(E2; 0_1^+ \rightarrow 2_1^+)$ values, are very sensitive to collectivity and deformation. These observables and the N_pN_n valence correlation scheme, for example [1], are useful for understanding the behavior of nuclei near and far from stability. Utilizing such schemes with new data from new regions of the nuclear landscape could quickly recognize deviant behavior and lead to understanding the nuclear structure in the new region.

Inverse kinematics allows for the removal of RNBs from the target area and for the use of compact particle detector geometries. Rutherford scattering events have the projectile peaked at $0°$ in the laboratory while inelastically scattered beam nuclei, such as from Coulomb excitation, are very forward focused in the laboratory frame. Scattered RNBs will then exit the target chamber and minimize background activity. Elastically and

CP638, *Mapping the Triangle: International Conference on Nuclear Structure*
edited by A. Aprahamian, J. A. Cizewski, S. Pittel, and N. V. Zamfir
© 2002 American Institute of Physics 0-7354-0093-8/02/$19.00

inelastically scattered target nuclei scatter predominantly at angles less than but near 90° in the laboratory frame. The measurement of the ejected target particle or scattered beam particle angle and energy determines the reaction kinematics that have occurred.

Coulomb excitation for $B(E2;0_1^+ \rightarrow 2_1^+)$ values

Coulomb excitation in inverse kinematics on a *low* Z target at energies safely below the Coulomb barrier is a highly selective population mechanism. In an even-even nucleus, single-step $E2$ excitation of the 2_1^+ state is the dominant process. The resulting γ-ray spectrum contains only a single γ-ray and allows for a simple analysis to extract the $B(E2;0_1^+ \rightarrow 2_1^+)$ value.

A high efficiency detector used for Coulomb excitation with RNBs is the Gamma-Ray Annulus for Inverse Kinematics (GRAFIK) detector [2]. The GRAFIK detector is an 8.3×7.6 cm cylindrical NaI detector in which a 3.2 cm diameter clearance hole has been made perpendicular to the cylinder axis. A thin target foil for Coulomb excitation is placed at the center of the detector through-well. The detector has an efficiency of 5.7% at 1.3 MeV for γ-rays emitted at the target position with an energy resolution of 6%. γ-rays emitted by the beam nuclei in flight are typically Doppler broadened by about 7%, but the simple spectrum and inherent resolution makes Doppler correction unnecessary.

The GRAFIK detector was used for Coulomb excitation of ^{130,132,134}Te on ~ 1 mg/cm^2 ^{nat}C at a beam energy of 350 MeV at the Holifield Radioactive Ion Beam Facility (HRIBF) [3]. The beams had significant isobaric contamination. For example, ^{134}Te nuclei comprised less than half of the total beam intensity on target. The beams of ^{130,132}Te were used for runs lasting 1.6 h and 7.3 h respectively, yielding integrated beam fluxes of 6.4×10^9 and 1.5×10^{10} incident Te nuclei. The closed neutron shell nucleus ^{134}Te, where Coulomb excitation is hindered by a small $E2$ matrix element and a high-lying 2_1^+ state, had a beam intensity of 10^5 nuclei/s on target for 15 h yielding 6.7×10^9 ^{134}Te nuclei.

A position sensitive tilted foil microchannel plate detector [5], was located downstream of the target for kinematic selection of scattered beam particles. Less than 0.5% of Rutherford scattered beam particles were scattered at angles $> 1.5°$ and Coulomb excited scattered beam particles were strongly peaked at a scattering angle of $\sim 5°$. A 1.5° elliptical hole was excised from the center of the tilted foil, allowing beam scattered in the extreme forward direction to pass through undetected while serving as a trigger for Coulomb excited scattered beam nuclei. Two thin foil microchannel plate timing detectors [4] were located upstream of GRAFIK to count the total number of incident beam nuclei for the determination of the Coulomb excitation yield and to provide a beam particle time signal as a coincidence requirement.

Background was negligible in the ^{130,132}Te runs but substantial in the ^{134}Te run. Gated spectra are shown in Fig. 1. In each spectrum the Te isotopes had the highest energy γ-ray expected from single-step Coulomb excitation compared with any Coulomb excitation of the isobaric contaminants. The spectra show a single high energy γ transition and associated lower energy Compton events and, to a lesser extent, background γ-rays.

The measured Coulomb excitation γ-ray yield depends upon the kinematic acceptance

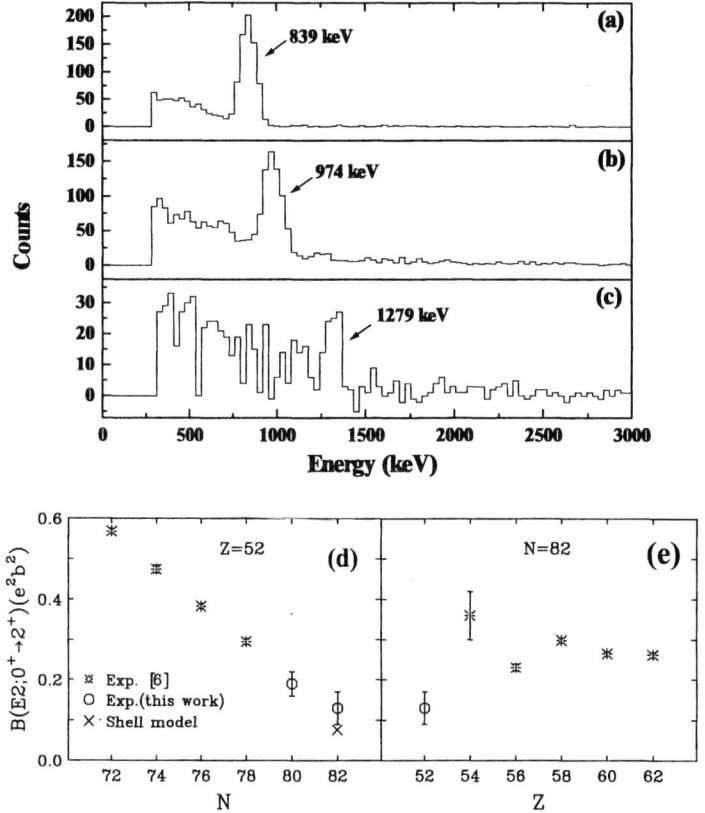

FIGURE 1. Gated NaI spectra, showing the $2_1^+ \rightarrow 0_1^+$ transition peak of interest, for all three experiments: (a) ^{130}Te, (b) ^{132}Te, (c) ^{134}Te (time random background subtracted). $B(E2; 0_1^+ \rightarrow 2_1^+)$ values from Raman [6] and the present work are shown in (d) with the predictions of the shell model even-even Te nuclei with $N > 72$. The isotonic systematics for the even-even $N = 82$ nuclei are shown in (e).

of the experiment. The $B(E2; 0_1^+ \rightarrow 2_1^+)$ value for ^{130}Te, well known from the literature [6] to be 0.295(6) e^2b^2, was used for normalization. The extracted $B(E2)$ values are 0.19(3) for ^{132}Te and 0.13(4) for ^{134}Te.

The results confirm the expectation that the $B(E2)$ values decrease toward the $N = 82$ magic number. A simple approach for a shell model calculation would be to expect the spectrum and transition rates of ^{134}Te to resemble those of two protons in the $50 - 82$ shell interacting with a surface delta interaction. The data is slightly more collective than this simple analysis would imply as can be seen in Fig. 1 (d). The isotonic systematics for $N = 82$ nuclei are also shown in Fig. 1 (e). Excepting ^{136}Xe, the increase and subsequent saturation of the $B(E2; 0_1^+ \rightarrow 2_1^+)$ values as valence protons are added is expected, as in the seniority scheme.

Coulomb excitation for $B(E3)$ values and quadrupole moments

Additional matrix elements can be obtained from Coulomb excitation by increasing the bombarding energy and/or the Z of the target.

The recent successful Coulomb excitation experiments in inverse kinematics using the CLARION array [8] has prompted a program to measure the $B(E3; 0_1^+ \rightarrow 3_1^-)$ value for ^{132}Te and additional neutron-rich nuclei investigating octupole collectivity in the $Z \sim 56$, $N \sim 88$ region where octupole deformation was predicted [9]. A beam of 370 MeV 10^7 ^{132}Te nuclei/s will be Coulomb excited on a ^{28}Si target [10]. With a CLARION γ-detector array efficiency of about 4% at 610 keV (the 3^- to 4^+ γ-ray transition), an expected γ-ray count rate of 100 counts/day is calculated. Si ions scattered out of the target will be detected in the HyBall CsI charged particle detector array.

A precision experiment to measure the quadrupole moment has been approved at REX-ISOLDE for the suspected oblate-deformed ^{70}Se nucleus[11]. The MINIBALL cluster detector array, 13% efficient, and the Edinburgh Compact Disc detector, covering angles from $10°$ to $40°$, will be used. A beam of 2×10^4 ^{70}Se/s on targets of nickel and silicon should complete the yield measurements in 2 days per target.

Inelastic scattering and transfer reactions with γ-ray spectroscopy and particle detection

Inelastic scattering and transfer reactions using γ-ray spectroscopy to resolve excited states will provide new sensitivity to the measurement of nuclear observables. Previously, experiments were limited to stable beam and target combinations and used magnetic spectrometers or NaI(Tl) detectors to resolve excited states. Modern γ-ray detector arrays will allow the excited states to be finely resolved while the angular distribution of scattered particles is measured.

Inelastic scattering above the Coulomb barrier involves the inclusion of nuclear interactions and a coupled-channel analysis becomes necessary. The benefit provided is additional off-diagonal matrix elements from which $B(E4; 0_1^+ \rightarrow 4_1^+)$ values, for example, can be extracted. The optical model parameters needed can be obtained from the elastically scattered particles and can be calibrated using electromagnetic matrix elements obtained from Coulomb excitation. Experiments with inelastic scattering will need particle detector placement similar to that for Coulomb excitation.

Transfer reactions in inverse kinematics, such as (p,d) and (d,p) reactions give information on the occupation probabilities for single particle valence orbitals of ground and excited states. The energy of single-particle states would verify changes in shell structure far from stability. Transfer reactions would need particle detection at forward angles for beam-like particles. Target-like particle detection is heavily weighted forward within $50°$ for (p,d) and (d,t) reactions and backwards at $110°$ for (d,p) reactions.[13]

TIARA is a charged particle detector for RNB experiments at GANIL with the ISOL facility SPIRAL. TIARA, shown schematically in Fig. 2, is a highly segmented array of Si detectors that covers forward, backward, and angles near $90°$ all with an angular resolution of $2°$. TIARA will be positioned at the center of the EXOGAM

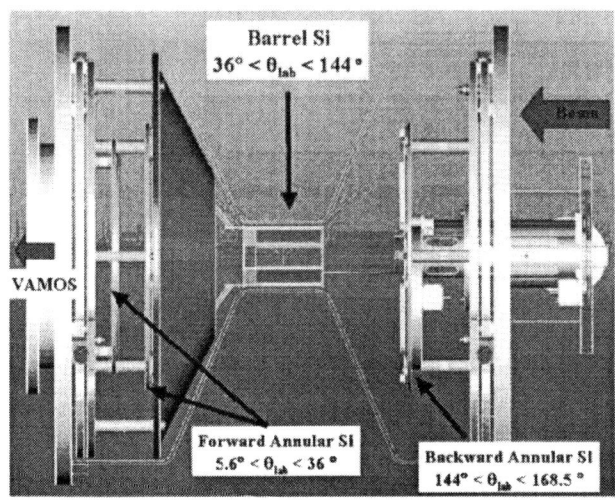

FIGURE 2.   Schematic view of TIARA showing the Si particle detectors and their angular coverage

γ-ray array, which in close geometry would have a photopeak efficiency of 20%. A recoil spectrometer, VAMOS, will be downstream of TIARA and could be used for the identification of beam and beam-like nuclei. Together, these powerful particle and γ-ray arrays will offer a versatile instrument for Coulomb excitation, inelastic scattering, and single particle transfer reactions in inverse kinematics. Initial proposed experiments cover the range of these reactions and hold the promise of supplying a great deal of data on the low-lying levels and nuclear structure of radioactive nuclear beams.

This work was supported by the US DOE under grant and contract numbers DE-FG02-91ER-40609, DE-FG02-88ER-40417, DE-AC05-00OR22725, and DE-CH02-98CH10886.

REFERENCES

1. R.F. Casten and N.V. Zamfir, J. Phys. G: Nucl. Part. Phys. **22** (1996).
2. C. J. Barton, R. L. Gill, R. F. Casten, D. S. Brenner, N. V. Zamfir, and A. Zilges, Nucl. Instrum. Methods **A391**, 289 (1997).
3. G. D. Alton and J. R. Beene, J. Phys. G **24**, 1347 (1998).
4. D. Shapira *et al.*, Nucl. Instrum. Methods A (submitted); D. Shapira *et al.* (private communication).
5. D. Shapira, T. A. Lewis, and L. D. Hulett, Nucl. Instrum. Methods **A454**, 409 (2000).
6. B. Singh, Nucl. Data Sheets **93**, 33 (2001).
7. S. Raman, C.W. Newsor, Jr., and P. Tikkanen, Atomic Data and Nuclear Data Tables **78**, 1 (2001).
8. D. C. Radford *et al.*, Proc. of 3rd Int. Conf. on Exotic Nuclei and Atomic Masses, July 2001, Eur. Phys J. A.
9. For a review, see P.A. Butler and W. Nazarewicz, Rev. Mod. Phys. **68**, 349 (1996).
10. N.V. Zamfir *et al.*, HRIBF proposal, (2001).
11. D.G. Jenkins, P. A. Butler *et al.*, ISOLDE proposal, CERN-INTC-2002-003, INTC/P 149 (2002).
12. S.M. Fischer *et al.*, Phys. Rev. Lett. **84**, 4064 (2000).
13. W.C. Catford *et al.*, TIARA proposal, (2000).

Nuclear level densities in Casten's triangle

Till von Egidy* and Dorel Bucurescu[†]

*Physik-Department, Technische Universität München, D-85748 Garching, Germany
[†]"Horia Hulubei" National Institute of Physics and Nuclear Engineering, Bucharest, Romania

Level density parameters of the back-shifted Fermi gas and of the constant temperature model were determined for 298 nuclei with a least squares fit to low energy levels and neutron resonances. The results are systematized with the valence correlation scheme and related to Casten's triangle.

INTRODUCTION

The energy dependence of nuclear level densities can be described with rather simple formulae using only two free parameters for each nucleus, such as those of the back-shifted Fermi gas (BSFG) [1, 2] or the constant temperature (CT) [1] models.

The level density parameters for the BSFG and CT models have been determined for 298 nuclides between ^{18}F and ^{250}Cf by fitting the two-parameter formulae to the complete level scheme at low excitation energies (≤ 3 MeV) [3] *and* the s-wave neutron resonance spacings at the neutron binding energies [4], with a procedure similar to that of [5]. The experimentally determined parameters are a and the back shift E_1 for the BFSG, and T and the back shift E_0 for the CT model, respectively. We correct the back shifts with the pairing energies (Δ) taken from mass tables [6]: $C_1 = E_1 - \Delta$ and $C_0 = E_0 - \Delta$. These largely accounts for the differences between even-even, odd-A and odd-odd nuclei and thus allows to put all the nuclei together [5]. For the parameters a and T the average, smoothly changing patterns have been determined, similarly to the previous work [5], as $a \sim A/9.5$ and $T \sim 18A^{-2/3}$, respectively. They also present oscillations around these smooth variations [5]. These are best evidenced by using the 'corrected' parameters, which we obtain by dividing their experimental values by the functions of A mentioned above. The parameters C_1 and C_0 are remarkably constant with A, their average values varying by only about 1 MeV over the whole mass range 18 to 250. More detailed discussions of nuclear level densities can be found in [7].

APPLICATION OF A VALENCE CORRELATION SCHEME

The evolution of many different structure parameters (observables) is governed by shell and residual interaction effects. This has been shown to be very efficiently systematized within the so-called valence correlation schemes (VCS) [8]. The idea is to represent these parameters versus simple functions of the numbers of the 'active' particles, N_p

CP638, *Mapping the Triangle: International Conference on Nuclear Structure*
edited by A. Aprahamian, J. A. Cizewski, S. Pittel, and N. V. Zamfir
© 2002 American Institute of Physics 0-7354-0093-8/02/$19.00

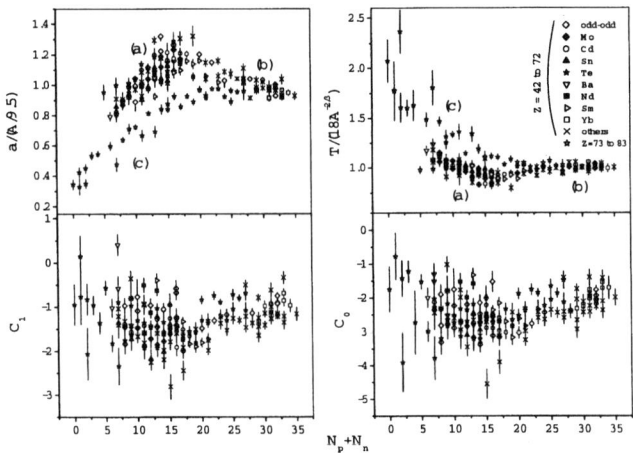

FIGURE 1. Level density parameters versus the total number of active particles for $Z = 42$ to 83 nuclei.

and N_n (the number of protons and neutrons, respectively, as particles or holes outside the closed shells). Since the main oscillations of our parameters are correlated with the shell fillings, we have investigated different VCS, based on the well-known quantities $N_p N_n$, $P = N_p N_n / (N_p + N_n)$, and also $N_t = N_p + N_n$. In the following we refer to results obtained with the numbers of active particles counted with respect to the 'classical' magic numbers 8, 20, 28, 50, 82, 126, 184. The corrected a and T parameters were found best correlated with N_t. The variations of the back-shift energies are somewhat more complicated, but still show some correlation with N_t.

Fig. 1 shows this type of representation for an extended nuclear region, $Z = 42$ to $Z = 83$. It is seen that, at least for the parameters a and T, the complicated patterns of the variation with A transform into rather simple and compact ones. One should remark that these plots contain together all nuclei from our collection, even-even, odd and odd-odd, therefore the coalescence of all the data into global patterns is rather remarkable. One can distinguish in Fig. 1 three classes of nuclei, which evolve on three distinct and rather compact patterns, denoted as (a), (b) and (c), better visible in the case of a and T.

Class (a) comprises nuclei with N_t below $\sim$17-20, for which a increases and T decreases, while C_i ($i = 0$ or 1) generally decrease. All these nuclei are generally of the 'transitional' type, between spherical and deformed ones. Actually, the 'upper' limit of this class at $N_t \approx 17 - 20$ roughly corresponds to $P \approx 4.0 - 5.0$, which is the point at which the nuclear deformation starts to set in [9]. This set of nuclei corresponds roughly to the 'universal anharmonic vibrator' type of nuclei [10].

Class (b) with almost constant (≈ 1) corrected parameters, therefore almost no influence of the shell structures, corresponds to the deformed nuclei. The nuclei with $Z \geq 88$ have also this pattern; they are all deformed, but not represented in Fig.1. There are a few isotopic chains with nuclei on both branches (a) and (b), e.g., Nd, Sm, Gd, which are known to evolve from spherical to well deformed (rotor) nuclei. ^{152}Sm realizes the coexistence between the two regimes and lies near this critical point [13].

Class (c) is formed by the nuclei in the mass region around 200 which strongly deviate

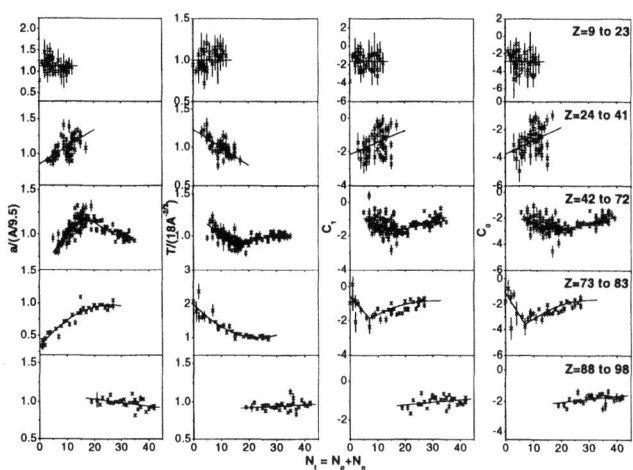

FIGURE 2. Same as Fig. 1 for all nuclei. The lines are linear or parabolic fits to the data.

from the smooth dependence on A: it contains some nuclei near closed shells (like Pb, Bi) and the nuclei with Z from 74 to 80.

The reason of the good scaling of the level density parameters with the total number of active particles (although not so perfect for the back shifts C_i) must be the statistical and combinatorial nature of the nuclear levels, which is reflected in the model parameters. Figure 2 shows the same representations of the corrected a and T parameters, and of C_1 and C_0, separated on 5 nuclear regions. Especially for the three regions of heavier nuclei, the well defined patterns may be effectively used to make predictions for other nuclei. Even predictions based on isotopic or isotonic sequences, are more reliable in the the the N_t representation. To briefly summarize the above results, the VCS type of representation of the level density model parameters (versus N_t) has revealed smooth variations which can serve practical purposes (prediction), and also highlight structural evolutions. In the following we will discuss in more detail the second aspect.

EVOLUTION IN CASTEN'S SYMMETRY TRIANGLE

The structural evolutions of the even-even nuclei are conveniently discussed within the Interacting Boson Model (IBA) [11], as evolutions between three dynamical symmetries: U(5) (anharmonic vibrator), O(6) (γ-unstable nucleus) and SU(3) (rotor). Recently there were introduced also transitional dynamical symmetries (at the critical transition points): E(5) - at the transition between U(5) and O(6) [12], and X(5) - between U(5) and SU(3) [13]. O(6) itself is both a dynamical symmetry and a critical point for the transition between prolate and oblate deformed (SU(3)) nuclei [14].

The way the real nuclei evolve with respect to these structural 'milestones' has been made very intuitive by the introduction of the Casten's symmetry triangle [15]. To picture Casten's triangle we have chosen our representation of ref. [16]: we plot the quantity $Q/\sqrt{B(E2)}$ as a function of $R_{4/2} = E(4_1^+)/E(2_1^+)$, where Q is the electric

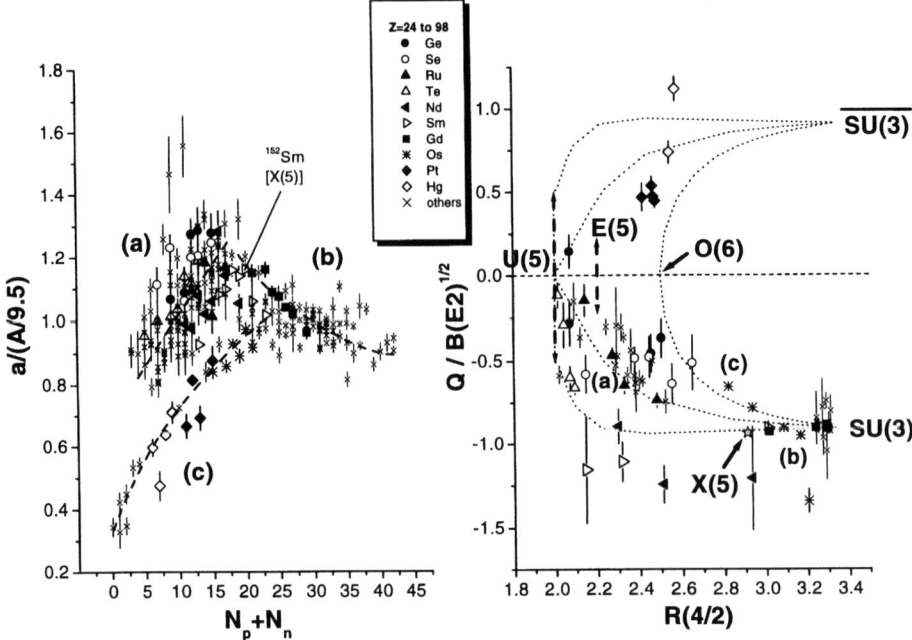

FIGURE 3. *Right part*: A particular representation of Casten's symmetry triangle (see text); the symbols represent even-even nuclei. The theoretical dynamic symmetry values are indicated. The dotted lines are certain IBA-1 calculated transition paths [16]. *Left part*: Correlation of the corrected *a* parameter of the BSFG model with N_t, for *all* nuclei of our set with $Z = 24$ to 98. The plot highlights the same isotope chains as the right side part. The dashed lines guide the eye for the three classes (a), (b), (c).

quadrupole moment of the 2_1^+ state and $B(E2)$ the $E2$ reduced probability for the $0_1^+ \rightarrow 2_1^+$ transition. It has the advantage that it separates between the prolate and oblate nuclei, thus showing the two possible SU(3) limits. The two observables have well defined values in the three limits: $Q/\sqrt{B(E2)}$ is ± 0.91, 0.0 and between 0.5 and -0.5 in the SU(3), O(6) and U(5) limits, while $R_{4/2}$ is 3.33, 2.5 and 2.0, respectively.

The right part of Fig. 3 shows this represention of Casten's triangle. The symbols show different nuclei for which the defined quantities have been experimentally determined. By looking, e.g., at different isotopic chains, one may describe their evolution with respect to the dynamical symmetries and the critical points. A few representative chains have been highlighted, to illustrate different structural evolutions. Thus, Ge, Se and Ru illustrate an evolution from U(5) towards SU(3) or O(6), along paths within the 'triangle'. The Te isotopes 122 to 128, remain close to the U(5) limit. Nd and Sm illustrate the U(5) to SU(3) transition, which goes practically on the lower leg of the triangle. The Nd, Sm and Gd symbols on the right side of the X(5) value are ^{150}Nd, ^{152}Sm, and ^{154}Gd, which have been associated to the X(5) symmetry [13]. The nuclei identified as near the E(5) symmetry are ^{134}Ba [12], ^{102}Pd [17] (second and third 'x' - symbols to the right of the E(5) bar) and ^{104}Ru [18] (the Ru symbol at $R_{4/2} = 2.48$). The Os isotopic chain (184 to 192) illustrates the SU(3) to O(6) transition, while the Pt

(192 to 198) and Hg (198 and 200) nuclei are near the O(6) dynamical symmetry, on the oblate side of the O(6) to SU(3) transition.

The left part of Fig. 3 is similar to Figs. 1 and 2. It shows the corrected a parameter of the BSFG model as function of N_t, for most of the nuclei of our set (Z from 24 to 98). The isotopic chains highlighted in the right side part of the figure, are represented by the same symbols in the left part of the figure. The two plots in Fig. 3 may not necessarily contain the *same* nuclei; the left side plot contains additional even-even nuclei (for which $Q(2_1^+)$ and $B(E2)$ are not measured), and also odd-mass and odd-odd nuclei. We find all nuclei which are near the U(5) limit, or transitional between U(5) and SU(3) or O(6), in the class that we have denoted with (a): their level densities increase with increasing N_t. The upper limit of this class corresponds to $N_t \approx 17\text{-}20$, which is near the nuclei that approximate the X(5) critical point. The class (b) of nuclei corresponds to the deformed nuclei, close to SU(3). Class (c) contains the isotopes of Tl, Pb and Bi which are the nearest to closed shell. It is continued with the O(6) nuclei Hg and Pt and, through the Os to W chains it smoothly joins class (b).

CONCLUSIONS

New level density parameters of the BSFG and CT models have been obtained for 298 nuclides between F and Cf. Plots of these parameters exhibiting rather simple and compact behaviours, have been obtained by using a valence correlation scheme, i.e., by representing the oscillating shell-dependent part of these parameters as a function of the total number of 'active' (outside the closed shells) particles. Three main global patterns have been ifentified. For small nuclear regions these allow a reliable extrapolation for nuclei far from stability. The evolutions along these three patterns could be understood as structural evolutions within the symmetry triangle of Casten.

REFERENCES

1. A.Gilbert and A.G.W.Cameron, Can. J. Phys. 43,1446(1965)
2. W.Dilg, W.Schantl, H.Vonach, and M.Uhl, Nucl. Phys. A217,269(1973)
3. Nuclear Data Sheets; R.B.Firestone et al., Table of Isotopes, John Wiley, New York, 1996.
4. G. Reffo, IAEA-TECDOC-1034, IAEA, Vienna, 1998, p.25.
5. T. von Egidy, H.H.Schmidt, and A.N.Behkami, Nucl.Phys.A481,189(1988)
6. A. H. Wapstra and G. Audi, Nucl.Phys. A432,55(1985); G. Audi, private comm. 2002.
7. A.V.Ignatyuk, IAEA-TECDOC-1034, IAEA, Vienna, 1998, p.65.
8. R.F. Casten, Nucl. Phys. A443,1(1985)
9. R.F.Casten, D.S.Brenner, and P.E.Haustein, Phys. Rev. Lett. 58,658(1987)
10. D.Bucurescu, N.V.Zamfir, R.F.Casten, and W.T.Chou, Phys. Rev. C60,044303(1999)
11. F. Iachello and A. Arima, The Interacting Boson Model, Cambridge Univ. Press, Cambridge, 1987
12. F. Iachello, Phys. Rev. Lett. 85,3580(2000); R.F.Casten and N.V.Zamfir, Phys. Rev. Lett. 85,3584(2000)
13. F.Iachello, Phys. Rev. Lett. 87,052502(2001); R.F.Casten and N.V.Zamfir, Phys. Rev. Lett. 87,052503(2001)
14. J.Jolie et al., Phys. Rev. Lett. 87,162501(2001)
15. R.F.Casten, Nuclear structure from a simple perspective, Oxford, N.Y., 1990
16. D.Bucurescu et al., Nucl. Phys. A672,21(2000)
17. N.V.Zamfir et al., Phys. Rev. C65,044325(2002)
18. A.Frank, C.E.Alonso, and J.M.Arias, Phys. Rev. C65,014301(2001)

Vibrational and Intruder Structures in the Stable Cd Nuclei

Steven W. Yates

Department of Chemistry, University of Kentucky, Lexington, KY 40506-0055 USA

Abstract. In (n,n'γ) experiments performed at the University of Kentucky 7.0 MV accelerator facility, the low-lying, low-spin level structures of several even-A stable Cd nuclei have been systematically studied. Level lifetimes have been determined with the Doppler-shift attenuation method. Initial results from [110-116]Cd generally support the interpretation of these nuclei as anharmonic quadrupole vibrators with low-lying intruder structures, although the details of the low-spin levels of [114]Cd and [116]Cd remain puzzling.

INTRODUCTION

The even-mass Cd nuclei are generally regarded as being among the best examples of vibrational or, using the language of the IBM, U(5) nuclei [1]; however, at excitation energies near the two-phonon triplet, additional 0^+ and 2^+ levels are typically observed. These "intruder" states have been interpreted as arising from proton excitations across the Z = 50 shell leading to 2p-4h configurations [2]. Higher-spin members of these structures have also been observed systematically [3,4].

In recent studies in our laboratory [5-9], we have focused on understanding the anharmonicities and ambiguities involved in describing the multiphonon structure of the stable Cd nuclei. While these studies have provided much new information, fundamental questions remain unanswered. How high in excitation do multiphonon states persist? To what degree do other excitations, such as the aforementioned intruder configurations or other collective structures, *e.g.*, mixed-symmetry states, affect the multiphonon structure?

EXPERIMENTAL

The methods employed in (n,n'γ) measurements performed at the University of Kentucky accelerator facility have been described in detail [10]. In our studies of the stable Cd nuclei, the targets were tens of grams of the enriched isotopes of [110,112,114,116]Cd. Gamma-ray angular distribution and excitation function spectra were obtained with a BGO Compton-suppressed HPGe detector, and γ-γ coincidence measurements were performed with four HPGe detectors in a close geometrical arrangement [11].

CP638, *Mapping the Triangle: International Conference on Nuclear Structure*
edited by A. Aprahamian, J. A. Cizewski, S. Pittel, and N. V. Zamfir
© 2002 American Institute of Physics 0-7354-0093-8/02/$19.00

A very exciting lifetime regime can be investigated in (n,n′γ) measurements by employing the Doppler-shift attenuation method (DSAM). While the recoil velocity imparted in neutron scattering reactions on heavy nuclei is small (v/c ≈ 0.001), it is sufficient to produce observable Doppler shifts, and lifetimes from a few femtoseconds to about one picosecond can be determined. The Doppler-shifted γ-ray energy, $E_\gamma(\theta)$, measured at a detector angle of θ with respect to the incident neutrons can be related to E_o, the energy of the γ ray emitted by a nucleus at rest, by the expression,

$$E_\gamma(\theta) = E_o \left[1 + F_{exp}(\tau) \, v_{cm} \cos \theta \, / c \right] \tag{1}$$

where v_{cm} is the velocity of the center of mass in the inelastic neutron scattering collision with the nucleus, and c is the speed of light. $F_{exp}(\tau)$ is the experimental attenuation factor determined from the measured Doppler shift and is compared with calculated attenuation factors to determine the lifetime. The techniques for measuring nuclear lifetimes using DSAM methods with the (n,n′γ) reaction have been developed to a high degree of precision in this laboratory in recent years [12]. The uncertainties of unknown feeding times can be mitigated by performing measurements just above the threshold for the level of interest. Relative uncertainties of γ-ray energies can typically be determined to < 10 eV, so the largest source of uncertainty now resides in our incomplete knowledge of the stopping powers of the residual nuclei recoiling at low velocities.

RESULTS AND DISCUSSION

Three-Phonon States in the Cd Nuclei

Ani Aprahamian and her coworkers, including Rick Casten, suggested that a complete isoscalar three-phonon quintuplet (0^+, 2^+, 3^+, 4^+, and 6^+ states) existed in ^{118}Cd; they based this identification of collective character primarily on level energies and γ-ray branchings [13]. From similar arguments, three-phonon structures have been identified in other Cd nuclei [1]; however, lifetime information, which is crucial in understanding the nature of excited levels, had typically been limited to only a few of the lowest levels in these nuclei. The B(E2) values of the transitions decaying from members of the 3-phonon quintuplets of ^{110}Cd and ^{112}Cd provided convincing evidence for collective enhancement consistent with that expected for 3-phonon states [5-7]. On the other hand, some members of the multiplet, usually the lower-spin (0^+ or 2^+) states, showed little enhancement in their decays. From our studies of ^{110}Cd and ^{112}Cd, we concluded that the delicate interplay between the phonon, intruder, and mixed-symmetry degrees of freedom can lead to states with quite mixed wave functions and, therefore, decay patterns and transition rates that are severely perturbed [5-7]. It was also found in ^{112}Cd that two-thirds of the expected B(M1) strength from the lowest 2^+ mixed-symmetry state was located in two states just above the 3-phonon 2^+ state and that these states mixed with intruder configurations [14]. This latter mixing leads to violation of selection rules of the U(5)-O(6) model [15], but it could still be treated reasonably well in IBM-2 calculations.

Intruder States

In Fig. 1, the low-lying 0^+ and 2^+ levels of the even-even $^{108-120}$Cd isotopes are shown. For clarity, the 2^+ members of the two-phonon triplets are omitted; the notation for the 0^+ states used by Kumpulainen *et al.* [3] and Juutinen *et al.* [4] is retained. The systematics of intruder levels reveal the expected V-shaped behavior in excitation energy, where the energies decrease to a minimum in the neutron mid-shell nucleus ^{114}Cd. Recent measurements have led to the determination of absolute transition rates for several of the crucial transitions, and B(E2) values, where known, are indicated. *The large $B(E2;2^+ \to 0^+)$ values for the transitions in the intruder bands are conspicuous and lend strong support to the identification of a strongly collective structure.*

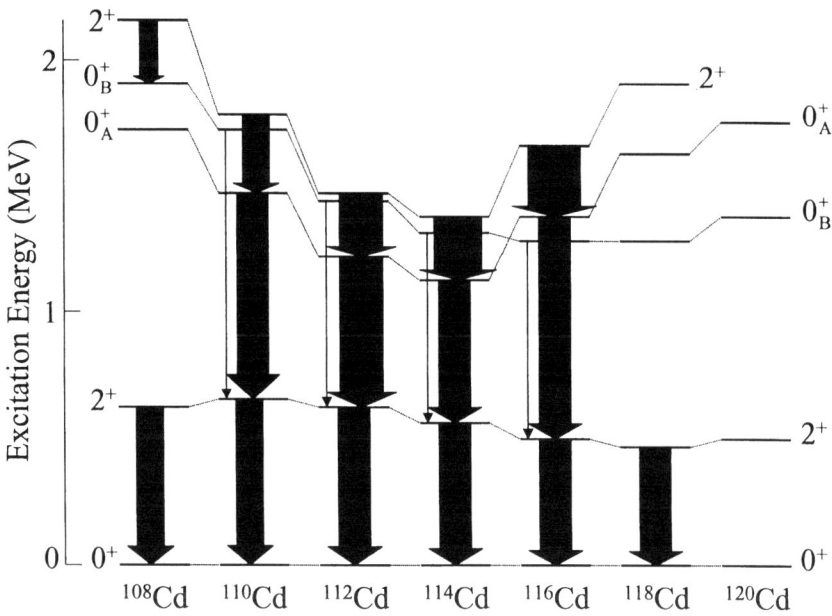

FIGURE 1. Systematics of the intruder configurations in the Cd nuclei. E2 transition strengths are indicated by the widths of the arrows. Only a lower limit for the transition in the intruder band of ^{108}Cd is available [16].

Support for the intruder state explanation for the additional 0^+ and 2^+ states found systematically near the two-phonon states in the Cd nuclei is clearly evident from the data presented in Fig. 1. As these intruder structures involve proton 2p-4h configurations, their behaviors might be expected to mimic those of the ground-state bands of their Ru and Ba isotones having 6 valence holes and 6 valence protons, respectively. Such correlations have been demonstrated previously for excitation energies [3,16], but not for transition rates. Table 1 illustrates the striking agreement between the $B(E2;2^+ \to 0^+)$ values of the Cd intruder bands and the ground-state bands of the corresponding Ru isotones. These observations further solidify the identification of the 0^+ states as intruder structures and raise an interesting question. Why does the intruder 0^+ state systematically decay strongly to the one-phonon state? Because of the strong mixing of the 0^+ intruder and phonon excitations that appear to be typical of the near mid-shell Cd nuclei, it might be argued that assignments of the character of these excitations may not be meaningful.

TABLE 1. B(E2; $2^+ \rightarrow 0^+$) Values in W.u. in "Six Valence Proton" Configurations.

Neutron Number	Cd Intruder Band	Ru Ground-State Band
64	56 ± 17	58 ± 5
66	61 ± 8	70 ± 5
68	86 ± 30	74 ± 7

Fig. 1 also illustrates a conundrum in ^{116}Cd that is not apparent in the lighter Cd nuclei. The 0^+ intruder band head, the second excited 0^+ state, which is populated by the enhanced decay of the 2^+ member of the intruder structure, decays by an enhanced E2 transition to the one-phonon state in ^{116}Cd. This behavior is contrary to that observed in the lighter Cd nuclei, where the lowest excited 0^+ state acts as the lowest-spin member of the two-phonon triplet. Configuration-mixing calculations show that it is the lowest excited 0^+ state that exhibits an enhanced decay and acts as the two-phonon member of the multiplet. In addition, the strong-mixing explanation applied to explain the lighter Cd nuclei leads a splitting between the 0^+ states of ^{116}Cd that is much too large.

Present And Future Studies

In the hope of clarifying the role of phonon-intruder mixing, we have turned to ^{114}Cd, the mid-neutron-shell Cd isotope. Rick Casten and his colleagues [17] measured lifetimes in the 3-phonon region of ^{114}Cd with the GRID (γ-ray induced Doppler) technique and commented on the unusual structure of this nucleus, which they described as an "enigma". They found that they could account for the B(E2) values they obtained with both mixing (between intruder and phonon states) and no-mixing approaches, although the latter led to large energy anharmonicities. They regarded this remarkable ambiguity as having origins in the basic properties of mixing between quantized excitations. In an extensive Coulomb excitation study with several projectiles, Fahlander et al. [18] determined a large number of E2 matrix elements in this nucleus but, despite this wealth of information, they were unable to choose meaningfully between two differing descriptions. They point out that the properties of the low-lying states can be explained partially in terms of mixing between a vibrational structure and a deformed intruder band; however, they raise serious arguments against this picture and conclude that these intruder states are interpreted best as vibrational structures.

There are a number of motivations for re-investigating ^{114}Cd. From systematic studies of the intruder excitations in the even-A Cd nuclei, it has been shown that, as expected, these states reach a minimum in energy at the mid-neutron-shell nucleus ^{114}Cd, while the ground-state properties of these nuclei are relatively constant [3,4]. The intruder 0^+ and 2^+ states are close to the two-phonon vibrational states, although not as close as in ^{116}Cd. Even the extensive GRID [17] and Coulomb excitation [18] measurements have left some important B(E2) values undetermined. We are measuring the lifetimes of the three-phonon and higher states (four-phonon candidates have been suggested [17]) in ^{114}Cd and hope to address the questions raised above. In addition, we have tentatively identified a single 2^+ mixed-symmetry state in ^{114}Cd at 2219 keV, an energy similar to the 2^+ mixed-symmetry states observed in ^{112}Cd [14]. Configuration mixing of this isovector quadrupole excitation with the intruder 2^+ state is anticipated [15,16].

ACKNOWLEDGMENTS

I wish to thank many colleagues who have contributed to these studies, especially Paul Garrett, Jan Jolie, Marcus McEllistrem, and Nigel Warr. This work was supported by the U. S. National Science Foundation under Grant No. PHY-0098813.

REFERENCES

1. Kern, J., Garrett, P. E., Jolie, J., and Lehmann, H., *Nucl. Phys.* **A593**, 21 (1995).
2. Heyde K., Van Isacker, P., Waroquier, M., Wenes, G., and Sambataro, M., *Phys. Rev.* **C25**, 3160 (1982).
3. Kumpulainen, J., Julin, R., Kantele, J., Passoja, A., Trzaska, W. H., Verho, E., Väärämäki, J., Cutoiu, D., and Ivascu, M., *Phys. Rev.* **C45,** 640 (1992).
4. Juutinen, S., Julin, R., Jones, P., Lampinen, A., Lhersonneau, G., Mäkelä, E., Piiparinen, M., Savelius, A., and Törmänen, S., *Phys. Lett.* **B386**, 80 (1996).
5. Lehmann, H., Garrett, P. E., Jolie, J., McGrath, C. A., Yeh, M., and Yates, S. W., *Phys. Lett.* **B387**, 259 (1996).
6. Corminboeuf, F., Brown, T. B., Hannant, C. D., Genilloud, L., Jolie, J., Kern, J., Warr, N. and Yates, S. W., *Phys. Rev. Lett.* **84**, 4060 (2000).
7. Corminboeuf, F., Brown, T. B., Hannant, C. D., Genilloud, L., Jolie, J., Kern, J., Warr, N. and Yates, S. W., *Phys. Rev.* **C63**, 014305 (2001).
8. Kadi, M., Warr, N., Garrett, P. E., Martin, A., Jolie, J., and Yates, S. W., to be published.
9. Reynolds, C. C., *et al.*, to be published.
10. Garrett, P. E., Warr, N., and Yates, S. W., *J. Res. Natl. Inst. Stand. Technol.* **105**, 141 (2000), and references therein.
11. McGrath, C. A., Garrett, P. E. , Villani, M. F., and Yates, S. W., *Nucl. Instrum. Methods Phys. Res.* **A421**, 458 (1999).
12. Belgya, T., Molnár, G., and Yates, S. W., *Nucl. Phys.* **A607**, 43 (1996).
13. Aprahamian, A., Brenner, D. S., Casten, R. F., Gill, R. L., and Piotrowski, A., *Phys. Rev. Lett.* **59**, 535 (1987).
14. Garrett, P. E., Lehmann, H., McGrath, C. A., Yeh, M., and Yates, S. W., *Phys. Rev.* **C54**, 2259 (1996).
15. Jolie, J., and Lehmann, H., *Phys. Lett.* **B342**, 1 (1995).
16. Gade, A., Jolie, J., and von Brentano, P., Phys. Rev. **C65**, 041305 (2002).
17. Casten, R. F., Jolie, J., Börner, H. G., Brenner, D. S., Zamfir, N. V., Chou, W.-T., and Aprahamian, A., *Phys. Lett.* **B297**, 19 (1992).
18. Fahlander, C., *et al.*, *Nucl. Phys.* **A485**, 327 (1988).

68

Q-phonons, Q-invariants, and company

P. von Brentano, A. Gade , N. Pietralla, and V. Werner

Institut für Kernphysik, Universität zu Köln, Köln, 50937 Köln, Germany

Abstract. The paper discusses the concept of Q-phonons and its connection to the concept of Q-invariants and the shape parameters of nuclei. It will also discuss some useful relations between Q-invariants and observables. These relations allow one to determine crucial nuclear observables such as the square of the quadrupole moment of the first 2^+ state from lifetime data of the gamma- and ground band which may have applications, in future measurements with rare isotopes beams. These concepts are discussed for the Barium, Xenon and Cerium nuclei with mass numbers around $A = 130$, because some of these concepts were either introduced or at least heavily used in the discussion of the nuclear structure of these nuclei.

INTRODUCTION

This paper is dedicated to the scientific work of Professor R. F. Casten on the occasion of his sixtieth birthday. We had the pleasure to work with him on many projects. On this occasion we remember many fruitful discussions, and discussion is very much at the heart of science. The paper discusses the Q-phonon scheme and its connection to the Q-invariants. The discussion will be made in the frame of the IBM model [1, 2, 3, 4] although it can be extended to other collective models [1, 2, 3, 4, 5, 6, 7] and to the shell model. Let us start by reminding a joint paper with Rick Casten [8], in which the dynamical O(6) symmetry [2] of the IBM was proposed for the Xenon and Barium region with mass numbers around A = 130. Of course prior to that paper Casten and Cizewski had identified ^{196}Pt as the O(6) nucleus but the Xe - Ba - Ce region is a much more extensive nuclear region.

What happened after that in the A=130 region and what do we still not know ? The collective bands were studied in great detail for A=124-132 in Cologne with " complete" gamma spectroscopy, that is with cold fusion reactions such as the (α,n), $(^3He, n)$, $(\alpha ,2n)$, $(^{13}C, 3n)$ reactions and also from the gamma decay following beta decay , which proved to be very powerful tools.

In many nuclei in this region 6 collective positive parity band heads were observed. As an example we cite recent Cologne work on ^{124}Xe [9], ^{126}Xe [10] and ^{132}Ba [11]. In these works both energies and B(E2) values from branched gamma decay were described very well by the IBM-1 as is shown for ^{124}Xe in fig.1

In the work by A. Dewald [12] absolute B(E2) values were studied. By the introduction of new methods he was able to obtain precision lifetime data for yrast bands. For the 2_1^+ states these new values follow very well the famous $N_p N_n$–law proposed by Casten [13]. In many cases the new and old values of the lifetimes differ by more than a factor 2 because the old data were from singles experiments where feeding and deorientation

CP638, *Mapping the Triangle: International Conference on Nuclear Structure*
edited by A. Aprahamian, J. A. Cizewski, S. Pittel, and N. V. Zamfir
© 2002 American Institute of Physics 0-7354-0093-8/02/$19.00

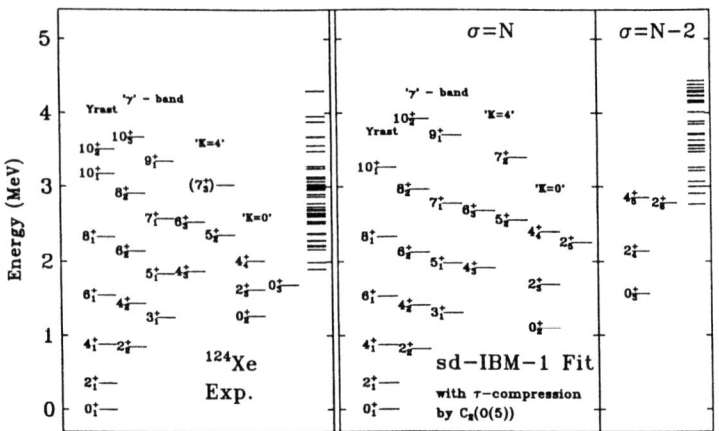

FIGURE 1. Collective bands in ^{124}Xe. Cologne data from V. Werner *et al.* [9]

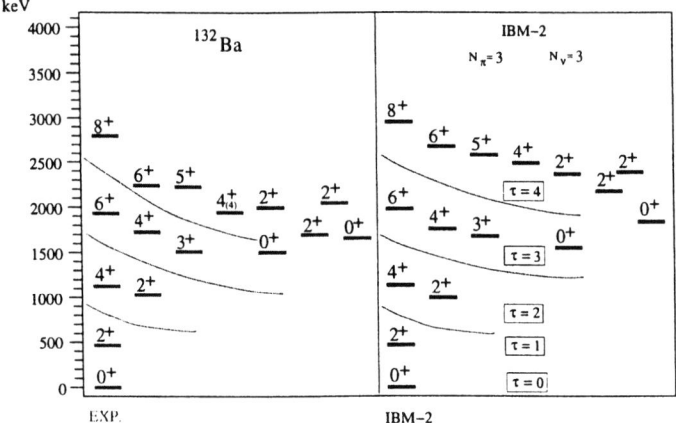

FIGURE 2. Collective bands in ^{132}Ba. Cologne data from A. Gade *et al.* [11]

is a problem.

THE Q-PHONON SCHEME

In the Q-phonon scheme the wave functions of different collective states are described by successive actions of the quadrupole transition operator Q on the ground state wave function. This scheme was suggested as an exact scheme for the O(6) dynamical symmetry by Otsuka and Kim [14] and was developed as an approximate scheme for all collective nuclei by the Cologne - Dubna - Tokyo - Yale collaboration [15, 16, 17, 18, 19, 20]. As an example the wave function of the first 2^+ state is given by :

$$|2_1^+\rangle = \mathcal{N}_Q Q |0_1^+\rangle + |R>$$

where the normalization constant $\mathcal{N}_Q$ is given by the expression

$$\frac{1}{\mathcal{N}_Q} = \sqrt{\frac{<0_1|(QQ)^{(0)}|0_1^+>}{1 - <R|R>}}$$

and where the rest wave function $|R>$ has a small length. This is shown in a paper by Pietralla *et al.* [16] where the value of the "length" of the "rest wavefunction" $<R|R>$ plotted from data for many nuclei is small. Clearly such description breaks down near magic shells. This finding has been recently supported by the results of microscopic calculations in terms of the shell model done by the Tokyo group [22]. The Q-phonon scheme suggests the existence of selection rules for the matrix elements of the Q operator which are supported by the vast amount of data in the A= 130 region, *e.g.* [9, 10, 11, 15, 23, 24, 25]. The "wavefunctions" of the Q-phonon scheme are of course, implicit as nothing is said about the structure of the ground state, which is highly complex, in general.

Q-INVARIANTS

The normalization constants in the wave functions of the Q-phonon scheme are expectation values of multiple moments of the quadrupole operator. The 2nd and 3rd moments of Q : q_2 and q_3 where extensively used by Kumar and Cline [26, 27, 28].

From the Q-invariants we can define [29] the dimensionless shape parameters K_n by normalizing relative to an appropriate power of q_2

$$K_n = \frac{q_n}{q_2^{n/2}} \text{ for } n \in \{3,4,5,6\} ; \quad q_n \propto \langle 0_1^+ | \overbrace{Q \dots Q}^{n-\text{times}} |0_1^+\rangle .$$

The shape parameters q_2 and K_3 are a generalization of the usual shape parameters beta and gamma to nuclei with a soft vibrating surface. In the rigid triaxial rotor one finds:

$$q_2 = \left(\frac{3ZeR^2}{4\pi}\right)^2 \langle \beta^2 \rangle \equiv \left(\frac{3ZeR^2}{4\pi}\right)^2 \beta_{\text{eff}}^2 ,$$

$$K_3 = \frac{\langle \beta^3 \cos 3\gamma \rangle}{\langle \beta^2 \rangle^{3/2}} \equiv \cos 3\gamma_{\text{eff}} ,$$

These two equations define effective values of the deformation parameters β_{eff} and γ_{eff}.

TABLE 1. Values of various parameters and observables for limit cases of IBA and for three nuclei, the boson number is $N_B = 8$ and the effective charge is $e_B = 0.13$ (from reference [9]).

χ	nucleus	q_2	K_3	K_4	K_5	K_6	γ_{eff}	σ_β	$\sigma_{\gamma\beta}$
0	O(6)	-	0	1	0	0.3	30^0	0	0.3
-1.33	U(5)	-	0	1.36	0	0.6	30^0	0.36	0.6
-1.33	SU(3)	-	1	1	1	1	0^0	0	0
-1.33	X(5)	-	0.93	1.17	1.2	1.33	21^0	0.17	0.47
0	E(5)	-	0	1.15	0	0.41	30^0	0.15	0.4
-0.26	^{124}Xe	1.55	0.27	1.11	0.34	0.42	25^0	0.11	0.35
-0.18	^{126}Xe	1.08	0.27	1.03	0.25	0.33	26^0	0.03	0.27
-0.04	^{134}Ce	1.14	0.20	1.03	0.20	0.30	26^0	0.03	0.26
0	O(6)	-	0	1	0	0.3	30^0	0	0.3

The shape invariants are measures of effective deformation parameters and their fluctuations. This is made more explicit by defining the following quantities as measures of the fluctuations of β^2 and of $\beta^3 \cos 3\gamma$:

$$\sigma_\beta = \frac{\langle \beta^4 \rangle - \langle \beta^2 \rangle^2}{\langle \beta^2 \rangle^2} = K_4 - 1,$$

$$\sigma_{\gamma\beta} = \frac{\langle \beta^6 \cos^2 3\gamma \rangle - \langle \beta^3 \cos 3\gamma \rangle^2}{\langle \beta^2 \rangle^3} = K_6 - K_3^2 .$$

We note that in the IBM-1 the theoretical values for K_4 are 1 for the SU(3) and the O(6), and 1.4 for the U(5) dynamical symmetry, distinguishing between β-rigid and vibrational nuclei, respectively. The fluctuation parameters $\sigma_\beta = K_4 - 1$ and $\sigma_{\gamma\beta} = K_6 - K_3^2$ are of special interest as they allow to distinguish between rigid and soft potentials in β and γ, respectively. We note that in previous papers we have referred to the fluctuation parameter $\sigma_{\gamma\beta}$ as σ_γ. This parameter, however, depends also very strongly on β. Q-invariants have been studied also in the GCM by the Yale group. In order to be concrete we give values of the K_n for some dynamical symmetries and for the nuclei ^{124}Xe, ^{126}Xe and ^{134}Ce in table 1.

APPROXIMATE EXPRESSIONS FOR THE SHAPE PARAMETERS Q_2 AND K_N IN TERMS OF NUCLEAR OBSERVABLES.

From the Q-invariants one can deduce information about nuclear deformations. The value of the Q-invariants and shape parameters q_2, K_n can be calculated directly from the IBM. Alternatively q_2 and K_3 have been obtained in the work of D. Cline and coworkers directly from the $E2$ amplitudes measured in Coulomb excitation [27, 28], namely

$$q_2 \approx B(E2; 0_1^+ \rightarrow 2_1^+),$$

TABLE 2. Comparison of $Q^2(2^+)$ from B(E2) relation with data (*cf.* reference [30]).

$Q^2(2^+)$	^{188}Os	^{190}Os	^{192}Os	^{194}Pt
relation	1.80(5)	1.14(10)	0.56(6)	-0.13(6)
exp.	1.72(35)	0.90(30)	0.84(22)	0.20(6)
$Q^2(2^+)$	^{196}Pt	^{114}Cd	^{106}Pd	^{160}Gd
relation	0.26(9)	0.31(4)	0.28(7)	4.14(6)
exp.	0.24(18)	0.13(6)	0.30(6)	4.33(17)

$$K_3 \approx \sqrt{\frac{7}{10}} \frac{<2_1||Q||2_1>}{<2_1||Q||0_1>} \; .$$

Recently some approximate relations, which allow one to determine K_3 and in addition K_4 from more easily accessible observables were given in the work of Jolos and coworkers [18, 30], namely

$$B(E2; 2_1^+ \to 2_1^+) + B(E2; 2_2^+ \to 2_1^+) \approx B(E2; 4_1^+ \to 2_1^+)$$

where we define $B(E2; 2_1^+ \to 2_1^+) \equiv 1/5 \langle 2_1^+||Q||2_1^+ \rangle^2$ and

$$K_4 \approx \frac{7}{10} \frac{B(E2; 4_1^+ \to 2_1^+)}{B(E2; 2_1^+ \to 0_1^+)} \; .$$

The above relation of $B(E2)$ values connects in a remarkably simple way three fundamental quantities of nuclear quadrupole collectivity: the square of the quadrupole moment of the 2_1^+ state, and the quadrupole transition strengths within the ground band and between ground and gamma band. This is a quite beautiful - albeit approximate - relation , which seems to agree with the data as is shown in table 2. It has been also checked in numerical calculations on the Casten triangle for N = 10 to an accuracy of 5-10% for reasonably big quadrupole moments [18, 30]. Clearly more extensive tests of this relation are needed. Such tests are important, because this relation may be the only feasible method to measure quadrupole moments in future work with rare isotopes beams.

Q-PHONON SCHEME WITH PROTONS AND NEUTRONS.

The Q-phonon scheme has been extended to the proton - neutron version IBM-2 in the work of Kim, Otsuka *et al.* [31]. The proper Q operator is again the $T(E2)$ operator : $T(E2) = e_\pi Q_\pi(\chi) + e_\nu Q_\nu(\chi)$ as this operator connects with the empirically observed $E2$ transitions. A considerable simplification occurs if we consider only states with good and maximal F-spin. In this case any of the operators $T(E2)$, $Q_\pi(\chi)$, $Q_\nu(\chi)$, $Q_\pi(\chi) + Q_\nu(\chi)$ can alternatively serve as the Q-operator of the Q-phonon scheme as

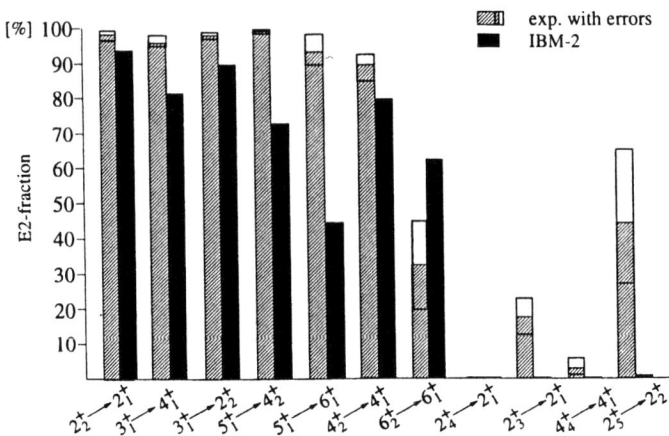

FIGURE 3. Comparison of the experimental and theoretical $E2$ fractions in the collective transitions in ^{126}Xe, Cologne data (from A. Gade *et al.* [10]).

their matrix elements in the $F = F_{max}$ space are all proportional to each other. The $M1$ transitions between collective states and the purity of the F-spin has recently been studied in the $A = 130$ region in the work of Alexandra Gade and Ingo Wiedenhoever [10, 11] at Cologne. We show a comparison of ^{132}Ba with IBM-2 in figure 2. A summary of their result is that below the 6^+_2 state there are only negligible $M1$ contributions to the dominant $E2$ decay, which indicates good F-spin. We show an example of this work for ^{126}Xe [10] in figure 3. But on the other hand there is an interesting paper by Jacob *et al.* [32], who found a severe breaking of F-spin in the g-factors of ^{132}Xe. Perhaps this is due to the near $N = 82$ shell, but it is certainly an important question.

To summarize: the Q-phonon scheme gives a global form of the wave functions of the low lying collective states in terms of special parameters namely the Q-invariants q_n [15, 16, 17, 18, 19, 20]. This scheme leads to approximate selection rules for $E2$ transitions, which are well fulfilled by data and generalize the selection rules, which are obtained from the quantum numbers of the dynamical symmetries. The Q-invariants q_2 and K_n characterize the fluctuating shape of the nucleus and thus generalize the usual shape parameters β and γ which apply to a rigid triaxial core only. The Q-phonon scheme gives a simple method to derive useful albeit approximate relations for various q_n and K_n in terms of direct observables. This is based on the fact that corrections from the non-commutativity of the components of the quadrupole moment operator in the IBM are small. The accuracy of these relations can be checked for the parameters of the Casten triangle. Finally the Q-phonon scheme allows to compare different nuclear models. A very interesting question concerns the relation of the Q-phonon scheme with the shell model and here we will hear about the work of the Tokyo group. A field in which we still know little is the proton neutron structure of the collective excitations. Clearly we need more data in this region. Particularly important would be g factors and quadrupole moments as well as lifetimes of the gamma band.

ACKNOWLEDGMENTS

For fruitful discussions and for joint work we thank Drs: R. F. Casten, A. Dewald, C. Fransen, A. Gelberg, R. V. Jolos, J. Jolie, R. Krücken, T. Mizusaki, T. Otsuka, I. Wiedenhoever and N.V. Zamfir. This work has been partly supported by the Deutsche Forschungsgemeinschaft under Contracts No. Br 799/11-1, Br 799/10-2, and Pi 393/1-2.

REFERENCES

1. A.Arima and F.Iachello, Phys. Rev. Lett. **35**, 1069 (1975).
2. F.Iachello and A.Arima, *The Interacting Boson Model* (University Press, Cambridge, 1987).
3. F. Iachello, P.Van Isacker, *The Interacting Boson-Fermion Model* (Cambridge University Press, Cambridge, 1987).
4. T. Otsuka, A. Arima,F. Iachello, I. Talmi, Phys. Lett. **B76**, 139 (1978).
5. A.S. Davydov and G.F.Fillipov, Nucl. Phys. **8**, 237 (1958).
6. G.Gneuss, W.Greiner, Nucl. Phys. **171**, 449 (1971).
7. R.V.Jolos, F. Doenau, D. Janssen, Nucl. Phys. **A224**, 740 (1974).
8. R.F. Casten and P. von Brentano, Phys. Lett. **152**, 22 (1985).
9. V. Werner, H. Meise, I. Wiedenhoever, A. Gade and P. von Brentano, Nucl. Phys. **A692**, 451 (2001).
10. A. Gade, I. Wiedenhoever, M. Luig, A. Gelberg, H. Meise, N. Pietralla, V. Werner and P. von Brentano, Nucl. Phys. **A665**, 268 (2000).
11. A. Gade , I. Wiedenhoever, H. Meise, A. Gelberg and P. von Brentano, Nucl. Phys. **A697**, 75 (2002).
12. J. Gableske, A. Dewald *et al.*, Nucl. Phys. **A 691** 551 (2001).
13. R. F. Casten, *Nuclear structure from a simple perspective*, 2nd edition , Oxford University Press, 2000).
14. T. Otsuka and K.-H. Kim, Phys. Rev. C **50**, R1768 (1994).
15. G. Siems,U. Neuneyer,I.Wiedenhover, S.Albers, M.Eschenauer, R.Wirowski, A.Gelberg, P. von Brentano,T. Otsuka, Phys. Lett. **320B**, 1 (1994).
16. N.Pietralla, P.von Brentano, R.F.Casten, T.Otsuka,N.V.Zamfir, Phys. Rev. Lett. **73**, 2962 (1994).
17. N.Pietralla, P.von Brentano, T.Otsuka, R.F.Casten, Phys. Lett. **B349**, 1 (1995).
18. R.V. Jolos, P. von Brentano, N. Pietralla, I. Schneider, Nucl. Phys. **A 618**, 126 (1997).
19. Yu. V. Palchikov, P. von Brentano, and R. V. Jolos, Phys. Rev. C **57**, 3026 (1998).
20. N. Pietralla, T. Mizusaki, P. von Brentano, R. V. Jolos, T. Otsuka, and V. Werner, Phys. Rev. C **57**, 150 (1998).
21. D. D. Warner and R. F. Casten, Phys. Rev. Lett. **48**, 1385 (1982).
22. N. Shimizu, T. Otsuka, T. Mizusaki, M. Honma, Phys. Rev. Lett. **86**, 1171 (2001).
23. C. Königshofen, K. Jessen, A. Gade, I. Wiedenhoever, H. Meise, P. von Brentano, Phys. Rev. C **64**, 037302 (2001).
24. A. Gade, H. Meise, I. Wiedenhoever, A. Schmidt, A. Gelberg, P. von Brentano, Nucl. Phys. **A686**, 3 (2001).
25. P. Petkov *et al.*, Phys. Rev. C **62**, 014314 (2000).
26. K. Kumar, Phys. Rev. Lett. **28**, 2 D (1972).
27. D. Cline, Ann. Rev. Nucl. Part. Sci. 36 (1986) 683, and references therein.
28. D. Cline, Acta Phys. Pol. B **30**, 1291 (1999).
29. V. Werner, N. Pietralla, P. von Brentano, R.F. Casten and R.V. Jolos, Phys. Rev. C **61**, 021301 (2000).
30. V. Werner, P. von Brentano and R. V. Jolos, Phys. Lett **B527**, 55 (2001).
31. K.-H.Kim, T. Otsuka , P. von Brentano, A. Gelberg, P. Van Isacker, R. F. Casten, in "*Capture gamma ray spectroscopy and related topics*" ed. A. Molnar (Spinger, Budapest, 1996), Vol.I, p. 195.
32. G. Jacob, N. Bencer-Koller *et al.*, Phys. Rev. C **65**, 024316 (2002).

What is the nature of $K^{\pi} = 0^{+}$ bands in deformed nuclei?

A. Aprahamian

Department of Physics, University of Notre Dame, Notre Dame, IN 46556

Abstract. New data is presented on $K^{\pi} = 0^{+}$ bands in ^{156,158}Gd and ^{178}Hf. Each of these nuclei exhibit a different aspect of the nature of $K^{\pi} = 0^{+}$ bands in deformed nuclei. In ^{156}Gd, there is evidence of a $K^{\pi} = 0^{+}$ band with significant two-phonon $\gamma\gamma$ component whereas in ^{178}Hf results are shown for a $K^{\pi} = 0^{+}$ band whose members exhibit two-phonon $\beta\beta$ character. New (p,t) reaction results point to the existence of thirteen excited 0^{+} states in ^{158}Gd below an excitation energy of 3.1 MeV. All the results combine to provide a challenge to our present understanding of the nature of these excitations.

INTRODUCTION

The nature of 0^{+} states is one of most important open questions in nuclear structure today. These states are crucial to the discussion of vibrational structure in spherical nuclei, of critical point symmetries, as well as the nature of excitations in well-deformed nuclei. My paper focuses on the nature of $K^{\pi} = 0^{+}$ bands in the rare-earth region of deformation. Traditionally, deformed nuclei are described as rotors with superimposed excitations where the first excited $K^{\pi} = 0^{+}$ bands along with the $K^{\pi} = 2^{+}$ bands were labeled as single-phonon "β" and "γ" vibrational excitations. The $K^{\pi} = 2^{+}$ excitations are well understood theoretically and shown to vary smoothly in collectivity across a given isotopic chain[1, 2, 3]. The nature of $K^{\pi} = 0^{+}$ excitations however remains enigmatic with an associated flurry of activity from both theoretical and experimental camps. Two recent review publications summarize the existing data on $K^{\pi} = 0^{+}$ bands and several possible interpretations in Ref.[4, 5]. The results are puzzling at best. In many deformed nuclei of this region, there are several excited $K^{\pi} = 0^{+}$ bands below the pairing gap. There are variations in collectivity amongst the $K^{\pi} = 0^{+}$ bands in the same nucleus, as well as enormous variations in collectivity of the first excited $K^{\pi} = 0^{+}$ bands in very narrow isotopic regions.

There has been an extensive discussion in the literature on the nature of $K^{\pi} = 0^{+}$ bands as excitations built on the first $K^{\pi} = 2^{+}$ γ-band. Examples include the ^{168}Er[1, 6], and ^{162,164}Dy nuclei[7]. In all these cases the first excited $K^{\pi} = 0^{+}$ band is above the $K^{\pi} = 2^{+}$ band by several hundred keV. In this paper, we explore nuclei where the $K^{\pi} = 0^{+}$ bands are very close in excitation energy or below the $K^{\pi} = 2^{+}$ bands. The goal here is not to present a definitive answer to the question that I pose in the title but to present new data on some aspects of the nature of $K^{\pi} = 0^{+}$ bands in deformed nuclei. These nuclei include ^{156,158}Gd and ^{178}Hf. Data on ^{156}Gd and ^{178}Hf were extracted from lifetime measurements carried out at the ILL in Grenoble, whereas the ^{158}Gd results

CP638, *Mapping the Triangle: International Conference on Nuclear Structure*
edited by A. Aprahamian, J. A. Cizewski, 3. Pittel, and N. V. Zamfir
© 2002 American Institute of Physics 0-7354-0093-8/02/$19.00

are from a (p,t) measurement carried out at the high-precision Q3D spectrometer of the University of Munich.

EXPERIMENTS

The lifetime measurements of levels in ^{178}Hf and ^{156}Gd were carried out using the GRID technique[8]. GRID allows lifetime measurements of levels populated in thermal neutron capture reactions from the Doppler broadening of a transition affected by the recoil of a previously emitted γ-ray. The recoil velocities are very small (typically 10^{-4} to 10^{-6} of c) with resulting Doppler shifts in the order of a few eV and very short slowing-down times in the target. The last point limits the optimum range of accessible lifetimes to a few picoseconds and lower. The essential ingredients that contribute to the line shape of a selected transition are the slowing-down process in the target material, the initial recoil velocity distribution feeding the level from which the transition is de-exciting, temperature of the target, and the lifetime of the state of interest. If the first three are known (or determined), the lifetime can be extracted from the line-shape. The main uncertainty in heavy nuclei comes from the feeding of the chosen level. The line shape of a particular transition is measured using a double flat crystal spectrometer (GAMS4)[9, 10] installed 15m from the core of the high flux reactor of the ILL in Grenoble.

The ^{158}Gd was studied at the high-precision Q3D spectrometer of the University of Munich MP tandem accelerator laboratory using a 27 MeV proton beam on a target of isotopically enriched ^{160}Gd with a Carbon backing. A 1.8m long focal plane detector provided the particle identification of the ejectiles of mass 1-4 in the Q3D spectrometer[11] with an energy resolution of approximately 4-6 keV for the energy range of interest from 1 to 3 MeV. Discussion of experimental results are given below for each nucleus.

^{156}Gd

Lifetimes from a total of twelve levels from four different excitation bands were measured in ^{156}Gd using the GRID technique. The four excited bands include three $K^{\pi} = 0^{+}$ bands beginning at 1049.5 keV, 1168.2 keV, and 1715.2 keV and a $K^{\pi} = 2^{+}$ band at 1154.2 keV. Nine of the measurements had previously been unreported and the other three lifetimes agree well with previously measured results. Figure 1 shows the relevant levels along with the extracted range of level lifetimes. Transitions from the $K^{\pi} = 0_2^{+}$ band at 1049.5 keV to the ground state band are strongly collective and therefore indicative of β phonon vibrational character. B(E2) calculations for transitions from the $K^{\pi} = 2^{+}$ band to the ground state band supports the assignment of this band as the γ band. Transitions from the $K^{\pi} = 0_3^{+}$ band to the ground state band are not collective while the $K^{\pi} = 0_4^{+}$ band at 1715 KeV is strongly connected to the $K^{\pi} = 2^{+}$ γ band and is evidence of a significant $\gamma\gamma$ multi-phonon vibrational component for this band. There is to date only one other example of a nucleus with significant $\gamma\gamma$ character and it is

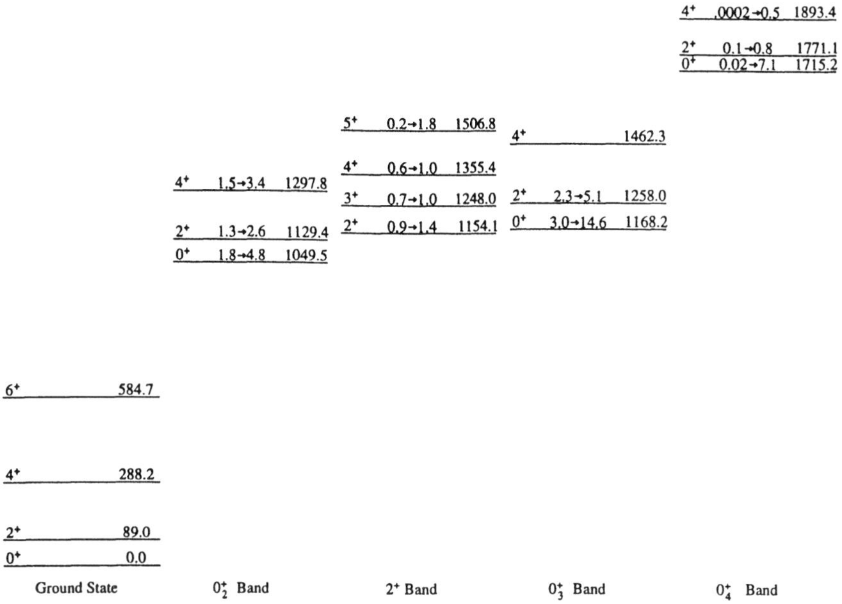

FIGURE 1. A partial level scheme of ^{156}Gd showing all levels with ranges of measured lifetimes.

in the ^{166}Er nucleus[12]. The main difference between the two is the observed level of anharmonicity for the vibrational bands. The excitation energy ratio of the two-phonon $E_{K^\pi=0^+}/E_{K^\pi=2^+}$ is 2.47 in ^{166}Er whereas the same ratio is 1.49 in ^{156}Gd.

^{178}Hf

Lifetimes of several $K^\pi = 0^+$ bands have been measured. The results reveal for the first time the existence of two excited $K^\pi = 0^+$ bands connected by strongly collective transitions. Figure 2 shows a partial level scheme of ^{178}Hf with all the known excited $K^\pi = 0^+$ bands in this nucleus. The B(E2) values from the 2^+ members of the second and third excited $K^\pi = 0^+$ bands to the g.s. band would perhaps raise a question regarding the identification of either band as the collective one-phonon β-vibrational excitation. However, the observed preference of decay from the 2^+ and 4^+ members of the $K^\pi = 0_5^+$ band at 1772.2 keV band is to the band at 1199.4 keV. This is compatible with the expected behavior of a collective vibrational excitation built on the 1199.4 keV band. The outstanding question is the collectivity of the first excited $K^\pi = 0^+$ band. Our new measurements are in excellent agreement with coulomb excitation results for all three states. Our results indicate some significant degree of collectivity for the first excited $K^\pi = 0^+$ band. The most important result is the indication for the first time of the existence of a collective $K^\pi = 0^+$ excitation built on an excited $K^\pi = 0^+$ band in any nucleus. The $K^\pi = 0_2^+$ band is less collective but in the same order of magnitude as

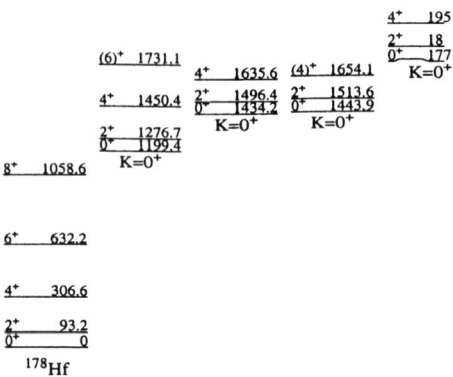

FIGURE 2. All the known $K^{\pi} = 0^{+}$ bands in the ^{178}Hf nucleus.

some of the single-phonon γ-vibrational bands in this region of deformed nuclei. If the collectivity of the first excited $K^{\pi} = 0^{+}$ band can be more forcefully established, then this work would also point to the first observation of a significant two-phonon $\beta\beta$ vibrational character. Five-band mixing calculations were carried out to determine if bandmixing gave rise to the observed collective transitions but we found it impossible to reproduce the collective transitions between the two excited bands by any $K=2$ or 0 mixing matrix elements[3].

^{158}Gd

The new high-precision (p,t) reaction on an isotopically enriched target of ^{160}Gd allowed the identification of thirteen excited $K^{\pi} = 0^{+}$ bands below 3.1MeV in excitation energy in the spectrum of ^{158}Gd. Three of these 0^{+} states were previously identified. Four of the thirteen had previous tentative assignments. We confirm three of these four to be 0^{+} states. In addition, there were seven new 0^{+} assignments. The new 0^{+} assignments are further strengthened by the placement of γ-rays that were clearly identified to belong to the ^{158}Gd nucleus with no previous level assignments. Such an abundance of 0^{+} states has not previously been seen in nuclei until the very present. A similar study of an N=84 nucleus with (p,t) had resulted in the observation of ten 0^{+} states[13] below an excitation energy of 4.1 MeV in the ^{146}Sm nucleus. In that particular case, a particle-core coupling model could provide a remarkable explanation for the nature of the observed 0^{+} states. In the case of a well-deformed nucleus, such as ^{158}Gd, it is not yet possible to decipher the nature of all the observed 0^{+} states. It is therefore essential to carry out lifetime measurements of the new 0^{+} states in order to get an insight into the nature of these states and their collectivity. With the present measurement, ^{158}Gd provides an unprecedented opportunity for the investigation of the nature of $K^{\pi} = 0^+$ bands. The observation of thirteen excited $K^{\pi} = 0^+$ excitations in one nucleus below an excitation energy of 3.1 MeV will be the strongest challenge yet to our understanding of $K^{\pi} = 0^+$ excitations.

CONCLUSIONS

The three nuclei presented here all show different aspects of the nature of $K^{\pi} = 0^+$ bands in deformed nuclei. In ^{156}Gd, the first excited $K^{\pi} = 0^+$ band is collective to the g.s. (even more so than the first excited $K^{\pi} = 2^+$ γ band) and behaves as one would expect for a vibrational excitation built on a deformed g.s. The excited $K^{\pi} = 0_4^+$ band at 1715 keV and its members ($0^+,2^+,4^+$)show enhanced collectivity in transitions to the first excited $K^{\pi} = 2^+$ band indicating the existence of significant two-phonon $\gamma\gamma$ components in this band. The example in ^{178}Hf shows a collective $K^{\pi} = 0_4^+$ band built on the first excited $K^{\pi} = 0^+$ band at 1199 keV. Furthermore, the abundance of 0^+ states observed in ^{158}Gd below an excitation energy of 3.1 MeV not only presents a unique challenge to our understanding of these states but also provides a true playground for further investigations into the nature of these states!

ACKNOWLEDGMENTS

Motivation for this project on the nature of 0^+ states and all types of vibrational structure along with the inspiration to pursue them comes from R.F. Casten to whom I owe a debt of gratitude. This research has been funded by the National Science Foundation under grant number 99-01133. Finally this work is the result of extensive collaborations with H. Boerner (ILL), R.C. de Haan (ND), S. Lesher (UK), P. Isacker (GANIL), A.M. Bruce (Brighton), G. Graw (Munich), and L. Trache (Texas A&M).

REFERENCES

1. Günther, C. et al., *Phys. Rev. C* **54**, 679 (1996).
2. Aprahamian, A. et al., *J. Phys. G* **25**, 685 (1999).
3. Aprahamian, A. et al., *Phys. Rev. C* **65**, R 031301 (2002).
4. Heyde, K. and Wood, J., *Nucl. Phys. A* **682**, 482c (2001).
5. Garrett, P.E., *J. Phys. G* **27**, R1 (2001).
6. Casten, R.F. and von Brentano,P. *Phys. Rev. C***50**, R1280 (1994).
7. Lehmann, H. et al., *Phys. Rev. C* **57**, 569 (1998).
8. Börner,H. and Jolie, *J. Phys. G: Nucl. Part. Phys.***19**, 217 (1993) and all the references therein.
9. Dewey, M.S. et al., *Nucl. Instrum. Meth. A* **284**, 151 (1989).
10. Kessler, E.G. et al., *J. Phys. G: Nucl. Phys.* **14**, 167 (1988).
11. Zanotti, E. et al., *Nucl. Instrum. Meth. A* **310**, 706 (1991).
12. Garrett, P.E., et al., *Phys. Rev. Lett.* **78**, 4545 (1997)
13. Oros, A.M., et al., *Nucl. Phys. A***613**, 209 (1997).

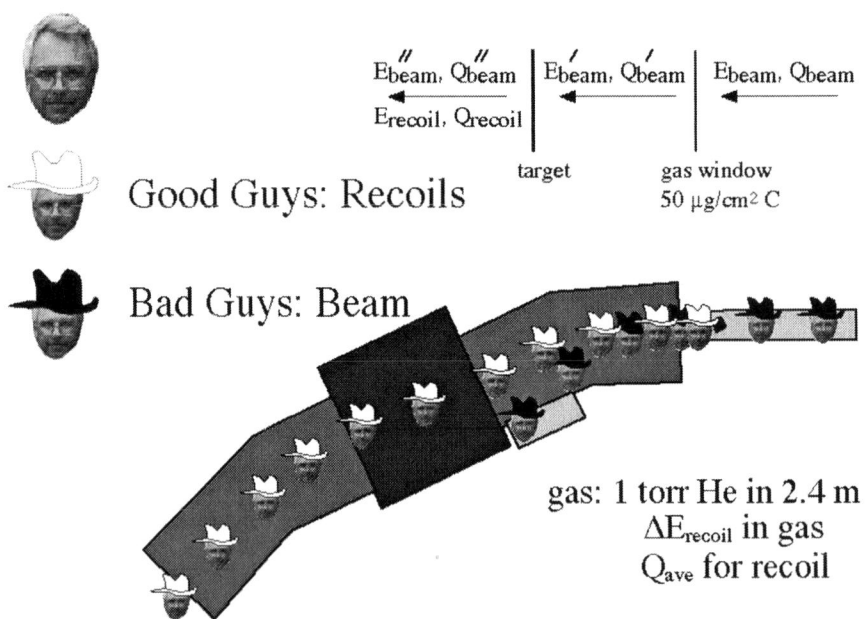

$$E''_{beam}, Q''_{beam} \leftarrow E'_{beam}, Q'_{beam} \leftarrow E_{beam}, Q_{beam}$$
$$E_{recoil}, Q_{recoil}$$

target

gas window
50 µg/cm² C

Good Guys: Recoils

Bad Guys: Beam

gas: 1 torr He in 2.4 m
ΔE_{recoil} in gas
Q_{ave} for recoil

High Resolution Gamma Spectroscopy at ILL: Past and Future

H.G. Börner, M. Jentschel, P. Mutti

Institut Laue Langevin, F38042 Grenoble, Cedex 9, France

Abstract. To obtain the highest possible resolution has always been a major challenge in high resolution γ–ray spectroscopy, using crystal diffraction at the Institut Laue Langevin (ILL), Grenoble. About a quarter of a century ago the very first measurements at ILL started with a resolution already sensibly better than the one which could be obtained with solid state detectors. To date γ-rays emitted after neutron capture can be recorded with unprecedented and unequaled resolving power. In the context of this meeting some key experiments, carried out in the last 25 years, will be discussed. Finally we will try to evaluate where the emphasis might be in the upcoming years.

INTRODUCTION

Originally, high resolution γ–ray spectroscopy at the Institut Laue Langevin (ILL) was developed to determine with high accuracy the energies of γ-ray transitions and to separate multiplet structures not resolvable by other means. Three decades ago the first generation *DuMond* type crystal spectrometers built at ILL allowed one to obtain a resolution $\Delta E/E$ on the order of 10^{-4}. Later on efforts were made to approach 10^{-5} [1]. In that period the data obtained were mainly used to construct nuclear level schemes with the aim to do that as completely as possible. Completeness could be achieved because the installation of such spectrometers at a high flux reactor allowed one to obtain a very high dynamic range in intensities. Up to 5 orders of magnitude between intensities of the strongest and weakest transitions could be obtained. This was also helped by the fact that spectra are scanned step by step and therefore one does not encounter dead time problems. A well known example is ^{168}Er [2] where - within the spin window which can be reached in neutron capture - a complete level scheme was established up to an excitation energy of close to 2.5 *MeV*. Due to the completeness achieved it became a famous testing ground for theoretical models. A bit more than a decade ago one succeeded in achieving a resolving power close to $\Delta E/E = 10^{-6}$. This ultra high resolving power allowed one to observe the tiny Doppler shifts observed when nuclei recoil after the emission of γ–rays and this led to the possibility of determining lifetimes of excited states populated after thermal neutron capture. This short outline demonstrates that the aim to obtain the highest possible resolution and precision in a measurement has always been one of the major challenges for experimental physicists at ILL.

CP638, *Mapping the Triangle: International Conference on Nuclear Structure*
edited by A. Aprahamian, J. A. Cizewski, S. Pittel, and N. V. Zamfir
© 2002 American Institute of Physics 0-7354-0093-8/02/$19.00

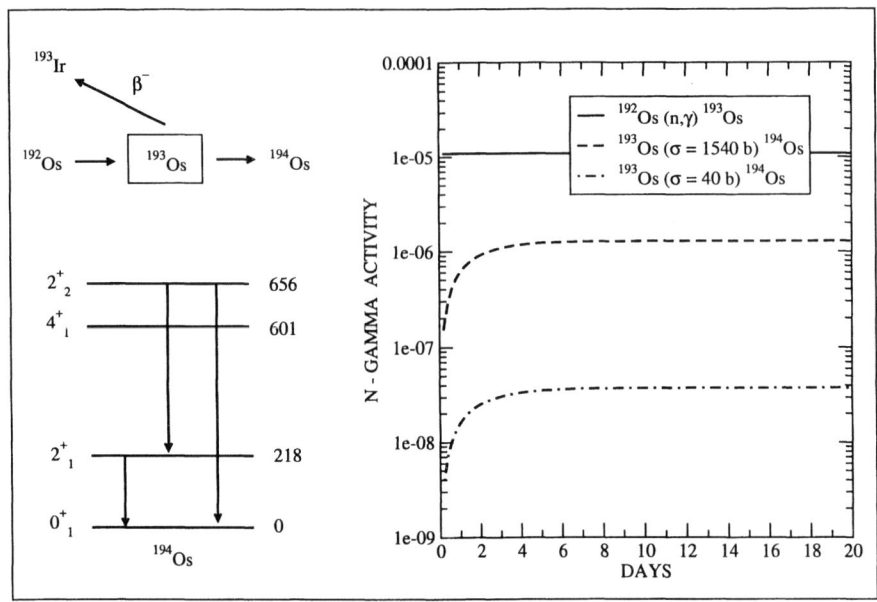

FIGURE 1. Level scheme for ^{194}Os (left part) and growth of intensities for ^{194}Os as a function of time of irradiation (right part). The flat curve describes the ^{192}Os (n,γ) activity. The growing curves describe the ^{193}Os (n,γ) activity for 2 cross sections of 40 b and 1540 b, respectively.

MEASURING TOGETHER WITH RICK

The first measurement together with Rick on these high resolution instruments was done in the region of Osmium. Of special interest was ^{194}Os [3]. As ^{193}Os is β unstable, ^{194}Os (when using neutron capture) can only be populated in successive capture of two neutrons (for simplicity we call it double neutron capture). Prior to this measurement only one double neutron capture experiment had been performed at ILL. The aim of that first measurement was to determine the thermal neutron capture cross section of unstable ^{147}Nd [4]. For ^{194}Os the aim was for the first time to obtain spectroscopical information on a nucleus which can only be reached by the capture of two neutrons. The half life of ^{193}Os is rather short - 30 hours - and the production of ^{194}Os competes with that of ^{193}Ir, the competition being determined by the relative sizes of the capture cross section and the half-life of ^{193}Os. The branching ratio between these two channels was determined by comparing the intensity of the 218.51 keV transition (see Fig. 1) assigned to ^{194}Os with that of the 219.17 keV line from ^{193}Ir. The transition intensity of the latter was known from previous work, while the value for the former was taken as $80 \pm 10\%$ per neutron capture. The result for the ratio (neutron capture / β decay) was 0.0039(6), yielding a thermal neutron capture cross section of ^{193}Os of 38 ± 10 b, much lower than the previously accepted one of 1540 b. It should be noted that the effective two neutron transfer cross section for capture into ^{194}Os was 8 mb. The criteria for identification of ^{194}Os γ–rays was that their intensities, like those of ^{193}Ir, grow with irradiation time whereas those for single neutron capture are constant. However, as can be seen from Fig. 1, the growth of intensity can only be observed during the first two days of irradiation. In contrary to that the typical time to scan a whole spectrum step by step was about 20 days. Therefore this manner of differentiation between single, twofold and higher order neutron capture was extremely tedious. To solve this problem a second target station was introduced which allowed one to expose the targets to half the neutron flux compared to the first target station. Single neutron capture then shows the same ratio for the relative intensities but double neutron capture scales with the square of this ratio, and so on. This made the assignment of capture γ–rays to single and/or multiple neutron capture very easy - by measuring the same spectrum twice. Rick's first visit had immediately resulted in a broadening of the experimental possibilities. This turned out to be an essential tool in the effort to establish level schemes as completely as possible, also because it allowed one to eliminate weak lines which originated from other nuclei, produced in subsequent neutron capture.

This measurement was soon followed by the study of ^{196}Pt [5] which became quite famous. It is not necessary to go into details here: ^{196}Pt became the first nucleus the level scheme of which was associated with the $O(6)$ limit of the IBA and many investigations of additional nuclei, important for IBA, then followed this measurement.

Finally, about 10 years ago, the GRID-method (**G**amma **R**ay **I**nduced **D**oppler broadening method) was introduced at ILL. The GRID method makes use of the fact that the neutron capture state will decay preferably by the emission of γ–rays. Each emitted γ–ray induces a recoil of the nucleus with velocities v in the order of 10^{-4} to 10^{-6} of the velocity of light. The measured Doppler shifts of subsequently emitted secondary γ–rays are quite small as they scale with v/c. Because there is no preferred direction for the emission of primary γ–rays, the measured profiles of secondary γ–rays will show a Doppler broadening rather than a Doppler shift. The Doppler profiles can be studied with ppm resolution using the crystal spectrometers $Gams4$ and $Gams5$ at ILL Grenoble [6, 7]. More details of the basic principles may be found elsewhere [8, 9]. The method was first tested in experiments with light nuclei [9] where the recoil energies are relatively large due to the small mass. But GRID is also well suited for the investigation of excitations in medium-heavy nuclei. One of the first examples studied in this region was ^{114}Cd where lifetime measurements of levels near 2 MeV gave quantitative evidence for collective 3-phonon vibrational states and an analysis of the level scheme revealed a remarkable ambiguity with its origin in the basic properties of quantum mechanical mixing [10]. The situation gets more difficult in heavy nuclei where the recoil energies are even smaller because $i)$ the intensity of feeding γ–transitions is generally much more fragmented and $ii)$ the mass of the recoiling nuclei much higher. The first heavy nucleus studied via GRID was ^{196}Pt [11]. Although its level structure had been shown to be associated with the $O(6)$ limit of the IBA, for a while this view was opposed to an interpretation in the vibrational $U(5)$ limit of the IBA. The GRID measurements in ^{196}Pt confirmed the interpretation in the frame of $O(6)$. This followed from a measurement of the lifetime of the 0_3^+ level at 1402 keV. This level decays mainly to the 2_1^+ state in ^{196}Pt. In an $O(6)$ interpretation this transition is forbidden ($\Delta\sigma = 2$) whereas it is allowed in $U(5)$ ($\Delta N = 1$). A lower limit of $\tau > 1.8$ ps was found which corresponds to a hindrance factor of more than an order of magnitude for the transition to the 2_1^+ state.

Soon after this measurement a level in ^{168}Er was associated with the 2-phonon γ– vibrational mode [12]. Other measurements followed and it turned out to be important that high resolution γ–ray measurements do not only allow for the determination of absolute transition rates but also the precise determination of the decay routes of a given level by complementary measurements of relevant stop over and/or cross over transitions. An example of such a test of placements is ^{158}Gd [13] and the potentially large and collective $K^\pi = 0^+$ to γ transitions from the $K^\pi = 0_3^+$ band at 1452 keV. However, precise energy measurements of the transitions and levels involved showed that previously assigned γ–ray lines from the 2^+ level of this second excited 0^+ band were incorrectly placed. This removed the existing evidence for a multiphonon character of this band.

The most recent measurements which we carried out together with Rick - in this now one quarter of a century of common projects - concerned the nucleus ^{152}Sm. Whereas ^{194}Os has been our first case to obtain spectroscopic information by double neutron capture, ^{152}Sm is to date the latest one. Whereas in the case of Osmium one wanted to obtain information about the level scheme, this time the challenge consisted in measuring lifetimes of excited states in ^{152}Sm. The new results [14] allowed to determine (amongst others) the lifetime of 3_1^+ state at 1234 keV. In Fig. 2 we show the relevant portion of the level scheme and the GRID profile for the 1112 keV transition depopulating this level. Whereas existing data for this 3^+ state presented serious un-certainties for any collective interpretation of this nucleus, the new results remove the previous potential discrepancies and support the interpretation of ^{152}Sm as exhibiting phase coexistence near the critical point of a sperical-deformed phase transition.

FUTURE TRENDS

It has been outlined above that high resolution γ spectroscopy at ILL started with the measurements of neutron capture γ ray spectra of secondary γ rays. The aim was to measure those as completely and as precisely as possible in order to obtain level schemes which would be as complete as possible. With time resolution was continuously improved and the investigations turned more and more away from the measurements of whole spectra towards the measurements of single line profiles which are characterized by Doppler broadening. Measurements of short lifetimes of nuclear excited states were discussed above. The GRID method also finds applications in other areas. So, for instance, if one wants to extract details about interatomic potentials from GRID measurements this can be achieved by introducing molecular dynamics (MD) simulations [15].

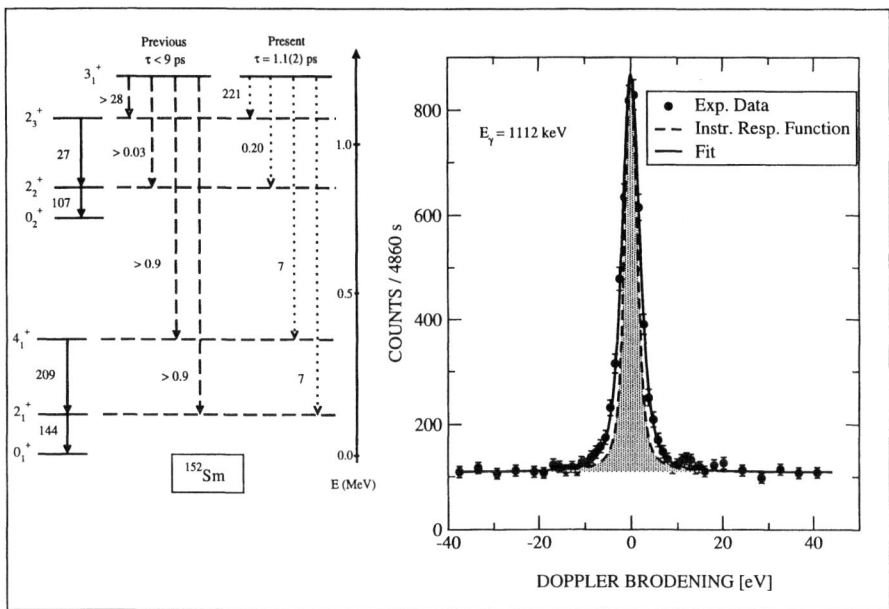

FIGURE 2. Partial level scheme for ^{152}Sm (left side) exhibiting the decay of the 3_1^+ state. The existing lifetime limit for the 1234 keV, 3_1^+ level is shown along with the new results. The right side of the figure shows the Doppler broadened line shape for the 1112 keV transition. The dotted line is the instrumental response. The solid line is a fit to the data that incorporates Doppler broadening due to the finite lifetime.

Gamma ray induced recoil energies typically range from eV to several hundreds of eV. At the low energy side they are comparable to the binding energies of the atoms in the solid (typically of the order of 10 eV). Recoils with less than the binding energy do not permit a permanent displacement of the recoiling atom but lead to oscillations around an equilibrium position. This regime has been studied in detail [15] by investigating electron-capture in ^{152}Eu. In this process ($p + e^- \rightarrow n + \nu$) the emitted neutrino causes a well defined recoil of about 3 eV. A short lived level (28 fs) is populated which decays in turn by the emission of γ rays. The study of the neutrino induced (NID) profiles of these gamma rays in conjunction with elaborate MD-simulations allowed to obtain a detailed description of the slowing down process at such sub barrier energies.

In β decay one obtains a continuous recoil energy distribution due to the emission of an electron and a neutrino. In recent GRID measurements, using natural Sm targets, we became aware that a huge amount of ^{156}Eu β decay activity was produced. In the β decay to ^{156}Gd recoils are induced due to the emission of electrons and neutrinos. Spin and parity of the decaying ground state in ^{156}Eu are $I^\pi = 0^+$. Levels populated in ^{156}Gd are either 0^+ or $1^{+/-}$ and are therefore either pure Fermi transitions or pure Gamov Teller transitions. The directional correlation coefficient a between the electrons

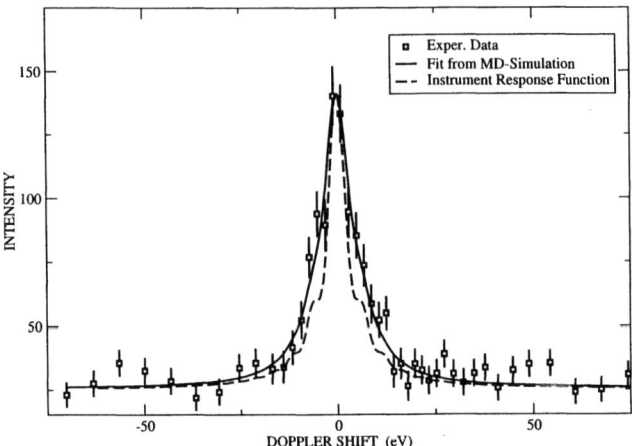

FIGURE 3. Doppler profile observed for the 1242 keV transition following the β–decay of the ^{156}Eu 0^+ ground state to the 1242 keV 1^- level in ^{156}Gd. The dotted line describes the instrumental response function. The solid line is a fit to the data that incorporates Doppler broadening due to the recoil caused by the emission of electrons and neutrinos in the decay of ^{156}Eu.

and neutrinos - and such the recoil distribution - is in principle known (in the frame of the standard *V-A* theory). This yields the possibility for an ideal test of the lifetime values deduced previously after neutron capture. Fig. 4 shows the line profile obtained for the decay of the 1242 *keV* level in ^{156}Gd. A value of $\tau = 65_{30}^{100}$ *fs* was obtained for the lifetime of this level (setting $a = -1/3$). This can be compared with $\tau = 55_{-16}^{+39}$ *fs* obtained from previous measurements [16]. For short known lifetimes (as is the case here) it is theoretically possible to invert the formulation of the problem and one can instead try to determine the correlation coefficient a from the measured BID profiles. This could, in principle, contribute to a test of the standard *V-A* theory. A χ^2 test yielded a minimum at $a = -1/3$, just as expected for a pure GT transition. This is very encouraging, however, the statistical uncertainty in this measurement is still too big to probe the validity of the *V-A* theory. Further refinement of the measurement procedure is needed.

CONCLUSION

Using some examples of measurements carried out in collaboration with Rick Casten we have briefly discussed how high-resolution measurements have evolved at ILL. Whereas in the past the main aim was directed to the construction of level schemes as completely as possible, the focus has slowly changed and is today - and most probably also in the next future - more directed to the study of Doppler broadened line profiles. The Doppler broadening can be caused by quite a variety of recoil mechanisms like GRID (**G**amma **R**ay **I**nduced **D**oppler broadening), NID (**N**eutrino **I**nduced **D**oppler broadening) and

BID (**B**eta neutrino **I**nduced **D**oppler broadening), opening such a big variety of subjects to be studied in the future.

REFERENCES

1. Koch, H. R., and et al., *Nucl. Instr. Meth.*, **175**, 401 (1980).
2. Davidson, W., and et al., *J. Phys.*, **G7**, 455 (1981).
3. Casten, R. F., Namenson, R. F., Davidson, W. F., Warner, W. F., and Börner, H. G., *Phys. Lett.*, **76B**, 280 (1978).
4. Heck, D., Börner, H. G., Pinston, J. A., and Roussiile, R., *Atomkernernergie*, **24**, 141 (1974).
5. Cizewski, J. A., and et al., *Nucl. Phys.*, **A323**, 349 (1979).
6. Dewey, M. S., and et al., *Nucl. Instr. Meth.*, **A284**, 151 (1989).
7. Kessler, E. G., and et al., *Nucl. Instr. Meth.*, **A457**, 187 (2001).
8. Börner, H. G., Jolie, J., Hoyler, F., Robinson, S. J., Dewey, M. S., Greene, G. L., Kessler, E. G., and Deslattes, R. D., *Phys. Lett.*, **B215**, 45 (1988).
9. Börner, H. G., and Jolie, J., *J. Phys.*, **G19**, 217 (1993).
10. Casten, R. F., Jolie, J., Börner, H. G., Brenner, D. S., Zamfir, N. V., Chou, W. T., and Aprahamian, A., *Phys. Lett.*, **B297**, 19 (1992).
11. Börner, H. G., Jolie, J., Robinson, S., Casten, R. F., and Cizewski, J. A., *Phys. Rev.*, **C42**, R2271 (1990).
12. Börner, H. G., Jolie, J., Robinson, S., Krusche, B., Piepenbring, R., Casten, R. F., Aprahamian, A., and Draayer, J. P., *Phys. Rev. Lett.*, **66**, 69 (1991).
13. Börner, H. G., Jentschel, M., Zamfir, N. V., Casten, R. F., Krticka, M., and Andrejtscheff, M., *Phys. Rev.*, **C59**, 2432 (1999).
14. Zamfir, N. V., Börner, H. G., Pietralla, N., Casten, R. F., Berant, Z., Barton, C. J., Beausang, C. W., Brenner, D., Caprio, M., Cooper, J. R., Krtcka, A. A., Krücken, R., Mutti, P., Novak, J. R., and Wolf, A., *submitted to Phys. Rev C* (2002).
15. Stritt, N., Jolie, J., Jentschel, M., and Börner, H. G., *Phys. Rev. Lett.*, **78**, 2592 (1997).
16. Helmer, R. G., *NDS*, **65**, 65 (1992).

Structures of Neutron-Rich Nuclei near ^{132}Sn

J.H. Hamilton[1], A.V. Ramayya[1], J.K. Hwang[1], Y.X. Luo[1,2,3], E.F. Jones[1], P. M. Gore[1], C.J. Beyer[1], X.Q. Zhang[1], J. Kormicki[1], J.O. Rasmussen[3], S.Z. Zhu[1,2,4], S.C. Wu[3], T.N. Ginter[3], C.M. Folden III[3], P. Fallon[3], P. Zielinski[3], K.E. Gregorich[3], A.O. Macchiavelli[3], I. Y. Lee[3], M. Stoyer[5], J.D. Cole[6], R. Donangelo[7], S.J. Asztalos[8], G.M. Ter-Akopian[9], Yu.Ts. Oganessian[9], A.V. Daniel[9], W.C. Ma[10], A. Covello[11], and A. Gargano[11]

[1]Department of Physics and Astronomy, Vanderbilt University, Nashville, TN 37235
[2]Joint Institute for Heavy Ion Research, Oak Ridge, TN 37830
[3]Lawrence Berkeley National Laboratory, Berkeley, CA 94720
[4]Physics Department, Tsinghua University, Beijing 100084, People's Rep. China
[5]Lawrence Livermore National Laboratory, Livermore, CA 94550
[6]Idaho National Environmental and Engineering Laboratory, Idaho Falls, ID 83415
[7]Universidade Federal do Rio de Janeiro, CP 68528, RG Brazil
[8]Massachusetts Inst. of Technology, Cambridge, MA 11830
[9]Flerov Laboratory for Nuclear Reactions, JINR, Dubna, Russia
[10]Department of Physics, Mississippi State University, MI 39762
[11]Dipartimento di Scienze Fisiche, Universita di Napoli Federico II and Istituto Nazionale di Fisica Nucleare, I-80126 Napoli, Italy

Abstract. We identified γ transitions in several neutron-rich nuclei near ^{132}Sn produced in the spontaneous fission of ^{252}Cf. These are of interest because they can test shell model calculations. These nuclei are ^{139}Ba, ^{133,135}Te, and ^{121,123}Cd. Excited states in neutron-rich ^{133}Te exhibit four types of particle-hole bands built on the known 334.3 keV isomer in ^{133}Te. The yrast and near yrast particle-hole states observed up to 6.2 MeV in ^{133}Te have characteristics quite similar to those in ^{134}Te. The band starting at 5.214 MeV is assigned as a neutron particle-hole excitation of the double magic core nucleus, ^{132}Sn and is a candidate for a tilted rotor band. Theoretical shell model calculations show good agreement with the experimental states up to 4.3 MeV. The new level scheme of ^{139}Ba includes 11 new transitions and 10 new levels and is compared with the 83-neutron isotone ^{135}Te. In ^{135}Te, a tilted rotor band, also, is proposed. Excited states built on the 11/2$^-$ isomeric states in neutron-rich ^{121,123}Cd are identified. The new data indicate that the 11/2$^-$ isomers in odd-A Cd isotopes with N>74 have spherical shapes similar to those in Te while the Cd ground states are likely deformed.

Particle and Hole States in ^{133,135}Te and ^{139}Ba

The extensive new information on the structures of neutron rich nuclei populated in spontaneous fission (SF) were reviewed in 1995 [1] and 1999 [2]. In 2000 with 102 detectors in Gammasphere we acquired 5.7×10^{11} triple and higher fold coincidences in ^{252}Cf SF. In this paper we present a few selected examples of the new physics to come from such higher statistical data.

CP638, *Mapping the Triangle: International Conference on Nuclear Structure*
edited by A. Aprahamian, J. A. Cizewski, S. Pittel, and N. V. Zamfir
© 2002 American Institute of Physics 0-7354-0093-8/02/$19.00

The high spin levels in ^{133,135}Te and ^{139}Ba have been independently extended, as shown in Figs. 1-3 [3,4]. These data provide new tests of particle-hole structures around double magic ^{132}Sn [3,4]. The ^{139}Ba energy levels upto 2091.72 keV were known [5]. The 10 new ^{139}Ba levels were identified in our SF work [4] (see Fig. 1). All these transitions in ^{133}Te were identified [3] as shown in Fig. 2. Similar work on ^{133}Te from ^{248}Cm [6] was published after our work was submitted. We identified 7 new levels and 9 new transitions not reported by Ref. [6]. These states are at 5600.8, 5501.5 (25/2$^-$),

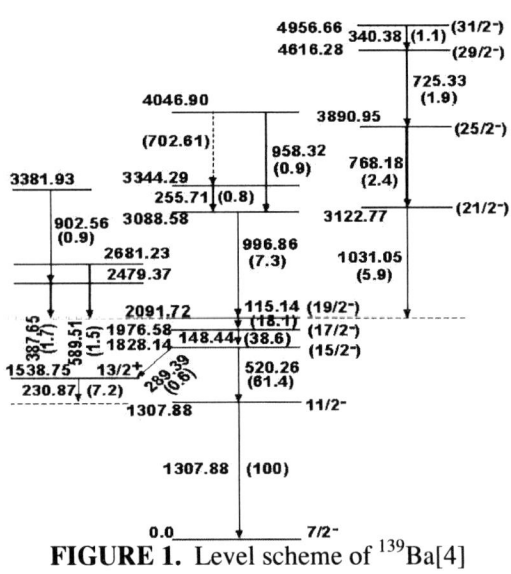

FIGURE 1. Level scheme of ^{139}Ba[4]

5214.7 (23/2$^-$), 4032.9 (23/2$^+$), 3934.5 (21/2$^+$), 3833.4, and 3825.4 keV in the level scheme as in Fig. 2. The ^{135}Te level scheme was previously proposed from ^{248}Cm [7] but with no intensities. Relative intensities of all transitions in ^{135}Te were carefully determined from compressed spectra.

Excited states in neutron-rich ^{133}Te exhibit four types of particle-hole bands built

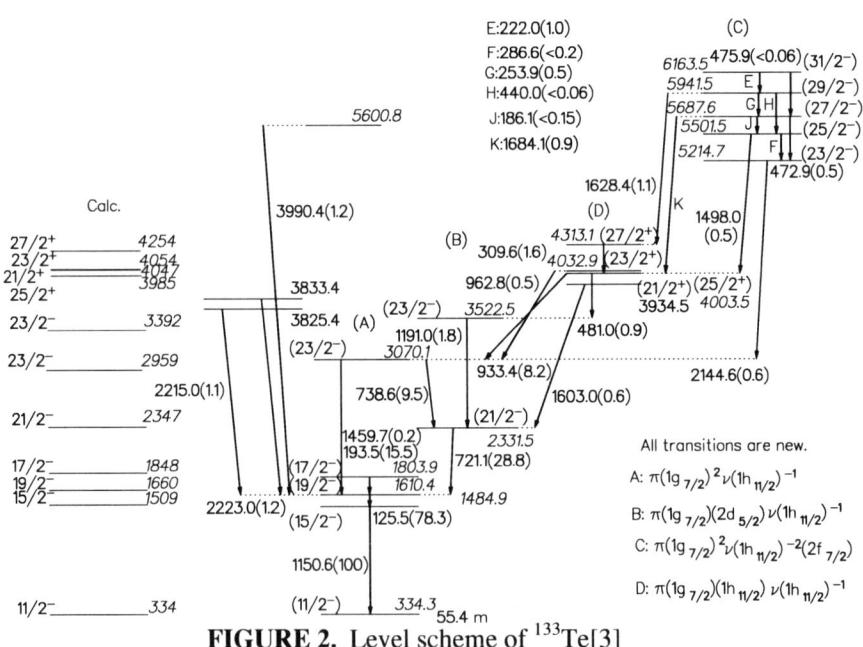

FIGURE 2. Level scheme of ^{133}Te[3]

92

on the known 334.3 keV isomer and have characteristics quite similar to those in [134]Te. These states are interpreted as a result of coupling a neutron $h_{11/2}$ hole to the [134]Te core. Groups of states lying at 4.023 MeV for [135]Te in Fig. 4 and 5.214 MeV for [133]Te in Fig. 3 related to a neutron particle-hole excitation of the double magic core nucleus, [132]Sn provide possible candidates for tilted rotor bands. Near the lower end of the band, the neutron total angular momentum can couple at near right angles to the proton angular momentum. Such bands are characterized by strong M1 cascade transitions with weak crossovers as we find in these [133,135]Te bands. We would expect on geometrical grounds that fission fragments would prefer populating prolate bands (see Fig. 4) as intermediates on the path to spherical ground states. For a prolate deformed potential in [135]Te, the two protons would be in the 1/2[431] orbital and the neutrons would have a pair in the 1/2[541] and a hole in the 11/2[505]. These are the first proposed tilted rotor bands in neutron rich nuclei and for prolate-spherical shape coexistence in nuclei around double magic [132]Sn.

The theoretical shell model calculations show very good agreement with the experimental data for the levels up to 4.3 MeV for [133]Te. Our additional 3934.5 and 4032.9 keV levels complete the agreement with theory for band D. The N=83 [135]Te and [139]Ba show marked differences associated with differences in their band structures related to one having proton particle and the other proton hole states both coupled to neutron particle states. For example, in [139]Ba an extra 17/2⁻ band is seen below the 19/2⁻ one and the E3 transitions to the [135]Te 19/2⁻ state are short circuited in [133]Te.

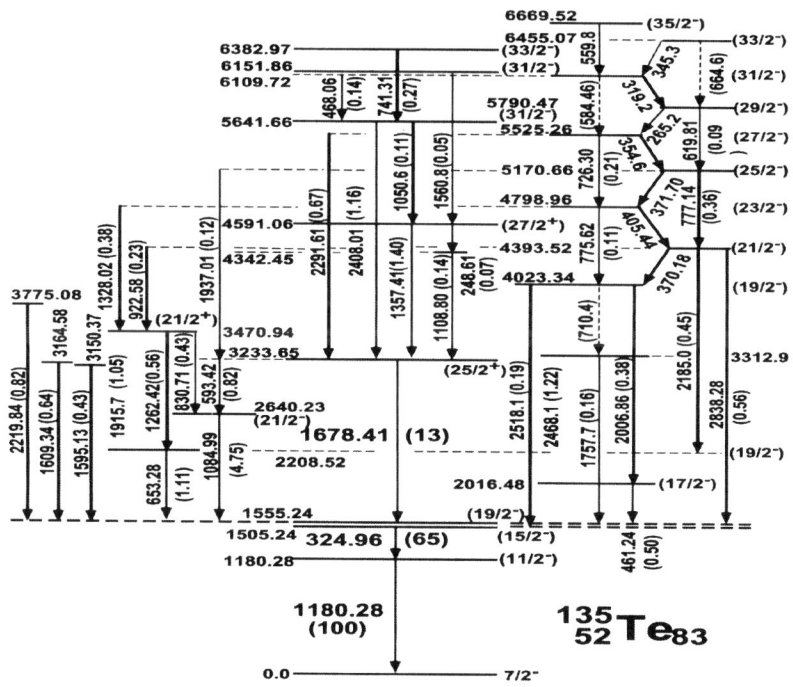

FIGURE 3. Level scheme of [135]Te[4]

93

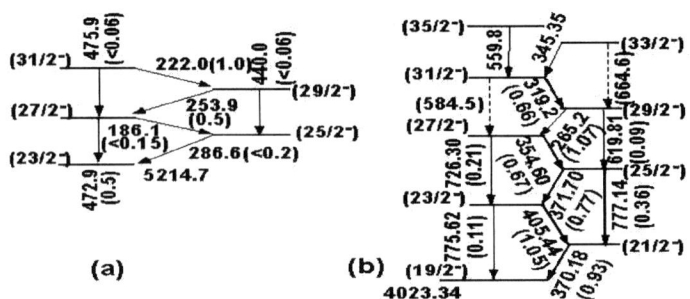

FIGURE 4. Proposed tilted rotor bands in (a) ^{133}Te and (b) ^{135}Te

New excited states in ^{121,123}Cd

The Z=48 Cd nuclei are close to the spherical Sn nuclei. The $h_{11/2}$ decoupled neutron bands are observed in $^{111-119}$Cd [8]. The A=120 region with Z < 50 and N < 82 is very important in astrophysics because of inconsistencies between the observed solar isotopic r-process abundances and the model calculations [9,10]. The center of the r-process path for Z=48 is ^{130}Cd [9,11]. However, nuclei to either side influence the r-process. Kratz et al. [9] suggested that presently inaccessible ^{120}Mo, $^{119-123}$Tc and ^{122,124}Ru may exhibit shape coexistence. Such shape coexistence will influence the r-process abundances in this region.

The near constant values of R = E(4$^+$)/E(2$^+$) $\cong$ 2.2 with β_2 ~0.12. [11] for N=78,80,

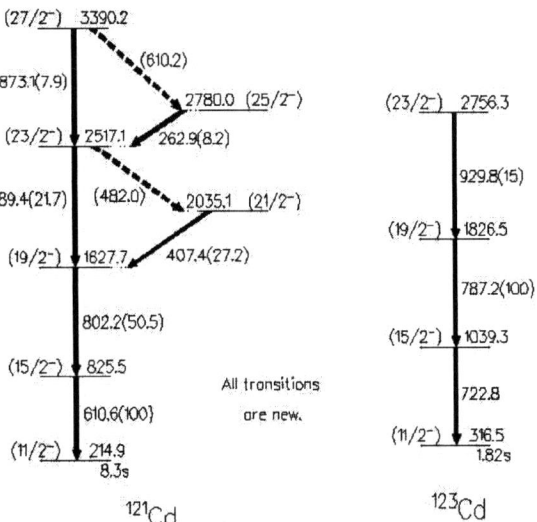

FIGURE 5. Level scheme of ^{121,123}Cd[12]

^{126,128}Cd isotopes indicate the possible quenching of the N=82 spherical shell gap. It is interesting to investigate the heavier odd -A Cd isotopes. From our SF work [12], the $h_{11/2}$ decoupled bands of ^{121,123}Cd are established (see Fig. 5). The ratios of R versus N for odd N Pd, Cd, Te and Xe are shown in Fig. 6. The even-even Cd isotopes show only a small decreasing trend for R with increasing N. The odd -A Te isotopes have nearly spherical shapes, where the value of R is $\cong$ 2.0-2.1.

The $h_{11/2}$ decoupled bands built on the isomeric states in odd-A Cd isotopes have a much larger value of $R \cong 2.5$ for $N \le 71$, characteristics of deformation. but strongly decreasing value of R to $R \cong 2.1$ for ^{123}Cd. This indicates a shape change with the 11/2⁻ isomer becoming spherical. The ground states of even and odd-A Cd isotopes may have non-zero deformations near A=130 as indicated by the near constant R for the even A isotopes out to ^{128}Cd. Our data suggest that the isomers in ^{125,127,129}Cd could be nearly spherical with β_2 much less than 0.1 compared to ^{126,128}Cd with β_2 =0.12 [11]. If this suggestion is correct, this shape coexistence effect can help remove the discrepancies between experiments and calculations in the solar r-process abundances in this region. It is suggested that the backbending observed at $\hbar\omega \approx 0.45$ MeV in ^{121}Cd is related to the alignment of the $h_{11/2}$ neutron pair as seen in the lighter A Cd nuclei [11].

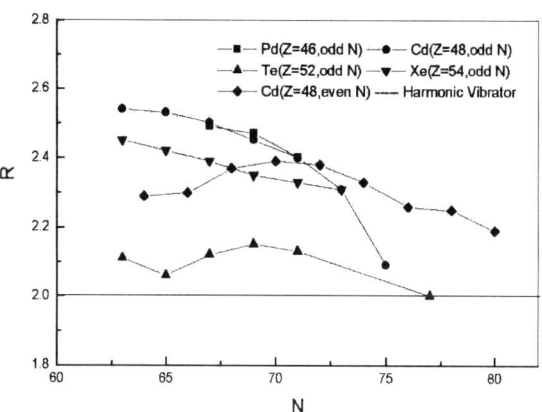

FIGURE 6. Ratios, R, of $(E_{19/2} - E_{11/2})/E_{15/2} - E_{11/2})$ for the odd N isotopes and $E(4^+)/E(2^+)$ for the even N isotopes[12].

The work at Vanderbilt University is supported by U.S. Department of Energy under Grant No. DE-FG05-88ER40407. The work (Y.X. Luo and S.J. Zhu) at Joint Institute for Heavy Ion Research is supported by U. of Tennessee, Vanderbilt University and U.S. DOE.

REFERENCES

1. J.H. Hamilton, et al., *Prog. in Part. and Nucl. Phys.* **35**, 635 (1995).
2. *Fission and Neutron Rich Nuclei*, edited by J. H. Hamilton et al., World Scientific (2000).
3. J. K. Hwang et al., *Phys. Rev.* **C65**, 034319 (2002).
4. Y.X. Luo et al., *Phys. Rev.* **C6405**, 4306 (2001).
5. H. Prade et al., *Nucl. Phys.* **A472**, 381 (1987).
6. P. Bhattacharyya et al., *Phys. Rev.* **C64**, 054312 (2001).
7. B. Fornal et. al., *Phys. Rev.* **C63**, 024322 (2001).
8 N. Fotiades et al., *Phys. Rev.* **C61**, 064326 (2000).
9. K.L. Kratz et al., *Astrophys. J.* **403**, 216 (1993).
10. B. Pfeiffer et al., *Z. Phys.* **A357**, 235 (1997).
11. T. Kautzsch et al., *Eur. Phys. J.* **A9**, 201 (2000)
12. J.K. Hwang etal., *J. of Phys.* **G28**, L9 (2002). Permission to reproduce granted by IOP Publishing Limited.

Microscopic Realization of O(6) and E(5) symmetries
- How nucleons are moving behind the Triangle -

Takaharu Otsuka*, Noritaka Shimizu†, Takahiro Mizusaki** and Michio Honma‡

*Department of Physics and Center for Nuclear Study, University of Tokyo, Hongo, Bunkyo-ku, Tokyo 113-0033, Japan
RIKEN, Wako-shi, Saitama 351-0198, Japan
†RIKEN, Wako-shi, Saitama 351-0198, Japan
**Institute of Natural Sciences, Senshu University, Higashimita, Tama, Kawasaki, Kanagawa 214-8580, Japan
‡Center for Mathematical Sciences, University of Aizu, Tsuruga, Ikki-machi, Aizu-Wakamatsu, Fukushima 965-8580, Japan

Abstract. Recent results of the Monte Carlo Shell Model and related studies will be presented for some Sn, Te and Ba nuclei. The anomalous $B(E2)$ value of ^{136}Te can be explained within the shell model description which describes the spherical-to-deformed phase transition in Ba isotopes. The microscopic realization of E(5) and O(6) symmetries are given with a shell model Hamiltonian which describes energy levels and $E2$ properties.

INTRODUCTION

The nuclear collective motion has been one of the most intriguing and characteristic problems of many-body physics. It is unique in the sense that the nucleus is a truly quantal system and, at the same time, one can discuss its "shape". Rick Casten has made and is still making important and unique contributions to this field of physics as we see in this conference.

The nuclear shell model has provided us with a very basic, sound and general basis for describing the nuclear structure. However, the description of the collective motion in heavy nuclei implies the diagonalization of tremendously huge Hamiltonian matrix, which has remained infeasible for decades. Only recently the shell model approach has become practical as the Monte Carlo Shell Model (MCSM) has been proposed [1, 2, 3, 4]. The MCSM has already played various roles in wide areas of the nuclear structure physics. We can refer details of the method to original papers [1, 2, 3, 4], and its applications to references [5, 6, 7, 8, 9, 10, 11]. A review article is also available [12]. In this talk, we would like to present very recent results of the MCSM calculations on some selected aspects of nuclear quadrupole collective motions to which Rick Casten has made significant contributions.

CP638, *Mapping the Triangle: International Conference on Nuclear Structure*
edited by A. Aprahamian, J. A. Cizewski, S. Pittel, and N. V. Zamfir
© 2002 American Institute of Physics 0-7354-0093-8/02/$19.00

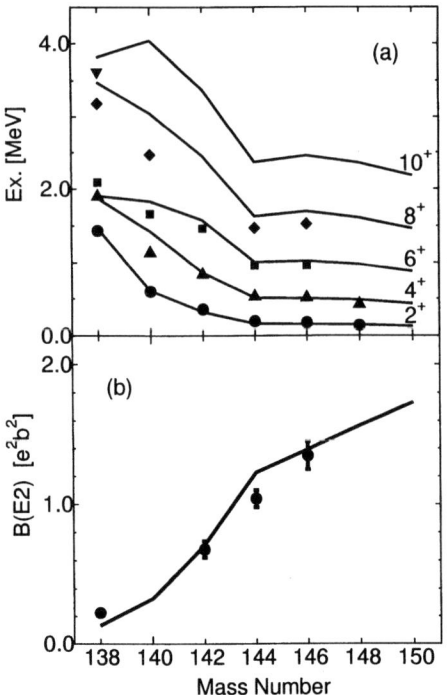

FIGURE 1. (a) Energy levels of Ba isotopes. The lines are calculations, while the symbols represent experiments [16]-[20]. (b) $B(E2;0_1^+ \rightarrow 2_1^+)$ values of Ba isotopes. Symbols are experimental [21], while the line is calculation.

SPHERICAL TO DEFORMED PHASE TRANSITION

We begin with the spherical to deformed phase transition [13]. This can be seen in many isotope chains as a function of neutron number (N) as well as in many isotone chains as a function of proton number (Z). We take the chain of Ba isotopes as an example. This phase transition can be seen quite well as N increases from a magic number 82. In other words, while the Ba isotope has a spherical shape in its ground state at N=82, the deformation becomes stronger as the N increases and the nucleus shows a rotational band around N=90. It is of much interest what is the basic mechanism of this phase transition.

We carried out an MCSM study for this purpose in [7], where we took a model space consisting of the Z=50-82 major shell for protons and the Z=82-126 major shell for neutrons. The shell model Hamiltonian H is comprised of the three parts as

$$H = H_p + H_n + V_{pn}, \tag{1}$$

98

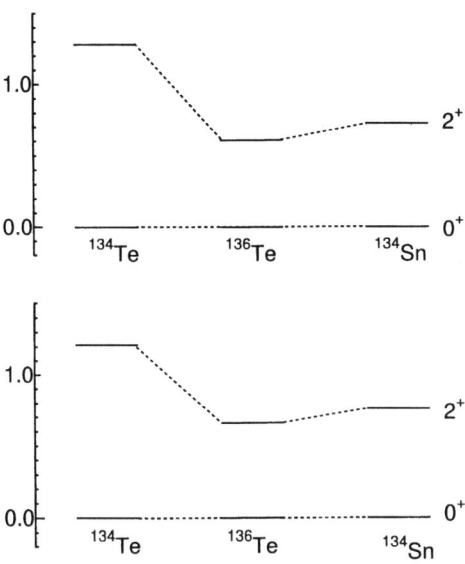

FIGURE 2. 2_1^+ levels of ^{134,136}Te and ^{134}Sn. The upper part is experiment, while the lower part calculated by the present work.

where $H_p(H_n)$ means the proton (neutron) Hamiltonian and V_{pn} denotes a proton-neutron interaction. The H_p includes single-particle energies and a two-body interaction. The single-particle energies are taken from experimental levels of ^{133}Sb [14]. The two-body interaction includes the monopole and quadrupole paring interactions and the quadrupole-quadrupole interaction. The monopole and quadrupole interactions are much stronger than the quadrupole-quadrupole interaction as can be seen in [7] with concrete strengths.

Likewise, the H_n includes single-particle energies and a two-body interaction, while the single-particle energies are taken from experimental levels of ^{133}Sn [15].

The proton-neutron interaction strength in V_{pn} is determined from the 2_1^+ level of ^{148}Ba. Details of the Hamiltonian can be found in [7].

Figure 1 shows yrast levels and $B(E2; 0_1^+ \rightarrow 2_1^+)$ values of $^{138-150}$Ba isotopes. One clearly sees that the spherical-to-deformed phase transition can be described to a good quantitative extent by this shell model calculation. We emphasize that the shell model Hamiltonian is fixed for all these nuclei, and the parameters of the Hamiltonian were determined mostly at semi-magic nuclei except for the proton-neutron interaction fixed at a well-deformed nucleus, without referring to how one can describe the phase transition. It is remarkable that such a simple theoretical framework can reproduce large changes of excitation energies, level structure and $E2$ transition strength. Since this kind of shell-model calculations require huge dimension of multi-nucleon Hilbert space [7, 12], one benefits from the capability of the MCSM calculation.

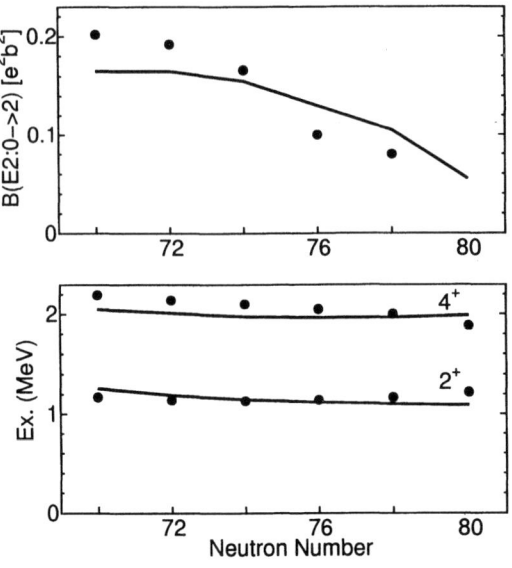

FIGURE 3. B(E2) of Sn.

ANOMALY IN ^{136}TE

Quite recently, a very interesting result was reported on the structure of ^{136}Te [22]. Figure 2 show 2_1^+ levels of ^{134,136}Te and ^{134}Sn. The shell model Hamiltonian and the model space is completely identical to those used in the previous section for describing the spherical-to-deformed phase transition. One sees a nice agreement between experiments and the present calculation. Since the number of valence nucleons is small, these calculations can be made by conventional calculations such as conventional shell-model calculations or extended TDA calculations.

We calculated $B(E2; 0_1^+ \rightarrow 2_1^+)$ values. The effective charges are also the same as those used in the previous section. Resultant values of $B(E2; 0_1^+ \rightarrow 2_1^+)$ are 0.098, 0.15 and 0.028 (e^2b^2), respectively. The first two values have been reported in [22] as 0.096 (12) and 0.103 (15) (e^2b^2). The agreement is good for ^{134}Te and reasonable for ^{136}Te. One can compare the latter value to a shell model result, 0.25 (e^2b^2) by Covello *et al.* [23] as cited in [22]. It should be investigated the origin of the difference between the two calculations.

The anomalously small $B(E2)$ value of ^{136}Te is due to the low-lying 2_1^+ level of ^{134}Sn, which is nothing but a 2$^+$ state of two neutrons, called D_ν usually. The ground state of ^{134}Sn (^{134}Te) is a 0$^+$ state of two neutrons (protons), called S_ν (S_π) likewise. The 2_1^+ state of ^{134}Te is a 2$^+$ state of two protons, called D_π. The 2_1^+ state of ^{136}Te is dominated by the product $S_\pi \times D_\nu$, because the other product has the excitation energy nearly twice

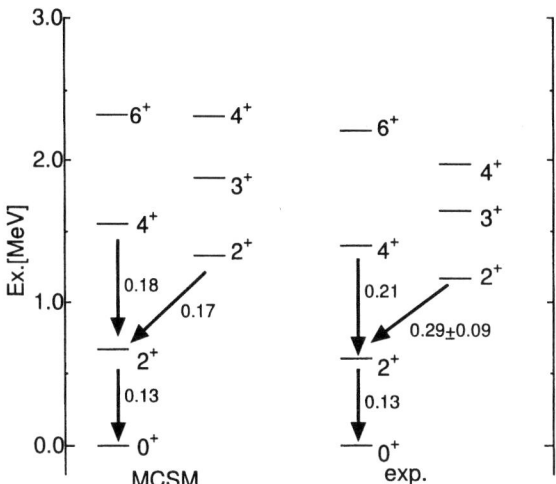

FIGURE 4. Energy levels of ^{134}Ba. Values of $B(E2)$ are also shown for low-lying collective transitions.

higher. In fact, the probability ratio between these two component is as large as about a factor four in ^{136}Te. This structure is the basic reason for the small $B(E2)$ value of ^{136}Te. In usual cases, $D_\pi \times S_\nu$ is equally important.

We note that the present proton-neutron interaction is only in the quadrupole channel, which is a simplification. In a more realistic study, the interaction, in particular the proton-neutron interaction has to be more complex, and such an interaction should have more prominent effects near closed shells where the collectivity does not dominate the structure. By including proton-neutron interaction beyond the quadrupole-quadrupole channel, the structure of ^{136}Te nucleus is affected more than other heavier Te isotopes, and its $B(E2)$ will become smaller.

Thus, the anomalously small $B(E2)$ value of ^{136}Te still remains anomalous and interesting, but can be explained in a consistent manner with the description of the spherical-to-deformed phase transition discussed in the previous section.

$B(E2)$ VALUES OF SN ISOTOPES

We now turn to nuclei with $N < 82$. The Hamiltonian is basically the same as the one used in previous sections. Since the neutron shell is changed, the parameters of the neutron Hamiltonian should be changed. The single particle energies are taken from the appropriate experimental data.

Figure 3 indicates $B(E2; 0_1^+ \rightarrow 2_1^+))$ values and 2_1^+ levels of Sn isotopes. Some of the $B(E2; 0_1^+ \rightarrow 2_1^+))$ values have recently been measured [22]. Although these

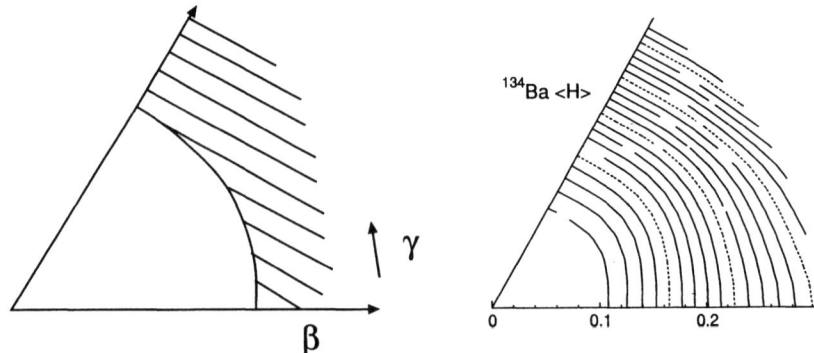

FIGURE 5. Potential energy surface of $E(5)$ symmetry (left) and that of ^{134}Ba (right).

$B(E2; 0_1^+ \rightarrow 2_1^+))$ values are rather unusually small, they can be described within the shell model. The theoretical description in 3 has been given in [24].

STRUCTURE OF ^{134}BA AND E(5) SYMMETRY

Rick has started and promoted experimental studies of O(6) dynamical symmetry extensively. He has shown with von Brentano that there are several good examples of O(6) symmetry in Xe and Ba region with $N \sim 76$ [25]. We would like to look at ^{134}Ba first. This is also because of more recent development of $E(5)$ symmetry scheme of Iachello [26] and its experimental realization [27] by Rick Casten and Zamfir. Energy levels of ^{134}Ba are shown in Fig. 4. The present shell model Hamiltonian reproduces experimental levels and $E2$ transitions quite well. One finds a basic pattern of the O(6) symmetry.

The potential energy surface (PES) of ^{134}Ba is then shown in Fig. 5. What is striking is that this PES has a flat bottom from the spherical limit to a certain deformation with a nearly perfect γ-instability. This situation can be compared well to the PES of the E(5) symmetry as indicated also in Fig. 5. From these comparisons, the structure of ^{134}Ba is quite consistent with the E(5) picture from a microscopic viewpoint, which is capable of describing the structure quantitatively. Although we do not show in this article, the same shell model Hamiltonian can describe neighboring nuclei as well.

STRUCTURE OF OTHER BA ISOTOPES

The potential energy surfaces (PES) are shown for three Ba isotopes in Fig. 6. The PES is obtained for angular momentum unprojected and projected ground states. One sees that ^{132}Ba is a quite spherical nucleus with slight tendency towards a weak oblate

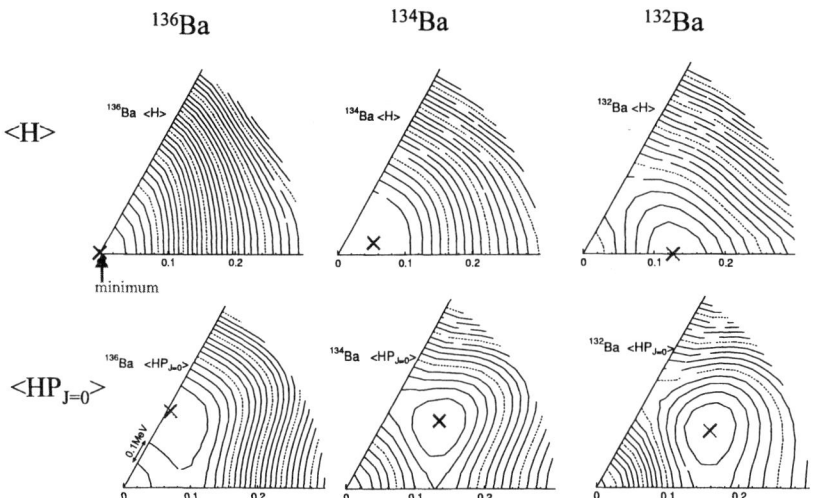

FIGURE 6. Potential energy surfaces of Ba isotopes.

deformation, ^{134}Ba is consistent with the E(5) symmetry, and ^{136}Ba is more like an O(6) nucleus. Thus, we can confirm that the appearance of E(5) and O(6) symmetries has a firm microscopic basis, and that the E(5) symmetry can appear in an even narrower region as expected from the origin of E(5). The angular momentum projected PES tends to depict a minimum with triaxiality in general. We just comment that this is true also in the O(6) limit [28, 29], and such a rigid triaxiality [30] should not be considered as a contradiction to the γ unstable nature of the O(6) limit [31, 32, 33].

ACKNOWLEDGMENTS

One of the authors (T.O.) acknowledges Rick Casten for his continuous and keen interests and also for his enthusiastic collaborations over past nearly 30 years. The MCSM calculations were performed partly by the Alphleet computer system in RIKEN, and by the Alphleet-2 system located in the CNS, the University of Tokyo. Some of the conventional shell calculations were made by the code OXBASH [34]. This work was supported in part by Grant-in-Aid for Scientific Research No. (A)(2) (10304019) and by Grant-in-Aid for Specially Promoted Research (13002001) from the Ministry of Education, Science and Culture.

REFERENCES

1. M. Honma, T. Mizusaki, and T. Otsuka, Phys. Rev. Lett. **75**, 1284 (1995).
2. T. Mizusaki, M. Honma, and T. Otsuka, Phys. Rev. C **53**, 2786 (1996).

3. M. Honma, T. Mizusaki, and T. Otsuka, Phys. Rev. Lett. **77**, 3315 (1996).
4. T. Otsuka, M. Honma, and T. Mizusaki, Phys. Rev. Lett. **81**, 1588 (1998).
5. T. Mizusaki, T. Otsuka, Y. Utsuno, M. Honma and T. Sebe, Phys. Rev. **C59**, 1846(R) (1999).
6. Y. Utsuno, T. Otsuka, T. Mizusaki, and M. Honma, Phys. Rev. **C60**, (1999) 054315.
7. N. Shimizu, T. Otsuka, T. Mizusaki and M. Honma, Phys. Rev. Lett. **86**, 1171 (2001).
8. T. Mizusaki, T. Otsuka, M. Honma, and B.A. Brown, Phys. Rev. **C63**, 044306 (2001).
9. Y. Utsuno, T. Otsuka, T. Mizusaki, and M. Honma, Phys. Rev. C **64**, 011301(R) (2001).
10. T. Otsuka, R. Fujimoto, Y. Utsuno, B.A. Brown, M. Honma, and T. Mizusaki, Phys. Rev. Lett. **87**, 082502 (2001).
11. M. Honma, T. Otsuka, B.A. Brown and T. Mizusaki, Phys. Rev. C **65**, 061301(R) (2002).
12. T. Otsuka, M. Honma, T. Mizusaki, and N. Shimizu, and Y. Utsuno, Prog. Part. Nucl. Phys. **47**, 319 (2001).
13. P. Ring and P. Schuck, *The Nuclear Many-Body Problem* (Springer-Verlag New York Inc. 1980).
14. L.A. Stone, *et al.*, Phys. Scr. **T56**, 316 (1995).
15. P. Hoff, *et al.*, Phys. Rev. Lett. **77**, 1020 (1996).
16. M. A. Islam, T. J. Kennett, and W. V. Prestwich, Phys. Rev. **C42**, 207 (1990)
17. L. J. Alquist, W. C. Schick, Jr., W. L. Talbert,Jr,. s. A. Williams: Phys. Rev. **C13** 1277 (1976)
18. S. M. Scott *et al.*, J. Phys. (London) **G 6**, 1291 (1980)
19. W. Urban *et al.*, Nucl. Phys. **A613**, 107 (1997)
20. J. C. Hill *et al.*, Phys. Rev. **C34**, 2312 (1986)
21. S. Raman, C. H. Marlarkey, W. T. Milner, C.W. Nestor, Jr., and P. H. Stelson, *At. Data Nucl. Data Tables* **36**, 1 (1987)
22. D.C. Radford, *et al.*, Phys. Rev. Lett. **88**, 222501 (2002).
23. A. Covello, *et al.*, in proceedings of the 7th Int. Spring Seminar on Nuclear Physics, edited by A. Covello (World Scientific, Singapore, 2002), p. 139; L. Coraggio, *et al.*, *Phys. Rev.* **C65**, 051306 (2002).
24. T. Mizusaki and T. Otsuka, Prog. Theor. Phys. Suppl. **125**, 97 (1996).
25. R.F. Casten and P. von Brentano, Phys. Lett. **B 152**, 22 (1985).
26. F. Iachello, Phys. Rev. Lett. **87**, 052502 (2001).
27. R.F. Casten and N.V. Zamfir, Phys. Rev. Lett. **87**, 052503 (2001).
28. T. otsuka and M. Sugita, Phys. Rev. Lett. **59**, 1541 (1987).
29. M. Sugita, T. Otsuka and A. Gelberg, Nucl. Phys. **A 493**, 350 (1989).
30. A.S. Davidov and G.F. Filippov, Nucl. Phys. **8**, 237 (1958).
31. L. Wilet and M. Jean, Phys. Rev. **102**, 788 (1956).
32. J. Meyer-Ver-Tehn, Phys. Lett. **B 84**, 10 (1979).
33. J. Dobes, Phys. Lett. **B 158**, 97 (1985).
34. A. Etchegoyen *et al.*, MSU-NSCL Report No. 524, 1985.

Ab Initio Large-Basis No-Core Shell Model and its Applications to Light Nuclei

B. R. Barrett*, P. Navrátil†, A. Nogga*, W. E. Ormand† and J. P. Vary**

*Department of Physics, P.O. Box 210081, University of Arizona, Tucson, AZ 85721
†University of California, Lawrence Livermore National Laboratory, Livermore, CA 94551
**Department of Physics and Astronomy, Iowa State University, Ames, IA 50011

Abstract.
We describe the *ab initio* No-Core Shell Model, in which the effective Hamiltonians are derived microscopically from realistic nucleon-nucleon potentials, as a function of the finite harmonic-oscillator basis space. We give results for the $A = 3$ nucleon system to demonstrate the viability of the technique. Extending our approach to p-shell nuclei, we present some of our latest results.

1. INTRODUCTION

The major outstanding problem in nuclear-structure physics is to calculate the properties of finite nuclei starting from the basic interactions among the nucleons. Such calculations have been performed so far only for light nuclei up to $A = 8$ [1]. We have developed a new *ab initio* technique for accurately computing nuclear properties, in which all A nucleons are taken to be active, interacting by realistic nucleon-nucleon (NN) interactions. We call this approach the No-Core Shell Model (NCSM) [2, 3, 4, 5, 6, 7, 8, 9].

The philosophy of the NCSM is to derive an A-body effective interaction for a given, finite model space from the solutions of an A-body bound-state system for a given Hamiltonian. In the standard formulation of this approach, the A-body effective interaction is usually truncated at the two-body level [3, 6], and more recently at the three-body level [8]. Utilizing a single-particle coordinate harmonic-oscillator (HO) basis, one determines the two-body effective interaction for the A nucleon system by solving a system of two nucleons bound in a HO well and interacting by the NN potential. We note that the use of a HO basis is crucial for ensuring that the center-of-mass (CM) motion of the nucleus does not mix with the internal motion of the nucleons. This approach is limited by the model space as well as by the fact that only a two-body or three-body effective interaction is used, despite the fact that higher-body effective interactions might not be negligible. Although the practical applications depend on the HO frequency and the model space, our results are guaranteed to converge to an exact solution once a sufficiently large model space is reached [4, 5].

CP638, *Mapping the Triangle: International Conference on Nuclear Structure*
edited by A. Aprahamian, J. A. Cizewski, S. Pittel, and N. V. Zamfir
© 2002 American Institute of Physics 0-7354-0093-8/02/$19.00

2. NO-CORE SHELL-MODEL APPROACH

2.1. Hamiltonian

In the no-core shell-model approach we start with the one- plus two-body Hamiltonian for the A-nucleon system,

$$H_A = \sum_{i=1}^{A} \frac{\vec{p}_i^2}{2m} + \sum_{i<j=1}^{A} V_{ij}^{NN} , \qquad (1)$$

where m is the nucleon mass and V_{ij}^{NN}, the NN interaction. In the next step we modify the Hamiltonian (1) by adding to it the CM HO potential $\frac{1}{2} Am\Omega^2 \vec{R}^2$, $\vec{R} = \frac{1}{A} \sum_{i=1}^{A} \vec{r}_i$. Later we can easily separate the effects of the CM interaction, because CM motion and internal motion do not mix. The added CM potential permits the use of the convenient HO basis and provides a mean field that facilitates the calculation of the effective interactions. The modified Hamiltonian, with a pseudo-dependence on the HO frequency Ω, can be cast into the form

$$H_A^\Omega = \sum_{i=1}^{A} \left[\frac{\vec{p}_i^2}{2m} + \frac{1}{2} m\Omega^2 \vec{r}_i^2 \right] + \sum_{i<j=1}^{A} \left[V_{ij}^{NN} - \frac{m\Omega^2}{2A} (\vec{r}_i - \vec{r}_j)^2 \right] . \qquad (2)$$

Since we solve the many-body problem in a finite HO model space, the realistic nuclear interaction in Eq. (2) will yield unreasonable results unless we employ a model-space dependent effective Hamiltonian. In general, for an A-nucleon system, an A-body effective interaction is needed. The effective interaction, utilized in the present calculations, is approximated by a two-body or a three-body effective interaction. Large model spaces are desirable to minimize the role of neglected effective many-body terms. In fact, large model spaces are desirable for the evaluation of any observable, *i.e.,* the larger the model space is, the smaller the renormalization contributions to any effective operator.

As the Hamiltonian H_A^Ω (2) differs from the Hamiltonian H_A (1) only by a CM dependent term, no dependence on Ω should exist for the intrinsic properties of the nucleus. However, because of the neglect of many-body terms in the effective-interaction derivation, a dependence on Ω appears in our calculations. This dependence decreases as the size of the model-space is increased.

2.2. Unitary transformation of the Hamiltonian

For the derivation of the effective interaction, we adopt approaches presented by Lee and Suzuki [10], Da Providencia and Shakin [11], and Suzuki and Okamoto [12] to find a unitary transformation, which yields an Hermitian effective Hamiltonian. We improved on these approaches by a scheme which allows us to calculate the unitary transformation non-perturbatively [3].

This is performed at a cluster level less than A, *i.e.,* at the two-body or three-body cluster level, for the effective interaction. The procedure for performing this cluster trun-

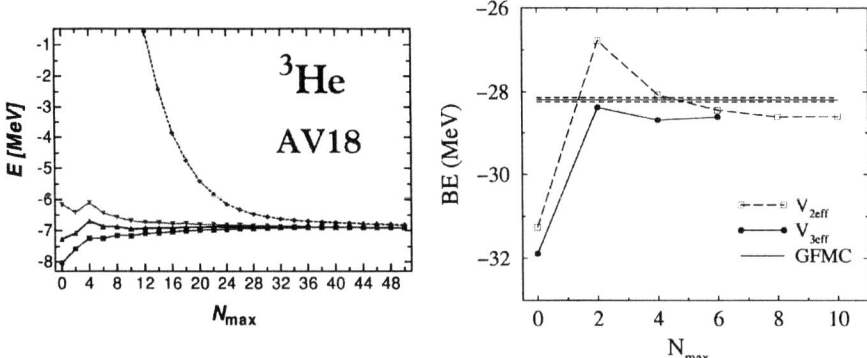

FIGURE 1. Left hand side: Ground-state energy dependence on the model-space size for ^{3}He interacting by the AV18 NN potential The dashed line shows the result based on the bare interaction. The solid lines with down-pointing triangles, up-pointing triangles and squares are results based on the effective interaction for $\hbar\Omega = 32$, 28 and 24 MeV,respectively. Right hand side: Same for ^{6}Li using the AV8' NN potential with Coulomb. The plotted energies occur at the HO frequency minima for the given value of N_{max} The results using both the two-body and the three-body effective interaction are compared with the GFMC results from Ref. [1]. The figure is from Ref. [8].

cation and computing the two-body and three-body effective interactions is described in detail in Refs. [6, 9].

The resulting two-body and three-body effective interactions depend on A, on the HO frequency Ω and on N_{max}, the maximum many-body HO excitation energy (above the lowest configuration) defining the model space. The effective Hamiltonians are translationally invariant by construction, and it follows that the effective two-body and three-body interactions approach their respective bare interactions for $N_{max} \to \infty$.

3. RESULTS

Our $A = 3$ results indicate the feasibility of our approach, showing that accurate values of physics properties can be obtained in sufficiently large model spaces. Even for the complicated AV18 NN potential [1], the $N_{max} = 50$ model space is sufficient for obtaining a converged result with an error less than 10 keV, as shown on the left hand side of Fig. 1. It is also seen that the utilization of the effective NN interaction speeds up the convergence significantly compared with the bare interaction.

For the ground state of ^{4}He with the two- and three body effective interactions we can obtain an accuracy of 200 keV or 0.7 % of the binding energy [5].

A bigger challenge for the NCSM is the p-shell, where model spaces increase rapidly in size with increasing N_{max}. Consequently, model spaces larger than $N_{max} = 10$ are not presently feasible for most p-shell nuclei. However, besides increasing N_{max} to improve convergence, one can also increase the cluster size of the effective interaction. This has been recently investigated by Navrátil and Ormand [8] for several p-shell nuclei. E.g. for

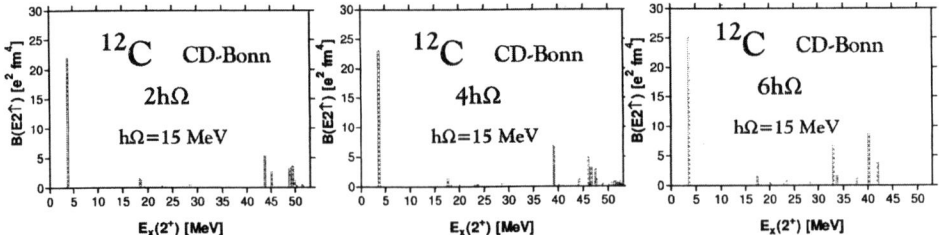

FIGURE 2. The B(E2) transition strengths in ^{12}C as a function of the model space size N_{max}. The ^{12}C spectrum was calculated utilizing the CD Bonn NN potential.

^{6}Li, it was demonstrated that three-body effective interactions accelerate convergence. This is is shown on the right hand side of Fig. 1.

The largest system that we have addressed so far is the vastly more complex ^{12}C. The ^{12}C nucleus has several pressing reasons for a more detailed investigation, such as the important role that it plays in neutrino studies using liquid scintillator detectors [13] and in parity-violating electron scattering, so as to measure the strangeness content of the nucleon [14, 15].

Our earlier results for ^{12}C appeared in Refs. [6, 16]. Here we present new calculations for the $B(E2)$ transition strengths in ^{12}C as a function of N_{max}, the size of the model space. Fig. 2 shows how the $B(E2)$ strength changes as N_{max} increases from 2 to 6. The main feature is the significant shift of the high-lying $E2$ strength from around 44-50 MeV to 33-42 MeV. This strength will move even lower as the excitation energies of the 2- or higher-$\hbar\Omega$-dominated states are brought down in energy, as the size of the model space increases.

4. CONCLUSIONS

In this contribution we described the *ab initio* NCSM approach and demonstrated its feasibility by applications to the $A = 3$ and 4 systems, for which we obtain well-converged results. For $A = 6$ we are also able to reach convergence. The acceleration of the convergence using three-body effective interaction has been demonstrated. In the case of ^{12}C, we are presently limited to model spaces up to $6\hbar\Omega$, where the dimensions already reach 32 million. We currently cannot reach full convergence for ^{12}C, but we do obtain a reasonable approximation for the lowest $0\hbar\Omega$-dominated states. On the other hand, the $B(E2)$ transition strengths clearly show that the 2- or higher-$\hbar\Omega$-dominated states have not yet converged but are steadily moving lower in energy as N_{max} increases. It will be interesting to employ three-body effective interactions for this system and to use consistently effective quadrupole operators for the calculation of the transition strengths. It should be noted that our calculations contain no adjustable parameters. The favorable comparison with available data that we obtain is a consequence of the underlying NN interaction.

ACKNOWLEDGMENTS

B.R.B. and A.N. acknowledge partial support by NSF grant No. PHY0070858. J.P.V. acknowledges partial support by USDOE grant No. DE-FG-02-87ER-40371. The work was performed in part under the auspices of the U. S. Department of Energy by the University of California, Lawrence Livermore National Laboratory under contract No. W-/405-Eng-48. P.N. and W.E.O. received support from LDRD contract 00-ERD-028.

REFERENCES

1. B. S. Pudliner, V. R. Pandharipande, J. Carlson, S. C. Pieper and R. B. Wiringa, Phys. Rev. C **56** 1720, (1997); R. B. Wiringa, Nucl. Phys. **A 631**, 70c (1998), S. Pieper and R. B. Wiringa, Ann.Rev.Nucl.Part.Sci. **51**, 53 (2001); Steven C. Pieper, V. R. Pandharipande, R. B. Wiringa, J. Carlson, Phys.Rev. **C 64**, 014001 (2001)
2. D. C. Zheng, B. R. Barrett, L. Jaqua, J. P. Vary, and R. L. McCarthy, Phys. Rev. C **48**, 1083 (1993); D. C. Zheng, J. P. Vary, and B. R. Barrett, Phys. Rev. C **50**, 2841 (1994); D. C. Zheng, B. R. Barrett, J. P. Vary, W. C. Haxton, and C. L. Song, Phys. Rev. C **52**, 2488 (1995).
3. P. Navrátil and B. R. Barrett, Phys. Rev. C **54**, 2986 (1996); Phys. Rev. C **57**, 3119 (1998).
4. P. Navrátil and B. R. Barrett, Phys. Rev. C **57**, 562 (1998), Phys. Rev. C **59**, 1906 (1999).
5. P. Navrátil, G. P. Kamuntavičius and B. R. Barrett, Phys. Rev. C **61**, 044001 (2000). E-print archive No. nucl-th/9907054.
6. P. Navrátil, J. P. Vary and B. R. Barrett, Phys. Rev. Lett. **84**, 5728 (2000); Phys. Rev. C **62**, 054311 (2000).
7. P. Navrátil, J. P. Vary, W. E. Ormand, and B. R. Barrett, Phys. Rev. Lett. **87**, 172502 (2001).
8. P. Navrátil and W. E. Ormand, Phys. Rev. Lett. **88**, 152502 (2002).
9. C. P. Viazminsky and J. P. Vary, J. Math. Phys., **92**, 2055 (2001).
10. K. Suzuki and S.Y. Lee, Prog. Theor. Phys. **64**, 2091 (1980); K. Suzuki, Prog. Theor. Phys. **68**, 246 (1982). K. Suzuki and R. Okamoto, Prog. Theor. Phys. **70**, 439 (1983).
11. J. Da Providencia and C. M. Shakin, Ann. of Phys. **30**, 95 (1964).
12. K. Suzuki, Prog. Theor. Phys. **68**, 1999 (1982); K. Suzuki and R. Okamoto, Prog. Theor. Phys. **92**, 1045 (1994).
13. A. C. Hayes, Phys. Rep. **315**, 257 (1999); P. Vogel, E-print archive nucl-th/9901027; E. Kolbe, K. Langanke and P. Vogel, Nucl. Phys. **A 652**, 91 (1999).
14. M. J. Musolf, *et al.*, Phys. Rep. **239**, 1 (1994).
15. W. E. Ormand, Phys. Rev. Lett. **82**, 1101 (1999).
16. B. R. Barrett, P. Navrátil, W. E. Ormand and J. P. Vary, in the Proceedings of the XXVII Marzurian Lakes School of Physics, Krzyże, Poland, September 2-9, 2001, Acta Phys. Pol. **B33**, 297 (2002).

Exactly Solvable Models Based on the Pairing Interaction

J. Dukelsky* and S. Pittel†

*Instituto de Estructura de la Materia, CSIC, Serrano 123, 28006 Madrid, Spain
†Bartol Research Institute, University of Delaware, Newark, Delaware 19716, USA

Abstract. The Pairing Model hamiltonian was first solved exactly by Richardson in the sixties. We give here the conditions of integrability for a wider variety of pairing hamiltonians for fermion and boson systems, together with the exact solution for their complete sets of eigenstates. The study of the exact solution for interacting boson models with repulsive pairing shows a new and unexpected mechanism for *sd* dominance.

INTRODUCTION

Exactly solvable models (ESMs) have played an important role in the understanding of several aspects of strongly correlated quantum problems, some not accessible from standard many-body methods. The main property of the ESMs is that they provide in analytic form the complete set of eigenstates, offering a unique tool with which to deeply understand the physics involved.

A common feature of ESMs in nuclear structure is that the model hamiltonian is written as a linear combination of the Casimir operators of a group decomposition chain, ideally representing the properties of a nuclear phase. Prototypical examples are the three dynamical symmetries of the Interacting Boson Model (IBM) [1], organized into the three vertices of the Casten Triangle. Others include the Elliott $SU(3)$ model and the $SU(2)$ one-level Pairing Model (PM) [2].

Superconductivity is a phenomenon common to both nuclear and condensed matter systems. It was first recognized in the latter field that superconductivity can be described by a microscopic pairing hamiltonian in the BCS approximation. While the pairing hamiltonian has been extensively used in both fields of physics in the BCS or more sophisticated approximations, the exact solution of the model given by Richardson in the sixties [3] passed almost unnoticed until very recently. The Richardson solution was rediscovered in an effort to describe the transition from superconductivity to the normal metal state in ultrasmall superconducting grains, which required a more accurate treatment than could be provided by Number Projected BCS [4]. Since then, much work has been done to generalize the model [5], while preserving its exact solvability, and to apply it to a variety of important physical systems. Applications have been reported to Bose condensates [6], interacting boson models [7], electrons in 2D lattices [5] and nuclear superconductivity [8].

In this contribution we review the recent generalization of the PM to three families of exactly solvable models. We then consider a model of repulsive pairing between bosons,

CP638, *Mapping the Triangle: International Conference on Nuclear Structure*
edited by A. Aprahamian, J. A. Cizewski, S. Pittel, and N. V. Zamfir
© 2002 American Institute of Physics 0-7354-0093-8/02/$19.00

which is part of the so-called rational family, and show that the model displays a quantum phase transition to a state with macroscopic occupation of the lowest two boson levels only. These results suggest a new mechanism for *sd* dominance in the IBM, which is triggered by the repulsion between bosons that arises from the Pauli Principle at the underlying nucleon level.

THREE FAMILIES OF EXACTLY SOLVABLE PAIRING MODELS

We begin our discussion of the three families of exactly-solvable Pairing Models by defining the elementary operators of the pair algebra,

$$K_l^0 = \frac{1}{2}\sum_m a_{lm}^\dagger a_{lm} \pm \frac{1}{4}\Omega_l \quad , \quad K_l^\dagger = \frac{1}{2}\sum_m a_{lm}^\dagger a_{l\overline{m}}^\dagger = \left(K_l\right)^\dagger , \qquad (1)$$

which in turn are the generators of the $SU(2)$ group for fermions, or the $SU(1,1)$ group for bosons and close the corresponding commutator algebras

$$\left[K_l^0, K_{l'}^+\right] = \delta_{ll'} K_l^+ \quad , \quad \left[K_l^+, K_{l'}^-\right] = \mp 2\delta_{ll'} K_l^0 . \qquad (2)$$

The operator $K_l^\dagger$ in (1) creates a pair of particles in time reversed states with $a^\dagger(a)$ the particle creation (annihilation) operator and $\Omega_l = 2l + 1$ the degeneracy of level l. Throughout the paper, the upper sign refers to bosons and the lower sign to fermions.

Assuming that there are L single particle states and taking into account that each $SU(2)$ or $SU(1,1)$ group has one degree of freedom, a model is integrable if there are L independent hermitian and global operators that commute with one another. These operators are the quantum invariants and their eigenvalues, the constants of motion of the system, completely classify their common eigenstates. To find these operators, we first define the most general set of hermitian and number-conserving one and two body operators in terms of the K generators:

$$R_l \;=\; K_l^0 + \left\{ 2g \sum_{l'(\neq l)} \frac{X_{ll'}}{2} \left(K_l^+ K_{l'}^- + K_l^- K_{l'}^+\right) \mp Y_{ll'} K_l^0 K_{l'}^0 \right\} . \qquad (3)$$

Up to this point, the matrices X and Y from which the R operators are defined are completely free. Here we fix them by imposing the condition that they must mutually commute to define an integrable model. The condition $\left[R_l, R_{l'}\right] = 0$ is fulfilled if the X and Y matrices are antisymmetric and satisfy

$$Y_{ij}X_{jk} + Y_{ki}X_{jk} + X_{ki}X_{ij} = 0 . \qquad (4)$$

An analogous condition was encountered by Gaudin [9] in a spin model now known as the Gaudin magnet. His model is based on R operators similar to (3), but without the one body term. Solving (4) leads to three families of solutions [10]:

112

I. The rational model

$$X_{ll'} = Y_{ll'} = \frac{1}{\eta_l - \eta_{l'}} \tag{5}$$

II. The trigonometric model

$$X_{ll'} = \frac{1}{\sin\left(\eta_l - \eta_{l'}\right)} \,, \quad Y_{ll'} = \cot\left(\eta_l - \eta_{l'}\right) \tag{6}$$

III. The hyperbolic model

$$X_{ll'} = \frac{1}{\sinh\left(\eta_l - \eta_{l'}\right)} \,, \quad Y_{ll'} = \coth\left(\eta_l - \eta_{l'}\right) \tag{7}$$

In all three families, the η_l are an arbitrary set of non-equal real numbers. Any choice of these parameters within any of the three families leads to an integrable model and any combination of the corresponding R operators produces an integrable hamiltonian.

The rational model was proposed in ref. [11] to demonstrate the integrability of the PM hamiltonian. Indeed the PM hamiltonian can be obtained as a linear combination of its R operators, viz: $H_{PM} = 2\sum_l \eta_l R_l^l$ plus an appropriate constant.

For all three models, the exact eigenstates in the seniority zero subspace can be expressed as

$$|\Psi\rangle = \prod_{\alpha=1}^{M} B_\alpha^\dagger |0\rangle \,, \quad B_\alpha^\dagger = \sum_l u_l(e_\alpha)\, K_l^+ \,, \tag{8}$$

where M is the number of pairs. The function u, which depends on a set of unknown *pair energies* e_α must fulfill the L eigenvalue equations $R_i|\Psi\rangle = r_i|\Psi\rangle$. A similar ansatz can be used for the states with other seniorities.

Here we summarize the results for the rational model, which is used in the application of the next section. Details on the other two families can be found in ref. [5].

$$u_i(e_\alpha) = \frac{1}{2\eta_i - e_\alpha} \,, \quad 1 + g\sum_j \frac{\Omega_j}{2\eta_j - e_\alpha} \mp 4g \sum_{\beta(\neq\alpha)} \frac{1}{e_\alpha - e_\beta} = 0 \,, \tag{9}$$

$$r_i = \pm\frac{\Omega_i}{4}\left[1 - \frac{g}{2}\sum_{j(\neq i)} \frac{\Omega_j}{\eta_i - \eta_j} \mp 4g\sum_\alpha \frac{1}{2\eta_i - e_\alpha}\right] \,. \tag{10}$$

For a given set of the $2L + 1$ free parameters η_i and g of the model, one has to solve the coupled set of M nonlinear equations (9) for the pair energies e_α. There are as many independent solutions as states in the Hilbert space.

A NEW MECHANISM FOR *SD* DOMINANCE IN THE IBM

Inherent in the success of the IBM [1] in describing the low energy properties of medium to heavy nuclei is the assumption of an effective decoupling of a subspace of collective bosons with angular momentum $L = 0$ (s) and $L = 2$ (d) from all other non-collective and higher spin bosons. Despite much effort through the years to derive the IBM from the underlying nuclear shell model, there has only been success in the spherical vibrational regime. It is known from Hartree-Fock-Bogoliubov (HFB) studies that deformed nuclei are also dominated by SD fermion pairs, though with no clear decoupling from the rest of the space. It is worthwhile noting here that HFB only takes into account fermion pair correlations. We will show in this section that there is a mechanism at the two-body boson interaction level (four fermion correlations) that further enhances sd dominance and that leads to a complete decoupling of the sd subspace in the limit of a large number of bosons. The origin is the repulsion between composite bosons that arises due to Pauli exchange of the constituent nucleons.

To represent this physics, we consider the rational model with high spin bosons and a repulsive pairing interaction. Exactly-solvable models of this type can be obtained from a linear combination of the constants of the motion in eq. (3),

$$H = \sum_l \varepsilon_l K_l^0 + g \sum_{l \neq l'} \frac{\varepsilon_l - \varepsilon_l}{\eta_l - \eta_{l'}} \left[K_l^+ K_{l'}^- - K_l^0 K_{l'}^0 \right] , \qquad (11)$$

where l here is the boson angular momentum and ε_l are the coefficients of the linear combination of R_l operators. The hamiltonian (11) contains a single boson term, a repulsive monopole pairing interaction and an attractive monopole-monopole interaction. To obtain the complete set of common eigenstates of the R_l operators and thus of the hamiltonian, it suffices to fix the parameters η_l and the pairing strength g and to look for solutions of the set of non-linear coupled equations (9).

We will present results for two choices of the model parameters. In Case I, we choose $\varepsilon_l = \eta_l = l$, which when inserted in (11) gives rise to a boson hamiltonian with equally spaced single boson energies and a constant pairing interaction. As discussed in [6, 7], this hamiltonian favors the scattering of boson pairs to high angular momentum states, a pathology that in fermion problems is often overcome by introducing a cutoff. Instead we also consider Case II with $\varepsilon_l = l$ and $\eta_l = l^2$, in which the pair scattering to high spin states has been softened. Once the set of parameters η_l is fixed, we can solve the set of equations (9) for given values of M and g. We can then calculate the occupation probabilities for the various boson states associated with the solution [6, 7].

For repulsive pairing and a large number of bosons the system shows a quantum phase transition from an s boson condensate in the weak pairing limit to a fragmented state in which only the two lowest boson states are macroscopically populated. In our case the two lowest states are precisely the s and d bosons. Quantum phase transitions take place in the thermodynamic limit, where the decoupling is complete. For finite systems the transition is softened and some non-sd bosons persist.

In Figure 1 we show the effect of the boson number on the fragmented phase for the two different cases, assuming a cutoff in boson angular momentum of $L = 12$. In the lower panel we display the occupation probabilities of the s and d bosons which in the large M limit both go to the well known $O(6)$ value of $1/2$. In the upper panel we plot the depletion from the sd subspace, or equivalently the summed occupation probability of the bosons with $L > 2$. In Case I, the depletion is small but non- negligible for small number of bosons (≈ 10) due to the unphysical properties of the constant

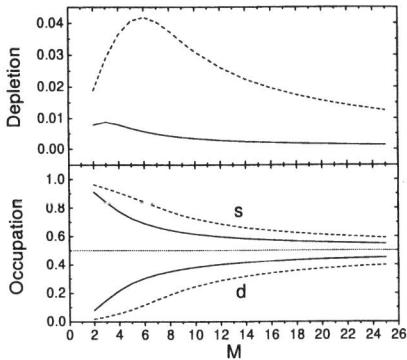

FIGURE 1. Occupation probabilities for $g = 0.5$ as a function of the number of boson pairs M. The dashed lines refer to Case I and the solid lines to Case II.

pairing interaction, but then goes to zero for increasing M values. In Case II the depletion is always small and goes to zero for moderate values of M.

Knowing the form of the ground state wave function, we can determine why only the s and d boson degrees of freedom survive in the thermodynamic limit. As discussed in ref. [7], specific phase relations must be satisfied by the different degrees of freedom for them to correlate under the influence of repulsive pairing, and these relations can be satisfied by only two at a time. Obviously they will be the lowest two, the s and the d.

ACKNOWLEDGMENTS

This work was supported in part by the the Spanish DGES under grant # BFM2000-1320-C02-02, by NATO under grant # PST.CLG.977000 and by the National Science Foundation under grant # PHY-9970749.

REFERENCES

1. F. Iachello and A. Arima, *The interacting Boson Model* (Cambridge University Press, 1987).
2. P. Ring and P. Schuck, *The Nuclear Many-Body Problem* (Springer-Verlag New York Inc. 1980).
3. R.W. Richardson, Phys. Lett. **3**, 277 (1963); R.W. Richardson and N. Sherman, Nucl. Phys. **52**, 221 (1964); R.W. Richardson, J. Math. Phys. **6**, 1034 (1965); R.W. Richardson, Phys. Rev. **141**, 949 (1966).
4. J. Dukelsky and G. Sierra, Phys. Rev. Lett. **83** (1999) 172.
5. J. Dukelsky, C. Esebbag and P. Schuck, Phys. Rev. Lett. **87**, 066403 (2001).
6. J. Dukelsky and P. Schuck, Phys. Rev. Lett. **86**, 4207 (2001).
7. J. Dukelsky and S. Pittel, Phys. Rev. Lett. **86**, 4791 (2001).
8. J. Dukelsky, C. Esebbag and and S. Pittel, Phys. Rev. Lett. **88**, 062501 (2002).
9. M. Gaudin, J. Phys. (Paris) **37**, 1087 (1976).
10. A new class of solutions was recently reported in R. W. Richardson, preprint # cond-mat/0203512.
11. M.C. Cambiaggio, A.M.F. Rivas and M. Saraceno, Nucl. Phys. A **424**, 157 (1997).

Transition Probability of Chiral Twin Bands

Jing-ye Zhang[*], F. Dönau[†], S. Frauendorf[**], P. Semmes[‡] and L. L. Riedinger[*]

[*]Department of Physics and Astronomy, University of Tennessee, Knoxville, TN 37996
[†]Forschungszentrum Rossendorf, 01314 Dresden, Germany
[**]Department of Physics, University of Notre Dame, Notre Dame, IN 46556, USA
[‡]Department of Physics, Tennessee Technological University

Abstract. The unique transition probability in chiral twin bands was predicted based on a simplified frozen alignment Particle plus Rotor (PR) model and a full PR model calculation.

The Tilted Axis Cranking (TAC) model predicted the possible existence of chiral twin bands (CTB) in rotating triaxial nuclei as a result of the spontaneous breaking of the chiral symmetry [1, 2]. Indeed candidates for the described structures, i.e. pairs of almost degenerate $\Delta I = 1$ rotational sequences with the same parity, were found in odd-odd N=75 isotones in the A=130 mass region [3, 4, 5, 6]. In order to further identify the CTB, the transition probability becomes crucial, since it reveals the structures of wave functions. The quasiparticle plus rotor (PR) models [7] possess the angular momentum I as a good quantum number, thus enabling one to account the electromagnetic ($M1$ and $E2$) transition probabilities without problems.

Based on a simplified frozen alignment PR model, in which, the two perpendicular quasiparticle spins $\vec{j}_p$ and $\vec{j}_n$ are frozen and coupled to the triaxial core [8], one can easily derive a relation for the intra band stretched $E2$ transition, $B(E2, I \rightarrow I-2) \propto \sin^2(\vartheta)$ where ϑ is measuring the angle between the rotational axis and the total quasiparticle

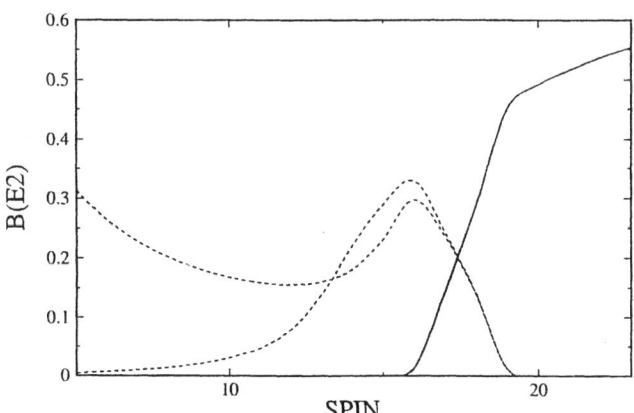

FIGURE 1. B(E2) values $[e^4b^2]$ for the lowest two bands calculated in the frozen alignment approximation to the PR model. Solid lines: inband transitions, dashed lines: interband transitions.

CP638, *Mapping the Triangle: International Conference on Nuclear Structure*
edited by A. Aprahamian, J. A. Cizewski, S. Pittel, and N. V. Zamfir
© 2002 American Institute of Physics 0-7354-0093-8/02/$19.00

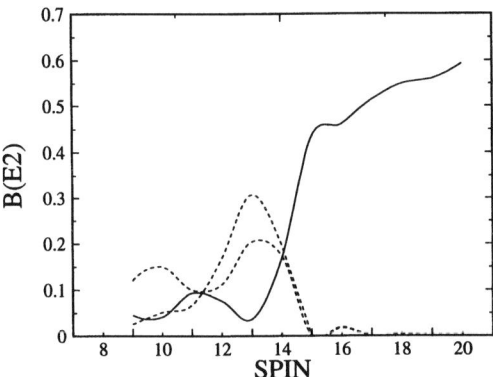

FIGURE 2. The stretched BE2's for the $\pi h_{11/2} \nu h_{11/2}$ band in ^{134}Pr from a full PR model calculation, with j-maxture and pairing. $\varepsilon = 0.2$, $\gamma = 30°$. the curve notation is the same as in Fig. 1.

alignment $\vec{j}_p + \vec{j}_n$. Hence, as shown on Fig.1 the BE2's for $I \le 15$ stay at zero before the CTB is formed ($\vartheta = 0$), and they steeply raise up once CTB is developed ($\vartheta > 0$). Fig.2 is the result from a full PR model calculation with j-mixture and normal pairing correlation for the $\pi h_{11/2}, \nu h_{11/2}$ band in ^{134}Pr.

The basic feature is the same as in the simplified model shown in Fig. 1. Therefore, such a sudden jump in BE2's might be taken as a helpful signal to identify chiral band structures in experiment.

ACKNOWLEDGMENTS

This work is funded by the U.S. Department of Energy through contracts no. DE-FG02-96ER40983 (University of Tennessee).

REFERENCES

1. Frauendorf, S., and Meng, J., *Nucl. Phys. A*, **617**, 131 (1997).
2. Dimitrov, V., Frauendorf, S., and Dönau, F., *Phys. Rev. Lett.*, **84**, 5732 (2000).
3. Petrache, C., et al., *Nucl. Phys. A*, **597**, 106 (1996).
4. Starosta, K., et al., *Phys. Rev. Lett.*, **86**, 971 (2001).
5. Bark, R. A., et al., *Nucl. Phys. A*, **691**, 577 (2001).
6. Hartley, D. J., et al., *Phys. Rev. C*, **64**, 031304(R) (2001).
7. Ragnarsson, I., and Semmes, P. B., *Hyp. Int.*, **43**, 425 (1988).
8. Dönau, F., Frauendorf, S., Zhang, J., Semmes, P., and Riedinger, L. (to be published).

Possible triaxial superdeformation in ^{174}Hf

D. J. Hartley*, M. Djongolov*, L. L. Riedinger*, F. G. Kondev†, R. V. F. Janssens**, K. Abu Saleem**, I. Ahmad**, D. L. Balabanski*, M. P. Carpenter**, P. Chowdury‡, D. M. Cullen§, M. Danchev*, G. D. Dracoulis¶, H. El-Masri‖, J. Goon*, A. Heinz**, R. A. Kaye††, T. L. Khoo**, T. Lauritsen**, C. J. Lister**, E. F. Moore**, M. A. Riley‡‡, D. Seweryniak**, I. Shestakova‡, G. Sletten§§, P. M. Walker‖, C. Wheldon§, I. Wiedenhöver**, O. Zeidan* and Jing-ye Zhang*

*Department of Physics and Astronomy, University of Tennessee, Knoxville, TN 37996
†Technology Development Division, Argonne National Laboratory, Argonne, IL 60439
**Physics Division, Argonne National Laboratory, Argonne, IL 60439
‡Department of Physics, University of Massachusetts-Lowell, Lowell, MA 01854
§Oliver Lodge Laboratory, University of Liverpool, Liverpool L69 7ZE
¶Department of Nuclear Physics, Australian National University, Canberra, ACT 0200
‖Department of Physics, University of Surrey, Guildford, Surrey, GU2 5XH
††Department of Chemistry and Physics, Purdue University Calumet, Hammond, IN 46323
‡‡Department of Physics, Florida State University, Tallahassee, FL 32306
§§The Niels Bohr Institute, Riso 4000, Roskilde

Abstract. Four regularly spaced rotational bands with large dynamical moments of inertia, which are consistent with known superdeformed bands in the Lu/Hf region, have been identified in ^{174}Hf. The states were populated in the ^{130}Te(^{48}Ca,4n) reaction at a beam energy of 194 MeV. The Gammasphere array detected the emitted gamma radiation. Ultimate cranker calculations predict substantial triaxial deformation ($\gamma \approx \pm 17°$) for highly deformed ^{174}Hf structures. However, ^{174}Hf is eight neutrons away from the previously established $N = 94$ triaxial superdeformed shell gap. Shell gaps at $N = 100$ and 106 with $\gamma \geq 15°$ are observed when $\varepsilon_2 \approx 0.45$, which may be responsible for the predicted TSD minima in ^{174}Hf.

The search for structures exhibiting properties of stable triaxial deformation has been ongoing for years. Although direct experimental evidence of triaxial shapes has been scarce, it has not kept the nuclear structure community from invoking triaxiality to describe various phenomena. In $^{163}_{71}$Lu, the rotational band based on the $\pi i_{13/2}$ excitation was found to be superdeformed [1, 2] and ultimate cranker calculations predict that this configuration has a substantial amount of triaxial deformation [3]. Recently, Ødegård et al. [4] presented evidence that a second superdeformed band (which lies higher in energy than the $\pi i_{13/2}$ band) may be the result of a wobbling excitation. Wobbling can only occur if the nucleus has a stable triaxial shape [5]; therefore, this is perhaps the best evidence for a non-axial shape to date. However, the calculations of Schnack-Petersen et al. [3] suggest that this is not unique to ^{163}Lu, but that similar structures likely have triaxial shapes throughout the $A \approx 165$ region. For this reason, sequences lying in the highly deformed minimum in this mass region are often referred to as triaxial

CP638, *Mapping the Triangle: International Conference on Nuclear Structure*
edited by A. Aprahamian, J. A. Cizewski, S. Pittel, and N. V. Zamfir
© 2002 American Institute of Physics 0-7354-0093-8/02/$19.00

superdeformed (TSD) bands. It was also concluded in Ref. [3] that a shell gap at $N = 94$ is responsible for the observed TSD structures. Thus, several experiments have been performed for Lu and Hf nuclei near $N = 94$ (see Ref. [4], and references therein). Although several cases have been observed in Lu, only [168]Hf [6] has shown evidence of TSD sequences in the $Z = 72$ nuclei. The lack of TSD bands in light Hf nuclei, and the fact that [174]Hf$_{102}$ is *eight neutrons* away from the $N = 94$ gap, make the recent discovery of four candidate TSD bands in this nucleus quite surprising.

High-spin states in [174]Hf were produced in the [130]Te([48]Ca,4n) reaction at a beam energy of 194 MeV. The target consisted of $\sim$0.5 mg/cm^2 of enriched [130]Te with a thin ($\sim$0.2 mg/cm^2) Au flashing, and the [48]Ca beam was provided by the ATLAS facility at Argonne National Laboratory. Decay γ radiation was detected with the Gammasphere [7] array, which contained 100 Compton-suppressed Ge detectors. A total of $\sim 5.5 \times 10^8$ three fold or higher coincidence events were recorded in approximately *one day of beam time*. It should be noted that the primary objective of the experiment was to observe rotational levels above the high-K isomers in [174,175]Hf [8]. As these sequences may be low in multiplicity, a relatively low trigger condition of only three or more suppressed Ge detectors in coincidence was used. Thus, this experiment was not optimized for a search of superdeformed bands, as higher multiplicity triggers and longer beam times are normally required to observe these weak structures. The transitions were Doppler corrected and then sorted into a coincidence $E_\gamma \times E_\gamma \times E_\gamma$ cube. Subsequent analysis of this cube was accomplished using the Radware package [9].

Representative spectra of the four sequences are displayed in Fig. 1. Combinations of double gates with the lowest five inband transitions were summed together in order to generate the coincidence spectra of bands 1 and 3. For bands 2 and 4, a limited number of clean gates were available, therefore these spectra were generated with double gates that included the 955 and 995-keV transitions, respectively. Unfortunately, none of the bands could be linked into the normal deformed structures of [174]Hf. However, strong coincidence relationships can be observed in the spectra between the bands and the ground-state structure in [174]Hf [8]. Band 1, shown in the top panel of Fig. 1, is the strongest of the four sequences with a relative intensity of $\sim$1% of the total population of [174]Hf. Note that this is perhaps four times stronger then the yrast TSD band observed in [168]Hf [6].

The dynamical moments of inertia, $\mathscr{J}^{(2)}$, of the [174]Hf TSD bands are plotted in Fig. 2(a), while the three sequences in [168]Hf [6] are shown in Fig. 2(b). Large dynamical moments of inertia are often associated with large deformation. The superdeformed nature of the yrast TSD bands have been confirmed with measured transition quadrupole moments of $Q_t = 11.4^{+1.1}_{-1.2}$ eb and $8.2^{+1.0}_{-0.6}$ eb for [168]Hf [6] and [163]Lu [2], respectively. (Normal deformed states have $Q_t \approx 6$ and 5 eb [2], respectively.) Since the bands in [174]Hf have similar $\mathscr{J}^{(2)}$ values to those in the lighter nuclei, they are likely superdeformed as well. However, lifetime measurements, that will determine quadrupole moments, are necessary to verify this assertion.

Total energy surfaces (TES) for [174]Hf were calculated using the Ultimate Cranker (UC) [10] code. The lowest energy configuration with $(\pi, \alpha) = (+,0)$ at $\hbar\omega = 0.55$ MeV is shown in Fig. 3. Several minima appear in the TES, and we will use the labeling convention of Bengtsson [11] to describe each of them. The remnants of minimum I, which

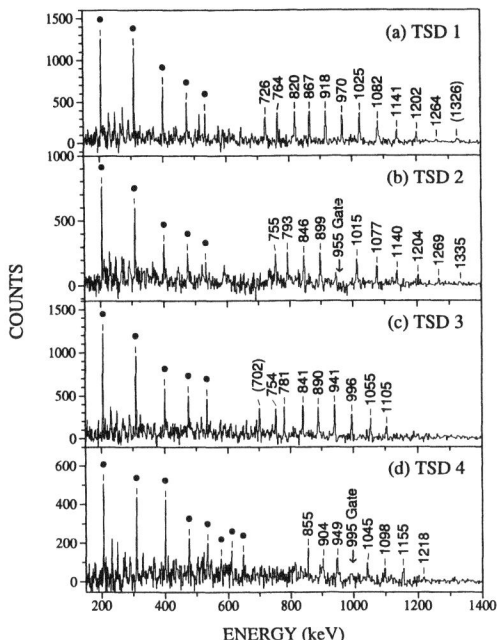

FIGURE 1. Coincidence spectra double-gated on all clean combinations of the lowest five transitions in the TSD bands of ^{174}Hf. For TSD bands 2 and 4, double-gates with the 955- and 995-keV transitions, respectively, were required. Transitions denoted with γ-ray energies are assigned to the TSD bands, while peaks marked with a filled circle are ground-state transitions in ^{174}Hf.

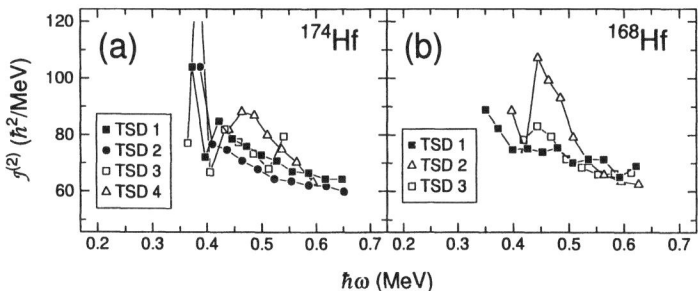

FIGURE 2. Dynamical moments of inertia of the TSD structures in (a) ^{174}Hf and (b) ^{168}Hf.

is lowest at frequencies less then 0.4 MeV, can still be seen in Fig. 3. It corresponds to the normal deformed well where $\varepsilon_2 \approx 0.25$ and $\gamma \approx 0°$. Minimum IA is lowest for the given frequency and has intermediate deformations of $\varepsilon_2 \approx 0.34$ and $\gamma \approx +7°$. This likely corresponds to the yrast sequence following an alignment of $h_{9/2}$ protons, which are known to drive the nuclear shape to higher deformations [12]. However, the structure of this minimum may be more complex as the predicted deformation is larger than that

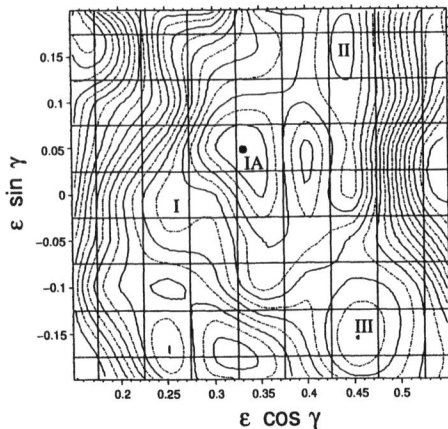

FIGURE 3. Total energy surface from ultimate cranker calculations for ^{174}Hf at a frequency of 0.55 MeV. A positive-parity, $\alpha = 0$ configuration was considered. Minima are labeled using the convention defined by Bengtsson [11].

measured for a nearby $h_{9/2}$ sequence [12].

Two superdeformed minima are observed in Fig. 3 as regions II and III. Minimum II is close in energy with respect to IA as it is only ~300 keV higher at this frequency. However, minimum III is located ~1 MeV higher in excitation energy than minimum II. This is consistent with calculations performed for nuclei with $N \approx 94$ [3, 11]. The deformations in the two TSD minima are $\varepsilon_2 = 0.453$ and $\gamma = +16°$ for II, and $\varepsilon_2 = 0.475$ and $\gamma = -19°$ for III. Therefore, the UC indicates that superdeformed structures in ^{174}Hf will also have significant triaxiality.

In order to determine the source of the TSD minima in ^{174}Hf, single-neutron energies were examined as a function of ε_2 and γ. One should first note the difference in predicted quadrupole deformations between the light $N = 92, 94$ Lu nuclei ($\varepsilon_2 = 0.389$ [3]) and ^{174}Hf ($\varepsilon_2 = 0.453$). A possible source for the larger deformation in the heavier nuclei may be explained in the left-hand portion of Fig. 4, which is the single-neutron energy as a function of ε_2. The hexadecapole deformation was set to zero for this and all following calculations. Orbitals that originate above the $N = 126$ spherical shell gap are strongly down-sloping in energy at higher deformations ($\varepsilon_2 > 0.25$) and highlighted with bold lines. These "intruder" orbitals are based on $i_{11/2}$ (solid lines) and $j_{15/2}$ (dashed lines), where significant deformation enhancement will occur if neutrons occupy these states. The light ($N = 92, 94$) Lu and Hf nuclei are less likely to involve the intruder states as their Fermi surfaces are further below these orbitals than $N = 102$ for ^{174}Hf. Thus, the larger deformation in ^{174}Hf may result from a higher occupancy of these $i_{11/2}$ and $j_{15/2}$ neutrons.

The right-hand panel of Fig. 4 displays the single-neutron orbitals as a function of γ with fixed deformation parameters $\varepsilon_2 = 0.453$ and $\varepsilon_4 = 0$. The ~15% increase in deformation from ^{163,165}Lu to ^{174}Hf shifts the relative placement of the neutron orbitals, which alters the single-neutron spectrum with respect to γ. Some notable

122

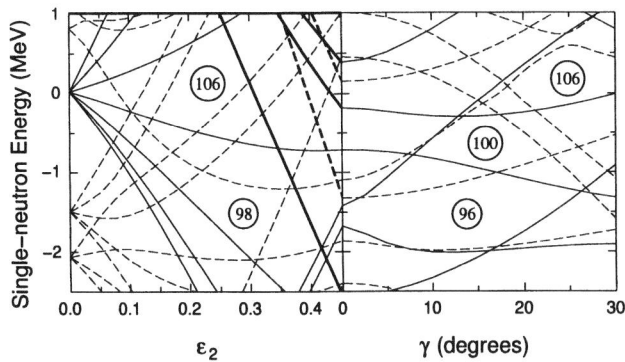

FIGURE 4. Left panel: single-neutron energy as a function of ε_2 from ultimate cranker calculations. Right panel: single-neutron energy as a function of γ, where $\varepsilon_2 = 0.453$ and $\varepsilon_4 = 0$.

differences occur in comparison with the calculations of Schnack-Petersen *et al.*, such as the evolution of the $N = 94$ gap to $N = 96$. The gap observed in Fig. 4 is nearly as large as the one previously calculated and spans a large range of γ. In addition, and perhaps more importantly, two other gaps are found at $N = 100$ and 106 (see Fig. 4) with well defined triaxiality, $\gamma \approx 15°$ and $\approx 25°$, respectively. As ^{174}Hf has $N = 102$, it is located near these gaps, which likely help create the TSD minima in Fig. 3. These gaps also suggest that TSD bands may be observed in $N = 100\text{-}106$ nuclei, which possibly opens a new region to explore triaxial superdeformation.

ACKNOWLEDGMENTS

This work is funded by the U.S. Department of Energy through contracts no. DE-FG02-96ER40983 (Tennessee) and W-31-109-ENG-38 (Argonne), as well as the National Science Foundation and the State of Florida (Florida State).

REFERENCES

1. Schmitz, W., et al., *Phys. Lett. B*, **303**, 230 (1993).
2. Schönwaßer, G., et al., *Eur. Phys. J. A*, **13**, 291 (2002).
3. Schnack-Petersen, H., et al., *Nucl. Phys. A*, **594**, 175 (1995).
4. Ødegård, S. W., et al., *Phys. Rev. Lett.*, **86**, 5866 (2001).
5. Bohr, A., and Mottelson, B. R., *Nuclear Structure*, vol. 2, Benjamin, New York, 1975.
6. Amro, H., et al., *Phys. Lett. B*, **506**, 39 (2001).
7. Janssens, R. V. F., and Stephens, F., *Nucl. Phys. News*, **6**, 9 (1996).
8. Gjørup, N. L., Walker, P. M., Sletten, G., Bentley, M. A., Fabricius, B., and Sharpey-Schafer, J. F., *Nucl. Phys. A*, **582**, 369 (1995).
9. Radford, D., *Nucl. Instrum. Methods Phys. Res. A*, **361**, 297 (1995).
10. Bengtsson, T., *Nucl. Phys. A*, **496**, 56 (1989).
11. Bengtsson, R., URL www.matfys.lth.se/~ragnar/TSD.html.
12. Joshi, P., et al., *Phys. Rev. C*, **60**, 034311 (1999).

SASSYER: A Gas-filled spectrometer at Yale

J. J. Ressler*, R. Krücken*, C. W. Beausang*, J. M. D'Auria[†], H. Amro*,
R. F. Casten*, M. A. Caprio*, G. Gürdal*, Z. Harris*, C. Hutter*,
A. A. Hecht*, D. Meyer*, M. Sciacchitano*, and N. V. Zamfir*

Wright Nuclear Structure Laboratory, Yale University, New Haven, CT
[†]*Simon Fraser University, Burnaby, British Columbia, Canada*

Abstract. The Wright Nuclear Structure Laboratory has recently acquired a gas-filled recoil separator previously used at Berkeley National Laboratory for heavy-element synthesis. The separator will be used to separate reaction recoils from primary beam particles and fission products following target bombardment. Commisioning of the separator has recently been completed, and the structure of ^{203}Rn investigated.

INTRODUCTION

Recoil separators [1, 2, 3] have been shown to be a powerful analytical tool for nuclear structure and decay studies. These devices provide an extremely clean method of channel selection and background suppression in heavy-ion fusion evaporation reactions. When coupled to Ge detector arrays, detailed spectroscopy of low cross-section ($\sim$nb) products may be accomplished.

The Wright Nuclear Structure Laboratory of Yale University has recently completed commissioning of a gas-filled magnetic separator, SASSYER (Small Angle Separator System at Yale for Evaporation Residues). This device will be used to study the structure of medium and heavy mass nuclei.

SEPARATOR AND GE ARRAY

SASSYER consistes of three magnets; two vertically-focussing dipoles and a single horizontally-focussing quadrupole. The quadrupole is placed between the two dipoles, in the configuration $D_v Q_h D_v$.

The magnetic separator was designed and implemented at Berkeley National Laboratory by A. Ghiorso *et al.* to specifically study element $Z = 110$. Descriptions of the magnets and experimental configuration at Berkeley may be found in Refs. [4, 5]. For Ghiorso *et al.*, the separator had an acceptance of 9 msr. At the WNSL, the target position has been moved further from the first dipole, reducing the acceptance to $\sim$5 msr. The target position was moved to accomodate a large germanium array, YRAST-Ball [6]. The array holds a maxmimum of 13 Compton-supressed clover detectors; nine at

CP638, *Mapping the Triangle: International Conference on Nuclear Structure*
edited by A. Aprahamian, J. A. Cizewski, S. Pittel, and N. V. Zamfir
© 2002 American Institute of Physics 0-7354-0093-8/02/$19.00

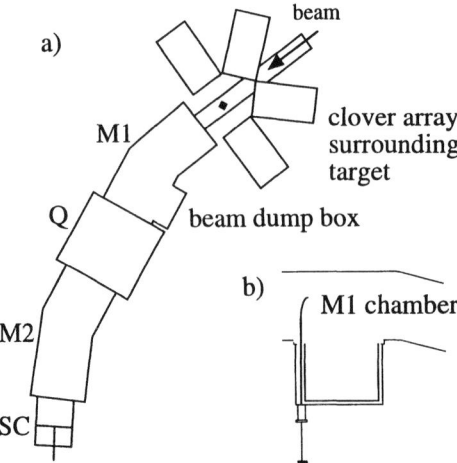

a)

beam

clover array
surrounding
target

M1

Q

beam dump box

b)

M1 chamber

M2

SC

FIGURE 1. Schematic of the SASSYER separator and dump box.

90° and four at 138.5° relative to the beam axis. Presently, the array consists of eight Compton-supressed clover detectors with an efficiency of ~ 2.5 % for 1.3 MeV γ-rays.

Together, the three magnets of SASSYER comprise an effective length of 2.4 m with a maximum rigidity of 2.2 T·m. The relatively compact length and high rigidity allows heavy nuclei with short lifetimes to be studied.

The magnet chamber is filled with ~ 1 torr of He gas. A thin carbon foil ($\sim 50 \, \mu g/cm^2$) is placed approximately 1.5 m in front of the first magnet to contain the gas. The gas enters the chamber near the entrance of the second dipole and is removed directly adjacent to the carbon window. The continuous flow minimizes contamimation and impurities. A schematic diagram of SASSYER is shown in Fig. 1a.

The purpose of SASSYER is to efficiently separate beam particles from recoiling products of heavy-ion fusion evaporation reactions. Fusion recoils will have a distribution of atomic charge states, q, due to the loss or pickup of electrons as they pass through the remainder of the target following production. If the flight path through the spectrometer is long relative to the mean free path of collisions between recoils and gas atoms, the multiple charge states of the recoils will coalesce into an average charge state q_{ave}. This clustering of many initial recoil charge states leads to much higher transmission through the separator relative to vacuum devices which typically accept only one or two charge states.

The presence of the gas induces a significantly lower average charge state than the most probable charge state in vacuum. For example, the ~ 24 MeV ^{150}Dy recoils produced from the ^{122}Sn(^{32}S,4n) reaction at 140 MeV have a calculated most probable charge state of 16^+ in vacuum. In 1 torr He gas, these recoils are calculated to have an average charge state of only 6.5^+, while the beam charge state is typically unaffected. When the dipole field is set to bend the recoil along the central axis in a gas-filled system, unreacted beam particles will have a significantly lower rigidity and therefore lower bending radius. To collect the beam, a dump box is placed on the low-ρ side of the first dipole magnet in SASSYER, see Fig. 1. Separation of the primary beam is then achieved shortly following the target.

The beam dump is a 34 x 34 cm^2 aluminum box, in which the interior is covered with thin lead shielding. A retractable Pb-covered arm may be extended into the first dipole to collect beam particles not fully bent into the dump. The beam dump box is shown schematically in Fig. 1b.

Recoil detectors are placed at the the exit of the SASSYER separator. For the commisioning and initial physics experiments, an array of solar cells has been used. Solar cells are small (1 cm^2), rugged silicon detectors of low cost.

HEAVY-ION REACTIONS

Initial commisioning of SASSYER was accomplished with standard alpha and fission radioactive sources. Light ($A < 50$) low energy stable beams were also used to test, calibrate, and align the magnet system.

Several heavy-ion reactions were used to test the system. The early experiments utilized only two clover detectors at 90° to the target position and a 3x5 cm^2 solar cell array behind the second dipole. Examples from two heavy-ion commisioning experiments are shown in Figs. 2 and 3. For the ^{32}S (140 MeV) + ^{122}Sn (0.5 μg/cm^2) reaction, PACE calculations [7] suggest the most populated channels to be ^{149}Dy (220 mb) and ^{146}Gd (110 mb). However, like many other nuclei in this mass region, both ^{149}Dy and ^{146}Gd have isomers with significant (>10 ns) lifetimes. Due to the presence of these isomers, only ^{150}Dy (28 mb) γ-rays were observed in the prompt gamma spectrum with significant intensity. The singles spectrum for the Ge detectors is shown in the upper portion of Fig. 2, with the recoil-gated spectrum below. Large background peaks at 511 keV (positron annhilation), 1174 keV, and 1332 keV (^{60}Co contamination) are observed in the singles spectrum. With the additional requirement of a recoil at the exit of SASSYER, these peaks are significantly reduced and a very clean spectrum of ^{150}Dy may be seen.

A similar comparison is shown in Fig. 3 for the ^{46}Ti (195 MeV) + ^{122}Sn (0.5 μg/cm^2). Large background peaks arising from positron annhilation and ^{60}Co contamination are seen in the singles spectra. The middle portion of Fig. 3 shows the resulting spectrum for a $\gamma - \gamma$ coincidence requirement. Increasing the required γ-ray multiplicity reduces the number of random events, but their presence remains significant. In the recoil-gated spectrum, bottom portion, peaks assigned to ^{164}Hf are easily observed.

The first significant physics study focussed on the structure of ^{203}Rn, which has been studied previously at the WNSL with YRAST-Ball [8]. The reaction ^{176}Yb(^{32}S,5n) at 165 MeV was used. The target was surrounded by the clover array, which consisted of four detectors at 90° and three at 138.5°. Recoiling reaction products were subsequently separated in SASSYER. Due to the high rigidity of the Rn recoils, ($B\rho = 1.9$ T·m), only ~0.7 torr He gas was used in the magnet chamber. Separated ions were implanted into a 3x10 cm^2 solar cell array located ~80 cm behind the second dipole.

Data was collected when an event was detected in one or more solar cell element(s). The resulting "prompt" gamma ray spectrum from the clover array surrounding the target is shown in Fig. 4b. Strong peaks assigned to ^{203}Rn and ^{202}Rn are evident. In Fig. 4a,

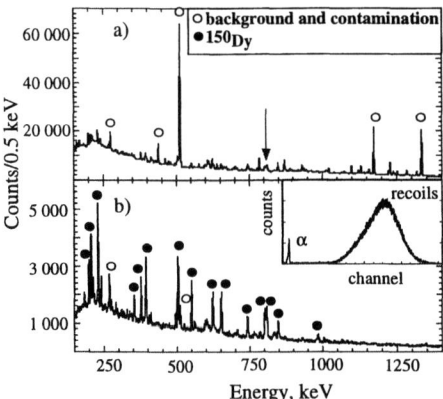

FIGURE 2. Results from the ^{122}Sn(^{32}S,xn) reaction. The upper portion (a) shows the γ-singles spectrum. The arrow points to the de-exitation energy of the first 2^+ state to ground in ^{150}Dy. The lower portion (b) shows the spectrum requiring a recoil at the exit of SASSYER. Numerous transitions assigned to ^{150}Dy are observed. The insert shows the energy spectrum for a single solar cell. A broad distribution of recoils are observed with sharp alpha peaks at lower energy.

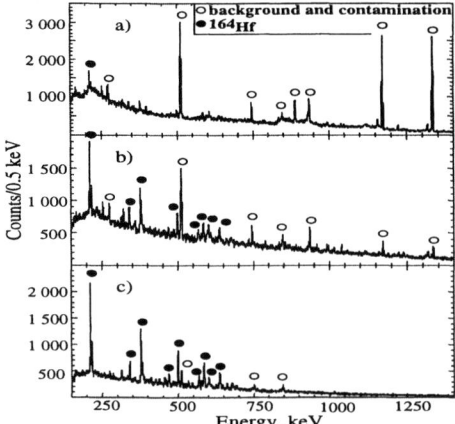

FIGURE 3. Results from the ^{122}Sn(^{46}Ti,xn) reaction. The upper portion (a) shows the γ-singles spectrum. The middle portion (b) shows the result when a $\gamma - \gamma$ coincidence is required. The bottom portion (c) shows the spectrum requiring a recoil at the exit of SASSYER. Transitions assigned to ^{164}Hf are seen.

$\gamma - \gamma$ coincidence data from the ^{174}Yb(^{34}S,xn) at 167 MeV of Newman *et al.* [8] is shown. This spectrum is dominated by a large fission background and intense peaks due to Coulomb excitation of the target. A omparison between these data and that obtained using SASSYER illuminates the improved background supression available with the magnetic separator.

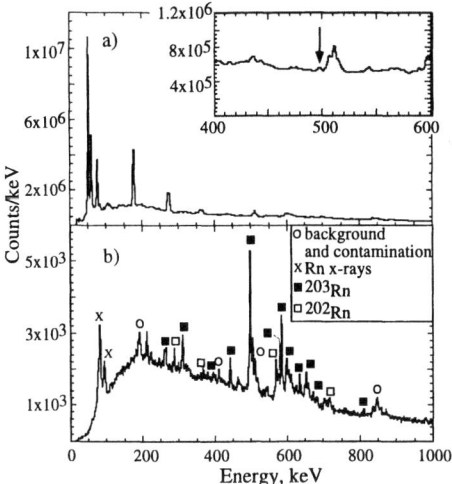

FIGURE 4. Gamma-ray spectra from the production of light Rn isotopes. In (a), results from the YRAST-Ball experiment [8] are shown. The spectrum is dominated by Coulomb-excitation of the Yb target. The inset shows the most intense transition in ^{203}Rn. In (b), the in-beam spectrum is shown with the requirement of a recoil exiting SASSYER. Transitions assigned to both ^{202}Rn and ^{203}Rn are observed.

ACKNOWLEDGMENTS

The SASSYER group would like to thank the Berkeley Gas-Filled Separator Group (notably A. Ghiorso and K. Gregorich) for their generosity and assistance with the magnets. We would also like to thank the RITU group of Jyväskylä for many useful discussions. The nuclear structure group would also like to gratefully acknowledge the hard work and tireless efforts of the accelerator operations staff – C. Miller, J. Ashenfelter, R. McGrath, W. Garnett, A. Ezeokoli, T. Barker, and R. Wagner. This work was conducted under U.S. DOE grant number DE-F602-91ER-40609.

REFERENCES

1. M. Leino *et al.*, Nucl. Instrum. Methods B **99**, 653 (1999).
2. C. N. Davids *et al.*, Nucl. Instrum. Methods B **70**, 358 (1992).
3. J. D. Cole *et al.*, Nucl. Instrum. Methods B **70**, 343 (1992).
4. A. Ghiorso, J. Radioanal. Chem.**124**, 407 (1988).
5. A. Ghiorso *et al.*, Phys. Rev. C **51**, R2293 (1995).
6. C. W. Beausang *et al.*, Nucl. Intrum. Meth. Res. A **452**, 431 (2000).
7. PACE computer code, A. Gavron, Phys. Rev. C **21**, 230 (1980).
8. H. Newman *et al.*, Phys. Rev. C **64**, 027304 (2001).

Nuclear Data Strategies for Mapping the Cosmos

Michael S. Smith* and Richard A. Meyer†

*Physics Division, Oak Ridge National Laboratory, Oak Ridge, TN, USA
†RAME', Inc., Teaticket, MA, & Chemistry Department, Clark University, Worcester, MA, USA

Abstract.
Significant advances are being made in understanding the structure of nuclei, especially those far from stability. The information from many such studies is vital to solving some important puzzles in astrophysics, such as the origin of the elements and the evolution of stars. However, dedicated efforts in data compilation, evaluation, dissemination, and coordination are needed to ensure that the latest nuclear measurements and theoretical calculations can be effectively utilized for astrophysics studies. A number of nuclear data strategies for astrophysics are presented.

NUCLEAR PHYSICS INFORMATION FOR COSMIC STUDIES

This conference featured many detailed descriptions of significant advances in understanding the underlying symmetries and behavior of nucleons in nuclei. Especially exciting are the efforts to probe nuclei far from stability - an area which represents a vital part of the future of the field of nuclear physics. Nuclear structure studies are important not only to understand the nature of the atomic nucleus, but also for "mapping the cosmos" - that is, for advances in astrophysics. For example, nuclei are produced via nuclear processes in astrophysical environments, so a knowledge of the structure of nuclei is needed to understand the cosmic origin of the elements. Furthermore, interactions between nuclei generate the energy we see as starlight and sunlight, and are responsible for changes in stellar composition and, therefore, for the evolution of stars. Measurements and theoretical descriptions of nuclei and their interactions, therefore, provide a foundation for sophisticated models of macroscopic astrophysical systems ranging from the Big Bang to exploding stars to the inner workings of our own Sun. In many instances, the ability of astrophysical models to accurately describe the latest, spectacular observations of the cosmos strongly depends on the input nuclear data. Thus, more extensive and precise nuclear data is required for advances in astrophysics.

It is not enough, however, to perform state-of-the-art nuclear measurements or theoretical calculations. To be utilized for astrophysical studies, this information has to be appropriately formatted for input into astrophysics simulation codes. This requires a *dedicated effort* in data compilation, evaluation, dissemination, and coordination. Currently, the results of many of the latest nuclear measurements or model calculations are not utilized in studies of the very astrophysical puzzles that - in some cases - motivated their generation. The situation is getting worse as more nuclear measurements are being made but not incorporated into reaction rate libraries and other astrophysical datasets.

CP638, *Mapping the Triangle: International Conference on Nuclear Structure*
edited by A. Aprahamian, J. A. Cizewski, S. Pittel, and N. V. Zamfir
© 2002 American Institute of Physics 0-7354-0093-8/02/$19.00

NUCLEAR DATA STRATEGIES FOR ASTROPHYSICS

The effort in nuclear astrophysics data is insufficient to meet the needs of the community, and there is now a recognition of its importance at the international level [1]. While some efforts in recent years have yielded important progress [2, 3], there are a number of strategies that will ensure a more effective utilization of nuclear physics information in nuclear astrophysics. First and foremost, more manpower is needed for evaluations. It is, however, also important to exploit the overlap between the nuclear data and nuclear astrophysics communities [4] to avoid duplication of efforts (e.g., with statistical model codes). It is crucial to coordinate plans to evaluate nuclear reactions or structure properties on a national and international basis in order to share expertise and avoid duplications. These, and other data activities, would be greatly facilitated by the establishment of a Nuclear Astrophysics Data Coordinator [5, 6] whose duties would be to:

- Establish, maintain, and update a central WWW archive of relevant datasets;
- Modify archive datasets for compatibility with astrophysical codes;
- Improve data accessibility via the creation of indices, search capabilities, graphical interfaces, bibliographies, error checking, plotting tools, and other enhancements;
- Other activities such as: help coordinate international data activities; establishing and maintaining a nuclear astrophysics email distribution list; publicize new nuclear astrophysics meetings, experimental results, and publications; and establish and maintain a priority list of important nuclear reactions and properties that require further study.
- Maintain an active research program using nuclear data to ensure the data activities truly fulfill the needs of data users.

These goals are all very realizable, and would have an impact on nuclear astrophysics research efforts worldwide with only a modest investment.

ACKNOWLEDGMENTS

ORNL is managed by UT-Battelle, LLC, for the U.S. Department of Energy under contract DE-AC05-00OR22725.

REFERENCES

1. Muir, D.W., and Herman, M., *IAEA Report* **INDC(NDS)-423**, 7 (2001).
2. Angulo, C. *et al.*, *Nucl. Phys.* **A656**, 3 (1999); http://pntpm.ulb.ac.be/nacre.htm.
3. M.S. Smith *et al.*, in *Proc. 10th Int. Symp. Capture Gamma-Ray Spectroscopy and Related Topics*, ed. S Wender, AIP Conf. Proc. **529**, 243 (1999).
4. M.S. Smith *et al.*, U.S. Nuclear Data Program Astrophysics Task Force Report, unpublished (1995); http://www.phy.ornl.gov/astrophysics/data/task/taskforce_report.html.
5. P.D. Parker *et al.*, Nuclear Astrophysics Data Project White Paper, unpublished; http://ie.lbl.gov/whitepaper.html
6. M.S. Smith, R.A. Meyer, *IAEA Report* **INDC(NDS)-**, in press (2002).

Nuclear Structure Studies with Rare Isotope Beams: Knockout Reactions

B. M. Sherrill

National Superconducting Cyclotron Laboratory, Michigan State University, East Lansing, Michigan 48824, USA

Abstract. This paper outlines the role of the Isospin Laboratory Steering Committee in the process that has lead to our current understand of the possibilities and challenges related to an advance rare isotope research facility. One of the challenges is to extract information using only weak rare isotope beams. Knock-out reactions are used as an example to illustrate that access to even weak beams can provide a large amount of nuclear structure information.

INTRODUCTION

Nuclei with a large excess of neutrons or protons are known to exhibit new features and teach us about aspects of nuclear structure not previously understood. In the past it was impossible to access many of the interesting nuclei because they could not be produced in sufficient quantities to study or our experimental techniques were not sufficiently sensitive. However, technology has advanced to the point where it is possible to build facilities, such as the proposed Rare Isotope Accelerator, which can produce a wide range of the interesting isotopes. Hence, there are a large number of exciting scientific possibilities in nuclear structure research that may be realized in the next few years.

In light of the opportunities, recently, the Nuclear Science Advisory committee has recommended the construction of the Rare Isotope Accelerator (RIA) as the highest priority for major new construction[1]. A schematic layout of the RIA facility is shown in Fig. 1. The goal of RIA is to provide the most intense source of rare isotopes for experimental studies. In order to do this, the facility allows for the optimization of the production mechanism for each desired isotope. These mechanisms can be projectile fragmentation or fission; target spallation, fragmentation and fission; and fusion. This requires the RIA driver be capable of accelerating all ions, up to Uranium, to at least 400 MeV/nucleon. Using superconducting linac technology it is possible to deliver very high beam intensities of more than 10^{14} ions/second for light ions and over 10^{13} uranium ions/s using current ECR ion source technology. The RIA concept will also requires a flexible array of target stations, including in-flight separation and ISOL. A full description of the RIA concept is provided in reference [2]. Included in the RIA concept is the capability of doing experiments with rare isotopes at a variety of energies from a few keV up to 100s of MeV/nucleon. The fast beams are critical since they allow for an increase in luminosity and allow for experiments with beams with extremely low intensity (even down to ions/week). This expands the scope of RIA to a wider range of the drip lines and to most of the r-process nuclei.

CP638, *Mapping the Triangle: International Conference on Nuclear Structure*
edited by A. Aprahamian, J. A. Cizewski, S. Pittel, and N. V. Zamfir
© 2002 American Institute of Physics 0-7354-0093-8/02/$19.00

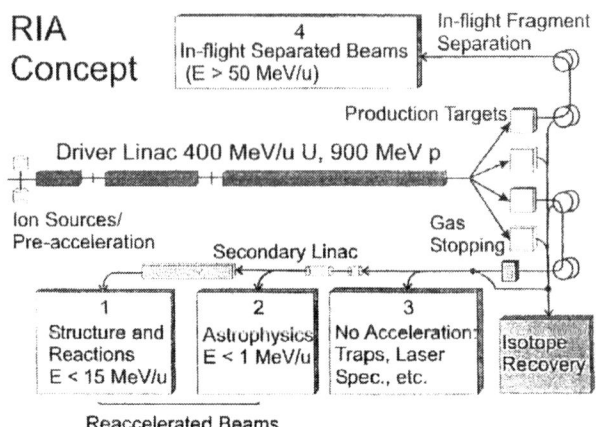

FIGURE 1. Schematic diagram of the Rare Isotope Accelerator facility. Rare isotopes are produced by beams from a superconducing linac, which can accelerate protons to 900 MeV and all ions up to uranium at 400 MeV/nucleon. Experiments will be possible at all secondary beam energies from 10 keV to 400 MeV/nucleon.

The RIA concept evolved over a long process which began with the 1989 NSAC Long Range Plan [3]. Enthusiasm for rare isotopes began to grow in the US in 1989 when there were a number of workshops devoted to the discussion of accelerated rare isotope beams, including a key meeting in Los Alamos [4]. The roots for the RIA project lie in the efforts in the mid 1980s to develop radioactive ion beam facilities including the Parksville conference in Canada, which has lead to the ISAC facility [5]. In the mid 1990s ORNL and ANL developed concepts for an advanced rare isotope research facility. Smaller scale, ISOL based radioactive ion beam facilities ISAC at TRIUMF and HRIBF at ORNL were build and operated in the later 1990s.

An NSAC subcommittee under the chairmanship of H. Grunder was constituted in 1998 to consider the optimal method for the production of accelerated rare isotope beams. This committee worked for more than one year evaluating the various options. Its deliberations lead to the current RIA concept [2]. The result is a facility which promises to allow most of the r-process nuclei to be studied as will as nuclei along nearly the entire proton drip line and the neutron drip line up to possibly Z=40.

HISTORY OF THE ISOSPIN LABORATORY COMMITTEE

At the time of the 1989 NSAC Long Range Plan it was decided by the nuclear science community that a group should form in order to push the idea of a dedicated rare isotope facility. It was hoped that the scientific opportunities would lead to future long range planning processes recommending the construction of an advanced rare isotope facility. The resulting, self-organized committee was called the Isospin Laboratory (ISL) Steering committee.

TABLE 1. Members of the Isospin Laboratory Steering Committee.

Member	Institution
R. Casten (Chairperson)	Yale
J. D'Auria	Simon Fraiser University
C. Davids	ANL
J. Garrett	ORNL
M. Nitschke	LBL
B. Sherrill	Michigan State University
M. Wiescher	Notre Dame University
E. Zganjar	Louisiana State University

TABLE 2. Summary of some of the scientific white papers that made the case for research with rare isotopes. NSAC Long Range Plan town meeting white papers are not included.

White Paper	Year	reference
Isospin Lab White Paper	1990	[6]
Isospin Lab White Paper Update	1995	[7]
Columbus White Paper	1997	[8]
Opportunities with Fast Beams	1999	[9]
RIA Physics White Paper	2000	[10]

The Isospin Laboratory Steering Committee original members are shown in Table 1. The committee solicited input from the community and organized a 400 person ISL users group. They issued newsletters, organized users meetings at DNP meetings, and wrote several scientific white papers, which are summarized in Table 2.

R. Casten was the chairperson of the committee and certainly one of its key members. Among other things he provided the glue for the committee and was responsible for coordinating and writing many of the scientific white papers, which justified the rare isotope beam science. A summary of some of the relevant White papers is shown in Table 2.

The process has been successful in that the 1996 Long Range Plan [11] placed a priority on rare isotope research and recommended going forward with the NSCL Coupled Cyclotron upgrade and plans for an advanced ISOL facility. Now that the advantages of a facility such as RIA are recognized, the 2002 Long Range Plan has endorsed this concept as the highest priority for major new construction.

KNOCKOUT REACTIONS

Some of the most interesting isotopes are difficult to produce because they are very far from stability. For example, S. Mizutori *et al.* [12] has made an estimate of the size of neutron skins over a wide range of nuclei. Compared to what RIA can make, the estimates are that RIA will be able to produce over a hundred isotopes with neutron

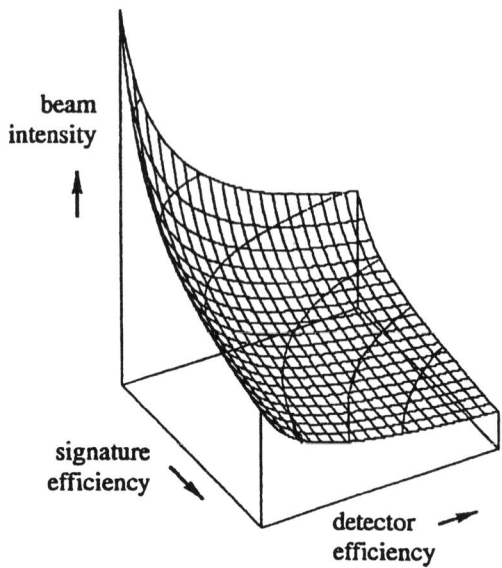

FIGURE 2. The Casten plot (see the footnote for an alternative name).

skins greater than 0.6 fm (that is nearly a full nucleon thick), but only at rates of around 100 ions per day.

The considerations of working with low-intensity rare isotope beams are illustrated by the Casten Plot [1]. The Casten Plot is illustrated in Fig. 2. The horizontal curved lines in the center of the plot represent lines of constant scientific knowledge. The plot illustrates that one can achieve a certain level of understanding at a certain beam intensity. However, the same level of understanding can be achieved with a lower beam intensity if the detector efficiency is higher or if a more sensitive technique is used to extract the information. An example of the latter is the use of N_pN_n systematics to look for changes in structure [13]. In fact, when beams of only 100 ions/day are available, it will be necessary to combine both high detector efficiency and efficient measures of information.

A relatively new technique (or at least new in its application to rare isotope beams) is nucleon knockout. Over the past few years, knockout reactions have been used to provide a high-energy equivalent to conventional low-energy transfer reactions of the type (p,d), (d,t), and (d,^{3}He), which are an important tool for the nuclear spectroscopy. Recent results suggest that the reaction theory based on the eikonal approximation provides accurate spectroscopic factors (perhaps good to 10%) with standard sets of

---

[1] A much more imaginative name was suggested by G.-E. Körner; the Casten Hammock.

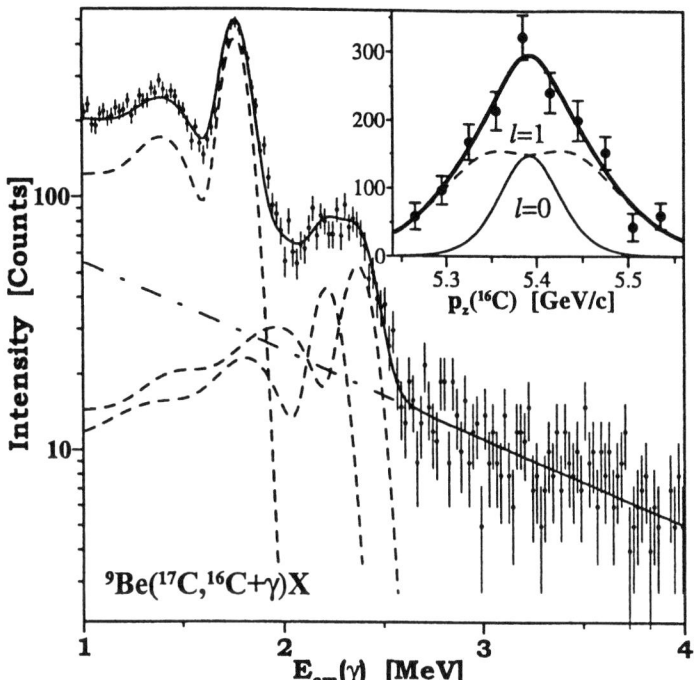

FIGURE 3. Gamma ray spectrum in the rest frame of coincident ^{16}C residues from the reaction ^{9}Be(^{17}C,^{16}C*)X at E_{beam}/A = 62 MeV. The strong peak at 1.77 MeV signals that approximately half of the cross section goes to the 2^+ first-excited level of ^{16}C. The momentum spectrum of coincident ^{16}C nuclei (inset) demonstrates that this partial cross section has two components, l=0 and l=2. From ref. [18]

input parameters. The most remarkable feature of the technique is its high sensitivity, which extends the possibility of obtaining spectroscopic information for beam intensities of only 100 ions/day. This is because the detection efficiency can be nearly 100 % and thick secondary targets can be used so that the interaction probability for the secondary ions is nearly 10 %.

Electron induced (e,e'p) knockout reactions have been used to study deeply bound states in nuclei and have found high momentum components in the wave functions and an interesting apparent lack of the full expected spectroscopic factors. These experiments are performed by measuring scattered electrons and protons and reconstructing the "missing momentum" for the recoil nuclei. The kind of knockout described in this paper is the same, but applied to rare isotope beams. In this case, the target is taken as a light nucleus. The "missing momentum" is directly determined by measuring the recoil momentum of the core. The final states of the core are determined by coincident γ-ray detection. An overview of the work done so far is given in [14]. Spectroscopic factors are determined by comparing the measured removal cross section to a calculated cross section assuming full spectroscopic strength. The angular momentum of the re-

TABLE 3. Comparison of spectroscopic factors determined from inverse kinematic proton knockout compared to those determined from (e,e'p) [19].

Nucleus	Spectroscopic factor, S	(e,e'p) Spectroscopic factor, S
^{12}C	0.57(2)	0.51(3)
^{16}O	0.68(4)	0.67(5)

moved nucleus can be determined from the width of the recoil momentum distribution [15, 16, 17]. Higher l-value removal has intrinsically larger momentum widths due to the stronger localization of the wave-function for high-l orbits. The results can be compared with spectroscopic factors obtained in various nuclear models, such as large basis shell-model calculations. For example in the case of ^{11}Li Simon *et al.* [15] were able to show the wave function is an almost equal mixture of s and p-waves.

As an example, Fig. 3 shows a spectrum of gamma rays measured [18] in coincidence with ^{16}C residues from the reaction ^{9}Be(^{17}C,^{16}C*)X at 5.4 GeV/c, for an incident ^{17}C beam intensity of 100-300/s. The gamma ray energies were transformed on an event-by-event basis into the ^{16}C rest-frame using the measured ^{16}C velocity and the emission angle and energy of the gamma rays recorded with a position-sensitive gamma-detector array. The calculated line shapes (dashed) are from a Monte-Carlo simulation, and the dot-dashed line represents a measured background, presumably associated with target fragmentation. The absolute intensities extracted from these simulations provide the cross sections to the ground state, to the 1.77 MeV 2^+ state, and to a group of states near 4.1 MeV, and the corresponding spectroscopic factors. The shape of the momentum spectrum shown in the insert allows the determination of the single-particle angular momentum l. For the present example, the partial cross section of 60 mb to the 2^+ state is explained by two contributions, approximately 44 mb for l=2 and 16 mb for l=0 – note the very different shapes of the two components. The implication is that the ground state has spin $3/2^+$ corresponding to a complex wave function of the form $|3/2^+> = c2\,|2^+ \otimes d5/2 > +c0\,|2^+ \otimes s1/2 > + \cdots$.

An interesting question is the extent to which this technique reproduces spectroscopic factors determined from more established methods, such as (e,e'p). The agreement is illustrated in Table 3 [19]. The central idea, as described in previous work [16, 18, 20], is in assuming that the theoretical partial cross section to a given final state nI^{π} of the residual nucleus (the core) can be written

$$\sigma_{th}(nI^{\pi}) = \sum_{j} S_{CM}(nI^{\pi}, lj)\sigma_{sp}(B_N, lj) . \tag{1}$$

Here $S_{CM}(nI^{\pi}, lj) = \frac{A}{A-1}S(nI^{\pi}, lj)$ is a spectroscopic factor with a center-of-mass correction. The quantity S expresses the parentage of the initial state with respect to a specific final state coupled to a nucleon with given angular-momentum quantum numbers (lj). In the knockout studies, approximately 10% of the wave function is samples and this factor is included in the quoted spectroscopic factors. The experimental spectroscopic factors for nuclear proton knockout on ^{16}O and ^{12}C are identical to those obtained in the (e,e'p) reaction, see the summary by Kramer *et al.* [21].

SUMMARY

This article has presented a summary of the history of the ISL steering committee. R. Casten played a major role in this committee and the effort to make the nuclear science community aware of the opportunities with rare isotope beams. He also introduced the concept of the Casten plot, which illustrates that the same level of science can be extracted from a rare isotope beam if either the beam intensity, detector efficiency, or signal efficiency is increased. The nucleon knockout technique is one example of a sensitive techniques that is able to provide spectroscopic information for extremely weak beams.

ACKNOWLEDGMENTS

The knockout reaction work described here is the collaboration of many people mentioned in the references. I would particularly like to acknowledge G. Hansen, A. Brown and J. Tostevin. The development of the Rare Isotope Accelerator Concept, RIA, has benefited from the contributions of many people. In particular Rick Casten has provided scientific leadership in the writing of the white papers listed here and in leading the Isospin Laboratory Steering Committee. The Casten plot picture is courtesy of R. Casten, its inventor.

REFERENCES

1. **2002 Long Range Plan for Nuclear Science**, DOE/NSF;
   http://www.er.doe.gov/henp/np/nsac/LRP_5547_FINAL.pdf .

2. ISOL Task Force Report, November 22, 1999, NSAC Report: a copy can be downloaded from http://srfsrv.jlab.org/isol/ISOLTaskForceReport.pdf .

3. **Nuclei, Nucleons, Quarks: Nuclear Science in the 1990's, A Long Range Plan for Nuclear Science**, NSAC (1989).

4. J.B. McClelland and D.J. Vieira, eds. Proc. of the Workshop on the Science of Intense Radioactive Ions Beams, LANL Reprot LA-11964-C (1990).

5. **The TRIUMF-ISOL Facilty: Proc. of the Acclerated Radioactive Beams Workshop**, Parsville, Canada, J.P.K. Lee and J.M. D'Auria eds., TRIUMF Report TRI-85-1 (1985).

6. **The IsoSpin Laboratory: Research Opportunities with Radioactive Nuclear Beams**, R. Casten *et al.*, LANL Report LALP 91-15.

7. **Overview of Research Opportunties with Radiactive Nuclear Beams: An Update–1995**, R. Casten *et al.*, a copy can be downloaded from:
   http://www.nscl.msu.edu/future/ria/process/whitepapers/isl95.pdf .

8. **Scientific Opportunities with an Advanced ISOL Facility**; a copy can be downloaded from:
   http://www.er.doe.gov/production/henp/isolpaper.pdf .

9. M. Thoennessen, **Scientific Opportunities with Fast Fragmentation Beams from RIA**, March 2000, a copy can be downloaded from: http://www.nscl.msu.edu/future/ria/process/whitepapers/opportunitiesffbeam.pdf .

10. **RIA Physics White Paper**, RIA 2000 Workshop in Durham, North Carolina, July 24-26, 2000, a copy can be downloaded from: http://www.nscl.msu.edu/future/ria/process/whitepapers/durham2000meeting.pdf .

11. **1996 Nuclear Science Long Range Plan**: the URL is http://pubweb.bnl.gov/~nsac/lrpmap.html.

12. S. Mizutori *et al.*, Phy. Rev. **C61** (2000) 044326.

13. R. Casten and B.M. Sherrill, Prog. in Particle and Nuclear Science **45** (2000) S171.

14. P. G. Hansen and B. M. Sherrill, Nucl. Phys. **A693**, (2001) 133.

15. H. Simon *et al.*, *Phys. Rev. Lett.* **83** (1999) 496.

16. A. Navin *et al.*, Phys. Rev. Lett. 81 (1998) 5089.

17. T. Aumann *et al.*, Phys. Rev. Lett. 84 (2000) 35.

18. V. Maddalena *et al.*, Phys. Rev. **C 63** (2001) 024613.

19. B.A. Brown, G. Hansen, B.M. Sherrill, and J. Tostevin, submitted to Phys. Rev. **C** (2002).

20. J.A. Tostevin, Nucl. Phys. A **682**, 320c (2001).

21. G.J. Kramer, H.P. Blok and L. Lapikas, Nucl. Phys. A **679**, 267 (2001).

Modification Of Shell Structure In p-sd Shell Nuclei Studied by Radioactive Beams

Isao Tanihata

RIKEN, 2-1 Hirosawa, Wako, Saitama 351-0198, Japan

Abstract Recent studies of interaction cross sections, fragments momentum distributions combined with standard quantities such as mass and beta-decay Q-values reveal new dynamic change of shell structures. In particular, special attention is paid on neutron halo nuclei in the p-sd shell. In halo nuclei and other neutron rich nuclei, the $2s_{1/2}$ orbital shows irregular behavior. One such extreme case is the new neutron magic number $N=16$. The global behavior of $2s_{1/2}$ shell is discussed in relation to the interaction cross-sections and fragment momentum distributions. Inconsistencies in understanding these data are pointed out. Peculiar changes of shells are discussed for ^{19}C and ^{23}O.

INTRODUCTION

Neutron halos appear in loosely bound nuclei near the neutron dripline. It is also known that the s-wave is strongly contributing to the formation of a halo. The effects of neutron halos are observed as a large enhancement of interaction (-or reaction) cross-section and a narrow distribution of a projectile-fragment momentum. In turn, these observations are a sensitive tool for the study of effect of s- and/or p-wave effects.

Changes of magic numbers in neutron rich nuclei have been discussed recently in many places [1]. The mixing of $1p_{1/2}$, $2s_{1/2}$, and $1d_{5/2}$ has been shown in Li and Be isotopes at $N=8$ shells showing the disappearance of $N=8$ magic number in neutron rich nuclei [2]. Also a disappearance of $N=20$ magic number has been confirmed below $Z=12$ (Mg) from γ-ray spectroscopy using radioactive beams [3]. Not only the disappearance, a new magic number $N=16$ has been observed in neutron rich nuclei [4]. Those changes of shells are most likely related to the movement of single particle levels and thus reflect the behavior of $2s_{1/2}$ and other orbital.

In the following, I show recent data on interaction cross-sections and fragment momentum distributions and then discuss the complicated change in the order of the orbitals. Also, I show the difficulty in explaining the interaction cross-sections and fragment momentum distributions simultaneously with present models.

CP638, *Mapping the Triangle: International Conference on Nuclear Structure*
edited by A. Aprahamian, J. A. Cizewski, S. Pittel, and N. V. Zamfir
© 2002 American Institute of Physics 0-7354-0093-8/02/$19.00

INTERACTION CROSS SECTIONS

An interaction cross section (σ_I) is the total probability for the removal of one or more nucleons from a projectile nucleus. It has been used to determine the sizes of unstable nuclei and has been a major tool leading to the discoveries of neutron halos and neutron skins [5,6]. Recent systematic measurements cover all nuclei in p shell and p-sd shell (proton in p-shell and neutron in sd shell) up to the drip line nuclei except ^{22}C [7]. Figure 1 shows the σ_I of ^{A}Z+C reactions measured at incident energy near 800A MeV.

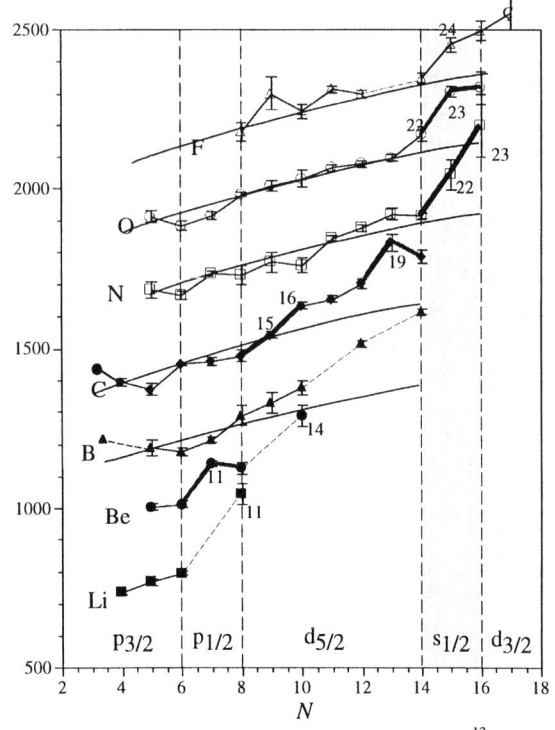

Fig. 1. Interaction cross sections of Li-F isotopes on ^{12}C target at ~800A MeV. Cross sections of different isotopes are shifted by (Z-3)*200 mb. Large increases indicate the contribution of s wave and a formation of neutron halo.

One can see easily in the ^{11}Li and ^{11}Be cases, the value of the σ_I increases abruptly when a halo neutron is added. In ^{11}Li, it is understood that the halo neutrons occupy $1p_{1/2}$ and $2s_{1/2}$ orbitals. The halo neutron in ^{11}Be is considered to mostly occupy the $2s_{1/2}$ state, though it is a more complicated mixed state. It has to be noted that such a large enhancement occurs only in p and s orbitals and in larger l orbitals because of the centrifugal barrier.

Smooth curves in Fig. 1 drawn for each isotope show the expected values of σ_I normalized at the stable isotope under the assumption that the interaction radius of an isotope is proportional to $A^{1/3}$. It is generally seen that the radius follows this curve in isotopes near the stable one but deviate from the curve in isotopes far from stability. The σ_I's of fluorine isotopes show rather normal tendency. The value increases in parallel with the curve up to 6 neutrons in the sd shell, the occupancy in $1d_{5/2}$ orbital. Then faster increase is seen for 7th and 8th neutrons, filling $2s_{1/2}$ orbital. The oxygen isotopes show similar behavior but no increase is observed from ^{23}O to ^{24}O.

In several of these cases, strong contributions of 2s orbital have been demonstrated. A recent measurement of $P_{//}$ distribution of ^{15}B fragment from ^{17}B has confirmed it for ^{17}B [8]. A large Coulomb dissociation cross section shows the strong contribution of $2s_{1/2}$ for ^{19}C. Therefore one can assume that the large increase of the $\Box_I$ is due to a contribution of 2s orbital. It is obvious that 2s orbital dominantly contributes near the neutron dripline for all isotopes. It seems that the neutron halos are formed in all isotopes free from the filling rule under normal shell ordering (near the stability line). An important conclusion is obtained from this fact; the halo is not formed just because the s-orbital is there for last neutron occupancy, but it is formed by bringing the s-orbital into the ground state.

MAGIC NUMBERS FAR FROM THE STABILITY LINE

Recent studies show considerable changes of magic numbers in neutron-rich nuclei. The first evidence was the disappearance of $N=20$ magic number in Na isotope from the isotope shift measurement [9]. Recently, B (E2) measurement of ^{32}Mg has also shown that $N=20$ is not a magic number for Mg isotopes [3]. It is now established that the $N=20$ magic disappears below $Z=12$. It is consistent with the fact that "expected" doubly magic nucleus ^{28}O is not bound.

Are magic numbers just weakened or do they disappear in neutron rich nuclei? It was a question until last year when Ozawa et al. analyzed the neutron separation energies and nuclear radii in neutron-rich region and found that a new magic number $N=16$ appears in very neutron rich nuclei [4].

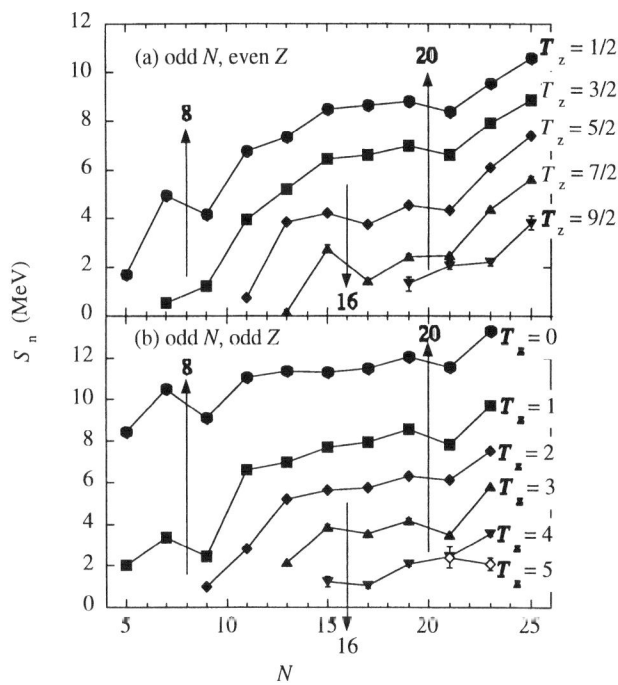

Fig. 2. Neutron separation energies of light nuclei. A break in the connected line indicates a neutron magic number. A new magic number $N=16$ appears for nuclei with isospin lager than 5/2.

It is clearly shown in Fig. 2 that the new magic number $N=16$ appears for nuclei with isospin larger than 5/2. It is also clearly seen that $N=8$ and 20 disappears for large isospin. The additional evidence of $N=16$ magic is seen as the neutron dripline for C, N, O. An important fact is the appearance of the magic $N=16$ in all nuclei with Z below 11. This magic number persists for a wide range of Z, probably from $Z=6$ to 11. It is an indication that this magic number is not due to deformation but a spherical configuration. It is also interesting to consider the double magicity of 24O. Kanungo et al. have extended such a systematic study to heavier nuclei [10]. The systematic behaviors of beta-decay Q-values, one- and two-neutron separation energies, excitation energies of the first excited states, and B (E2) values were studied. They found additional candidates of magic numbers at $N=30$, 32 and $Z=16$ in neutron-rich region of the nuclear chart.

All evidence points to the persistence of shell structure in neutron-rich nuclei with a change in the orbital ordering. There are many possible reasons for this and we list some of them below:

1. Deformation
2. Weak (or strong) binding
   Stronger binding for smaller l orbital (Halo)
3. Change of a potential shape
   Change of diffuseness
   Long density tail
4. Change of *ls* coupling
   Separation of p and n distributions (skin)
   Change of diffuseness
   Change of coupling strength
5. Residual p-n interaction and tensor coupling
6. Paring into continuum

Deformations play an important role in the disappearance of the $N=20$. In such nuclei, protons occupy $d_{5/2}$ orbitals and therefore have prolate deformed wave function for nuclei in this region. If orbitals are in their normal order, neutrons are in $d_{3/2}$ shell and have oblate-shaped wave functions. Therefore the overlap between protons and neutrons is small.

A weak binding energy can also influence the order of orbitals considerably. The binding energy of neutrons in neutron-rich nuclei could be extremely small. It is very often below 1 MeV in a nucleus near the dripline. In such a case, orbitals with smaller l have larger binding energies compared with larger l orbitals. In particular, the 2s orbital is much more strongly bound and helps to form a neutron halo. Similar situations may occur for the p orbital. This effect is seen in many cases of neutron halo nuclei as the lowering of the s-orbital below d-orbital. Neutron halos in Be, B, C near the dripline and they are consistent with this mechanism.

The combined effects of deformation and weak binding are clearly seen in ^{11}Be. This nucleus has a well-known abnormal ground-state spin-parity, $1/2^+$. The "normal" $1/2^-$ state appears as an excited state. The transition from this $1/2^-$ excited state to the $1/2^+$ ground state is known to have the strongest E1 transition probability among all other E1 transition in nuclei. If this $1/2^+$ state is a branch of deformed orbital extended out from $d_{5/2}$, 1/2[220], the E1 transition are forbidden because $1/2^-$ has quantum number 1/2[101]. On the other hand, the weak binding brings $s_{1/2}$ orbital down but it still is difficult to bring it below $p_{1/2}$. The spin-parity and strong E1 strength are explained only when two effects, deformation and weak binding, are combined.

The other possible effects listed above are not yet clearly identified in nuclei discussed in this paper. Therefore no further discussion is given.

SPECTROSCOPIC INFORMATION OBTAINED BY FRAGMENT MOMENTUM DISTRIBUTIONS AND INTERACTION CROSS SECTIONS

Momentum distributions of projectile fragment have been used to identify and to yield the spectroscopic information of halo nuclei successfully. The $P_{//}$ distribution of projectile fragment reflects directly the internal momentum distribution of the removed nucleon(s). The width and the shape of the $P_{//}$ distribution depend strongly on the l of the orbital that was occupied by the nucleon(s). For example, s wave gives a narrow distribution and d wave gives wider distribution. The behavior is just opposite to the density distribution that has extended tail in s wave and smaller tail in d wave.

A recent example is the measurement of ^{16}C fragments from ^{17}C at MSU [11]. They observe the $P_{//}$ distribution of ^{16}C fragments in coincidence with γ-rays from the excited states of ^{16}C. As shown in Fig. 3, they observed the dominance of the d wave in the last neutron and some amount of mixing of s wave from the narrow component observed in the 2^+ final state of ^{16}C. It is consistent with the assignment of spin-parity of ^{17}C ground state as $3/2^+$.

On the other hand, the momentum distribution of ^{18}C from fragmentation of

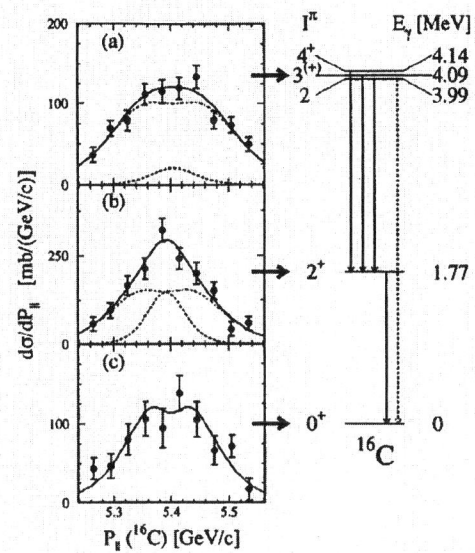

Fig. 3. $P_{//}$ distribution of ^{16}C fragments from ^{17}C in coincidence with γ rays from the excited state of ^{16}C. Wide and top-flat distributions show the dominance of d wave. Some contribution from s-wave is seen in the transition to 2^+ state.

^{19}C is narrow and consistent with the dominance of the s wave with small amount of admixture of d wave. It is thus considered to be a good tool to use $P_{//}$ distribution for spectroscopic information of a neutron halo.

Independently from $P_{//}$ distribution, interaction cross sections has been also used to obtain spectroscopic information of halos. An interaction cross section measured at high energy (~800A MeV) was fitted by changing the mixing ratio of different waves [7]. Consistent results were obtained between this analysis and $P_{//}$ analysis in ^{11}Li and ^{11}Be. Therefore the method is considered to be reasonably accurate.

Inconsistency in ^{19}C and ^{23}O

The measurements of the interaction cross-section and the Coulomb dissociation, a dominance of $2s_{1/2}$ wave is suggested for ^{19}C. In contrast, large amplitude of $1d_{5/2}$ wave is suggested from the $P_{//}$ distribution. The first quantitative study to see the consistency of mixing amplitudes was reported by Kanungo et al. for ^{19}C[14]. They have assumed the ground state with 0^+ and an excited state with 2^+ for the core (^{18}C). Because the spin-parity of ^{19}C is not known, it is assumed to have a value either $1/2^+$, $3/2^+$, or $5/2^+$. They have found that the any values of mixing ratio between $2s_{1/2}$ and $1d_{5/2}$ cannot reproduce the momentum distribution and the interaction cross-section, simultaneously. Therefore it is considered either some part of this model is improper or a considerable modification of the core is occurring from the bare ^{18}O.

A similar problem has been observed in ^{23}O [15]. Under the assumption that ^{23}O is represented by a ^{22}O core and a neutron. The $P_{//}$ distribution of ^{22}O fragment can be explained well by an s-wave dominance. However, the interaction cross section can not be reproduced even if one assumes 100% $2s_{1/2}$

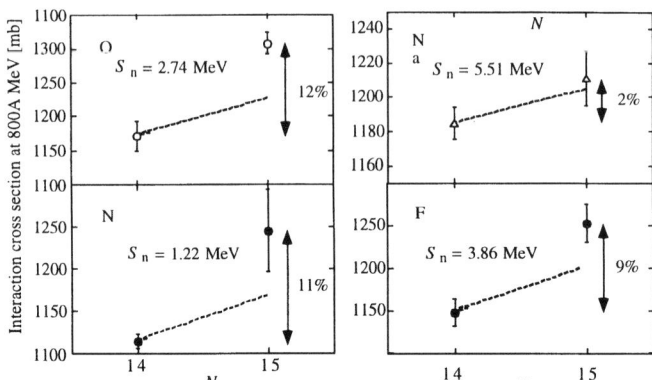

Fig. 4. Interaction cross-sections at 800A MeV on a C target. Lines indicate the prediction of the cross section based on 100% s-wave neutron added to cores.

wave, which is considered to give the largest cross section under the assumed configuration. Figure 4 shows the same difficulty in several nuclei. This anomaly always occurs at $N=15$. Increases of the interaction cross sections are much larger than the model calculation even if 100% $2s_{1/2}$ is assumed for the last neutron. Such a problem occurs for $Z=7, 8, 9$ but not for larger Z. Carbon case is not known yet because of the lack of the data. Again, a modification of the core is suggested.

Two-neutron removal experiment of ^{23}O

To understand the modification of the core or to find a reason for the inconsistency, a new type of experiment has been done for ^{23}O [16]. It is a $P_{//}$ distribution measurement for the two-neutron removal from ^{23}O. In this experiment, the combination of a high resolution Time-of-Flight detector ($\square_{tof}$~50 ps) and a high-resolution energy detector ($\square_E$~0.3 % at 80A MeV) was used to measure the one- and two-neutron removal simultaneously.

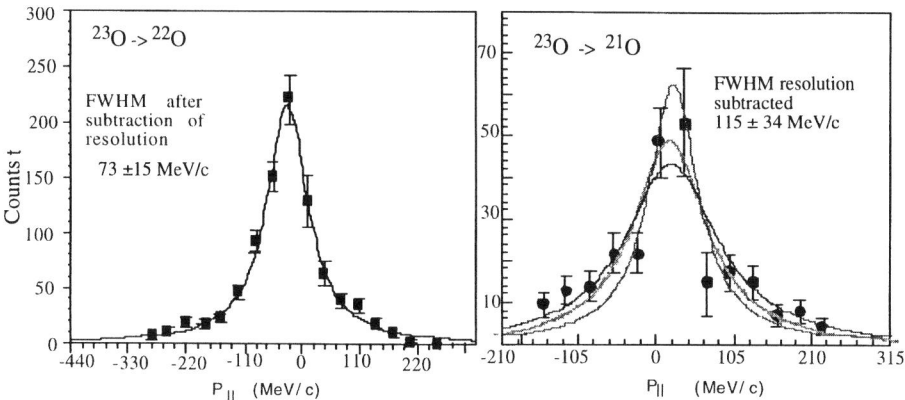

Fig. 5. $P_{//}$ distributions of ^{22,21}O fragment from ^{23}O at 70A MeV [16].

The obtained $P_{//}$ distribution is shown in Fig. 5. The FWHM of ^{22}O is 73 ± 15 MeV/c and is consistent with the s-wave. The FWHM of ^{21}O is 115 ±34 MeV/c. It is considered that the ground state of ^{21}O has $(\square 1d_{5/2})^5$ configurations. Therefore it is expected that the ^{21}O momentum distribution should be as wide as the d-wave, if the second removed neutron is kicked out directly by the collision. A possible combination of the removed nucleon orbitals and core states were considered but failed if one assume the configuration of ^{23}O be $(\square 1d_{5/2})^6 (\square 2s_{1/2})^1$. Instead, the configuration $(\square 1d_{5/2})^5 (\square 2s_{1/2})^2$ can reproduce both ^{21}O and ^{22}O distributions, simultaneously. The other possibility is the removal of the second neutron as evaporation. The calculation, however, shows that this distribution is wider than the observed distribution. It is intuitively understood as follows. The evaporation of the second neutron requires a highly excited intermediate state. Therefore, the removal of a deeply bound neutron is required for the first neutron. Then this is not the narrow $2s_{1/2}$ wave and moreover the effective separation energy is large. The momentum width is wider.

Therefore, this $(\square 1d_{5/2})^5 (\square 2s_{1/2})^2$ configuration also helps to solve the interaction cross-section problem. In the original model, up to one neutron occupy $2s_{1/2}$ orbital, but two neutrons can occupy the $2s_{1/2}$ orbital and enhances the cross section. The simple model calculations support this qualitative idea but the obtained cross section is not large enough and need an increase of an additional few percent. This change of orbital may be seen as a precursor. If one follows the change of the interaction cross-section carefully along the oxygen isotopes, the increase from ^{21}O to ^{22}O is already larger than the increase seen at the smaller neutron numbers. Therefore it can easily be considered that the $2s_{1/2}$ wave will start to contribute considerably already in ^{22}O nucleus. Unfortunately we do not have a good model that can treat the two-neutron removal from many-body configurations with correlations.

147

SUMMARY

Changes of the shell structures were discussed from the studies of neutron rich nuclei in p-sd shell. Dynamic changes of the shell have been observed as shell quenching at $N=8$ and 20, and as new magic numbers at $N=16$, 30, and 32 and $Z=16$. It is also found that the neutron halo is formed not just because of near s- or p- wave occupation, but due to bringing these small l orbitals to lower the total energy.

It was pointed out that the spectroscopic information obtained from $P_{//}$ distributions and interaction cross sections are not consistent each other in ^{19}C and ^{23}O nuclei with the conventional shell occupancies. The first simultaneous measurement of $P_{//}$ distributions of one- and two-neutron removal from ^{23}O has been reviewed. An abnormal configuration is suggested for ^{23}O. This result also may give interesting possibility in understanding the ^{19}C data consistently. It is desirable to have the $P_{//}$ distribution of two-neutron removal from ^{19}C experimentally, as well as, a model to calculate the interaction cross-sections and momentum distributions of one- and two-neutron removal from the more than two valence neutron configuration with particle correlations.

REFERENCES

[1] See proceedings of recent conferences, for example, I. Tanihata, Proceedings of the Seventh International Conference on Nucleus-Nucleus Collisions, Strasbourg, France July 3-7, 2000. Nucl. Phys. A **685** (2001) 80c, and other articles in the proceedings. I. Tanihata, Proceedings of the Conference on Nuclear Structure 2000, East Lansing, Michigan, USA, August 15-19, 2000. Nucl. Phys. A **682** (2001) 114c, and references therein.

[2] A. Navin et al., Phys. Rev. Lett. **85** (2000) 266.

[3] T. Motobayashi et al., Phys. Lett. B **346** (1995) 9.

[4] A. Ozawa et al., Phys. Rev. Lett. **84** (2000) 5493.

[5] I. Tanihata et al., Phys. Rev. Lett. **55** (1985) 2676.

[6] T. Suzuki et al., Phys. Rev. Letters **75** (1995) 3241.

[7] A. Ozawa, T. Suzuki, and I. Tanihata, Nucl. Phys. A **693** (2001) 32.

[8] T. Suzuki et al., Phys. Rev. Lett. **89** (2002) 12501-1.

[9] G. Huber et al., Phys. Rev. C **18** (1978) 2342.

[10] R. Kanungo et al., Phys. Lett. B **528** (2002) 58.

[11] V. Maddalena et al., Phys. Rev. C **63** (2001) 024613.

[12] Y. Ogawa et al., Nucl. Phys. A **543** (1992) 722.

[13] M. M. Obuti et al., Nucl. Phys. A **609** (1996) 74.

[14] R. Kanumgo et al., Nucl. Phys. A **677** (2000) 171.

[15] R. Kanungo et al., Phys. Lett. **512** (2001) 261.

[16] R. Kanungo et al., Phys. Rev. Lett. **88** (2002) 142502.

Fine structure in proton emission

K. P. Rykaczewski[1,2], R. K. Grzywacz[1,2], J. C. Batchelder[3],
C. R. Bingham[1,4], C. J. Gross[1,5], D. Fong[6], J. H. Hamilton[6], D. J. Hartley[4],
P. Hausladen[1], J. K. Hwang[6], M. Karny[2,4], W. Królas[6,7,8], Y. Larochelle[4],
T. A. Lewis[1], K. H. Maier[1,7], J. W. McConnell[1,7], A. Piechaczek[9],
A. V. Ramayya[6], K. Rykaczewski[10], D. Shapira[1], M. N. Tantawy[4],
J. A. Winger[11], C. -H. Yu[1], E. F. Zganjar[9], A. T. Kruppa[3,7,12],
W. Nazarewicz[1,4,13], T. Vertse[3,7,12], and K. Hagino[14]

[1]*Physics Division, Oak Ridge National Laboratory, Oak Ridge, TN 37831*
[2]*Institute of Experimental Physics, Warsaw University, Pl-00681 Warsaw, Poland*
[3]*UNIRIB-ORAU,Oak Ridge, TN 37831*
[4]*Dept. of Physics and Astronomy, University of Tennessee, Knoxville, TN37996*
[5]*ORISE, Oak Ridge, TN 37831*
[6]*Dept. of Physics and Astronomy, Vanderbilt University, Nashville, TN 37235*
[7]*Joint Institute for Heavy Ion Research, Oak Ridge, TN 37831*
[8]*H.Niewodniczański Institute of Nuclear Physics, Pl-31342 Kraków, Poland*
[9]*Dept. of Physics and Astronomy, Louisiana State University, Baton Rouge, LA 70803*
[10]*Oak Ridge High School, Oak Ridge, TN 37830*
[11]*Dept. of Physics and Astronomy, Mississippi State University, MS 39762*
[12]*INR, Hungarian Academy of Sciences, H-4001 Debrecen, Hungary*
[13]*Institute of Theoretical Physics, Warsaw University, Pl-00681 Warsaw, Poland*
[14]*Yukawa Institute for Theoretical Physics, Kyoto University, Kyoto 606-8502, Japan*

Abstract. Deformations and wave functions of proton-radioactive nuclei are studied using measured fine structure properties of proton emission and microscopic theoretical models. The experimental data are available for ^{131}Eu and ^{145}Tm decays, as well as for ^{141gs}Ho, where an observation of fine structure in proton emission is reported for the first time.

The first proton-radioactive decay of an isomeric state ^{53m}Co [1] was observed over thirty years ago at Berkeley. Today, we know about thirty proton radioactive isotopes, with about forty proton-emitting ground- and isomeric-states. While the energies of the emitted protons are within 1 MeV range, the halflives vary over six orders of magnitude, from a few microseconds to seconds - see a recent compilation [2].

At the beginning of the eighties, the first ground-state proton emitters ^{151}Lu [3] and ^{147}Tm [4] were detected using the recoil velocity filter SHIP and on-line mass-separator at GSI (Darmstadt), respectively. Proton emission rates for these rare-earth nuclei, and for a number of proton emitters discovered later, were well explained within the spherical approach - see [5, 6, 7]. The identification of first shape-transitional emitters in Munich, the ^{109}I and ^{113}Cs [8], showed that a deformation of the nuclear potential can play an important role in the proton emission process. Complex structure of the wave function can substantially reduce the decay rates estimated within the simple spherical picture. Theoretical studies on deformed proton emitters were intensified in the late

CP638, *Mapping the Triangle: International Conference on Nuclear Structure*
edited by A. Aprahamian, J. A. Cizewski, S. Pittel, and N. V. Zamfir
© 2002 American Institute of Physics 0-7354-0093-8/02/$19.00

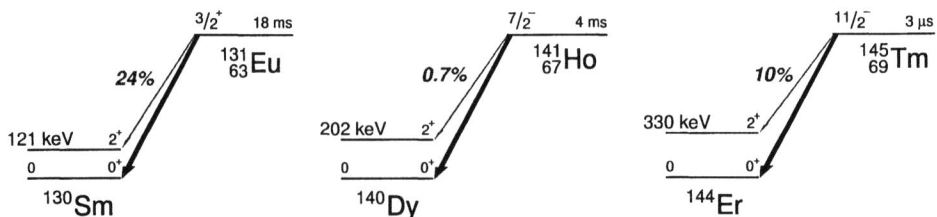

FIGURE 1. Summary of proton emission properties for the decay of ^{131}Eu, ^{141gs}Ho and ^{145}Tm.

nineties after the identification of ^{131}Eu and ^{141gs}Ho [9] at Argonne and ^{141m}Ho [10] at Oak Ridge.

Particularly interesting for the analysis of a wave function of proton-radioactive nuclei is the observation of a fine structure in proton emission. There are three odd-Z even-N proton radioactivities known to date, where proton transitions to both, the 0^+ ground-state and to the 2^+ excited state in an even-even daughter nucleus were detected, see Figure 1. These experimental data include the first observation made at Argonne for ^{131}Eu [11], and the results of Oak Ridge experiments on a 3-μs activity of ^{145}Tm [12, 13] and, very recently, on ^{141gs}Ho decay. The energies of respective 2^+ states in deformed even-even daughter nuclei were estimated before the experiments by using the N_pN_n valence particle coupling scheme [14, 15, 16]. The value of 120 ± 20 keV predicted for ^{130}Sm [17] is in a perfect agreement with observed value of 122 ± 3 keV [11]. The 2^+ energy of 344 keV could be expected for the ^{144}Er from a direct comparison to the "N=82 mirror" nucleus ^{156}Er, close to the experimental value $E_{2^+}=330(10)$ keV deduced from the proton energy spectrum [12, 13]. However, for ^{140}Dy, the N_pN_n scheme predicts the energy of 160 ± 20 keV [17]. The measured value of 202 keV [18, 19] is considerably larger indicating deviations from the simple valence scheme.

Proton radioactivity studies at Oak Ridge are performed at the Recoil Mass Separator (RMS) at the Holifield Radioactive Beam Facility (HRIBF). A detailed description of the RMS-based experiments is given in [20]. Recently, a detection of recoiling ions was improved by exchanging the gas avalanche counter for a Microchannel Plate detector(MCP) [21]. MCP was originally designed to monitor the position and intensity of postaccelerated radioactive beams. This detector offers essentially 100% efficiency for recording the position and time signals of the recoils at the RMS final focus. All signals from the MCP as well as from the Double-sided Silicon Strip Detector (DSSD) and single Si veto-detector behind the DSSD are recorded using digital pulse processing electronics from XIA [22] - see [23, 24] and references therein for detailed description of the HRIBF digital data acqusition. Digital signal processing based on the Digital Gamma Finder (DGF) modules was crucial for the discovery of a fine structure in proton emission from very short-lived ^{145}Tm. The 25-μs wide traces of the preamplifier signals were recorded allowing us to analyze the recoil-proton pile-up signals. The observation window for recording proton decay events was open from 500 nanosec to 10 μs after the recoil implantation into the DSSD. It was a major factor for the increase of counting rate by about an order of magnitude, as compared to the first experiment using analog electronics [25]. Very recently, the acquisition system based on the DGF modules

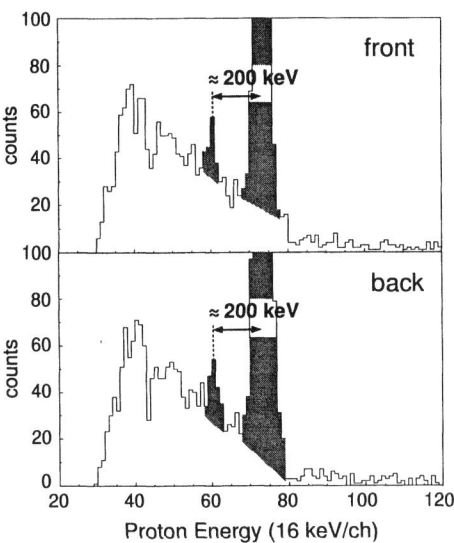

FIGURE 2. Evidence for fine structure in proton emission from ^{141gs}Ho. The 0.97 MeV proton lines were observed in the energy spectra collected at the front and at back strips of the DSSD.

allowed us to observe first ground-state two-proton radioactivity, of ^{45}Fe, produced and studied at the Fragment Separator at GSI Darmstadt [26].

For the study of 4-millisecond decay of ^{141gs}Ho, the time and amplitude of the detectors signals were analyzed on-board by the DGF modules. Good selectivity of the RMS allowed us to run with over 20 pnA intensity of 300 MeV ^{54}Fe beam on a 1mg/cm^2 thick ^{92}Mo target, without overloading the detectors and data acquisition. However, some degradation of the DSSD energy resolution was observed, and a noise level increased at low energies (below 0.7 MeV). During six days of experiment, the energy resolution of DSSD has changed from about 18 keV to about 25 keV for 1.17 MeV proton signals. A total of 7000 counts were collected at 1.17 MeV and about 50 counts in a line about 200 keV below the main transition, see Figure 2.

The energy of 202.2(2) keV for the first 2^+ state in ^{140}Dy was measured at Oak Ridge shortly before the ^{141}Ho study [18], and is also presented at this meeting [27]. The energy of the 2^+ state was deduced from the decay pattern of ^{140m}Dy interpreted as a new 7-μs, I$^{\pi}$=8$^-$, [ν9/2$^-$[514]⊗ν7/2$^+$[404] K-isomer at 2166 keV in ^{140}Dy. The properties of ^{140m}Dy decay [18] were already confirmed by an independent experiment at Argonne [19]. The studies on the ^{140m}Dy and ^{141gs}Ho decays suggest the interpretation of the observed 0.97 MeV proton line as the 0.7% transition to the 2^+ state in the daughter nucleus ^{140}Dy, with over 99% branching for 1.17 MeV protons populating the 0^+ ground state. The weak branching of 0.7% translates into about 2 nanobarn cross section for the fine structure line.

The measured energy of the 2^+ state of an even-even nucleus is commonly used to estimate the deformation [28, 29]. Following the recent publication of Raman *et al.* [29], we can derive values of β_2=0.33-0.34 for ^{130}Sm, β_2=0.23-0.24 for ^{140}Dy, and β_2=0.18

for ^{144}Er. The result for ^{140}Dy indicates somewhat smaller quadrupole deformation than earlier anticipated β_2 values of 0.27-0.28 [10, 30, 31]. However, the β_2 value of 0.23-0.24 is close to β_2=0.25 derived for ^{141gs,m}Ho [32] from the observed level scheme. This constitute first time experimental evidence that there is no dramatic shape change during proton emission process from ^{141}Ho parent to the ^{140}Dy daughter. The latter is a commonly used assumption during theoretical analysis of proton radioactivity. However, one should remember that there are model assumptions made during the conversion of measured 2$^+$ energies values into the quadrupole deformation parameters [28, 29]. The measurement of the B(E2) γ-transition probabilities for discussed 2$^+$ states could provide more solid experimental basis for the determination of the shape of the potential tunneled by the protons.

The interpretation of the structure of proton emitting states in ^{131}Eu and ^{141gs}Ho and their decay probabilities was recently made within the non-adiabatic coupled-channel method [33, 34]. The respective composition of the wave function and corresponding decay width were obtained. The wave function of emitting states, the 3/2$^+$[411] for ^{131}Eu and 7/2$^-$[523] for ^{141gs}Ho, is expressed as a sum of spherical components lj coupled to the I^π=0$^+$, 2$^+$, 4$^+$,6$^+$,8$^+$ rotational states of the ground state band in the daughter even-even nucleus, with respective c_{lj}^2 coefficients. The corresponding decay width Γ_{lj} for each component is also calculated. One finds the proton emission from ^{131}Eu to be dominated by 1% of the total wave function, the small $\pi d_{3/2}$ component ($c_{0,2,3/2}^2$=1%). The fine structure can be explained by the presence of $\pi d_{5/2}$ orbital in the wave function, with the $c_{2,2,5/2}^2$ value of 60%. The observed partial halflives are reproduced within 50% of the measured values. For ^{141gs}Ho, the main part ($\approx$80%) of the wave function is composed of the $\pi h_{11/2}$ orbital, but the observed decay is governed by a few percent admixture of the $\pi f_{7/2}$ component. However, the fine structure branching ratio is overestimated by a factor of three, and the total decay probability is underestimated by a factor of ten. Very likely, this disagreement is caused by the more complex shape, in comparison to the considered [β_2,β_4] deformation space. There are preliminary indications that calculations done with a triaxial shape of the tunneling potential improve greatly the agreement between the observed and theoretical decay rates [35].

For the transitional nucleus ($\beta_2 \approx$0.18) ^{145}Tm, the calculations of K. Hagino [36] indicate over 97% of $\pi h_{11/2}$ component in the wave function. Most of $\pi h_{11/2}$ (73%) is coupled to the ground-state of ^{144}Er, i.e., to the 0$^+$ core, while $\pi h_{11/2} \otimes 2^+$ accounts for 24% component. The observed 10% fine structure branching ratio results from the presence of a small 1% component, of the $\pi f_{7/2}$ coupled to the 2$^+$ excited core of ^{144}Er.

The results on ^{131}Eu, ^{145}Tm, and particularly on the pair of parent ^{141}Ho and daughter ^{140}Dy nuclei pave ground for the directions for future studies at the proton drip line. The experimental investigations will aim at the complete spectroscopy, including the proton emission rates, and the excited levels in the proton emitters (obtained via recoil Decay Tagging methods) and in the daughter nuclei. Hopefully, with the help of the Rare Isotope Accelerator, we will be able to reach and study new regions of proton- and two-proton radioactivities.

ORNL is managed by UT-Battelle, LLC, for the U.S. Department of Energy under Contract No. DE-AC05-00OR22725. The Joint Institute for Heavy Ion Research is supported by its members, University of Tennessee, Vanderbilt University and Oak Ridge National Laboratory. This work was also supported by the U.S. D.O.E. through Contracts No. DE-FG02-96ER40983, DE-FG05-88ER40407, DE-FG02-96ER40978, DE-FG02-96ER41006, DE-AC05-76OR00033, DE-FG02-96ER40963, by the Polish KBN Grants No. 2P03B 08617 and 2P03B 04516, by the Hungarian OTKA Grants No. T026244 and T029003 and by NATO Grant No. PST.GLG.977613. ATK and TV acknowledge support from the UNIRIB Consortium. KR was an honor student at Oak Ridge National Laboratory.

REFERENCES

1. Jackson, K.P. et al., *Phys. Lett.* **33B**, 281 (1970).
2. Sonzogni, A.A., *Nucl. Data Sheets* **95**, 1 (2002).
3. Hofmann, S. et al., *Zeit. Phys.* **A305**, 111 (1982).
4. Klepper, O. et al., *Zeit. Phys.* **A305**, 125 (1982).
5. Hofmann, S., *Radiochim. Acta* **70/71**, 93 (1995).
6. Aberg, S.,Semmes, P.B. and Nazarewicz,W., *Phys. Rev.* **C56**, 1762 (1997).
7. Aberg, S.,Semmes, P.B. and Nazarewicz,W., *Phys. Rev.* **C58**, 3011 (1998).
8. Faestermann, T. et al., *Phys. Lett.* **B137**, 23 (1984).
9. Davids, C.N. et al., *Phys. Rev. Lett.* **80**, 1849 (1998).
10. Rykaczewski, K. et al., *Phys. Rev.* **C60**, 011301 (1999).
11. Sonzogni, A.A. et al., *Phys. Rev. Lett.* **83**, 1116 (1999).
12. Rykaczewski, K. et al., *Nucl. Phys.* **A682**, 270c (2001).
13. Karny, M. et al., to be published.
14. Casten, R.F., *Phys. Rev.* **C33**, 1819 (1986).
15. Casten, R.F., and Zamfir, N.V., *J. Phys.* **G22**, 1521 (1996).
16. Casten, R.F., "Nuclear Structure from a Simple Perspective", 2nd ed., Oxford, NY, p. 297-330 (2000).
17. Zamfir, N.V., private comm. (1998).
18. Królas, W. et al., *Phys. Rev.* **C65**, 031303R (2002).
19. Cullen, D.M. et al., *Phys. Lett.* **B529**, 42-49 (2002).
20. Gross, C.J. et al., *Nucl. Instr. Meth. in Phys. Res.* **A450** 12 (2000).
21. Shapira, Lewis,T.A. and Hulett, L.D., *Nucl. Instr. Meth. in Phys. Res.* **A454** 409 (2000).
22. Hubbard-Nelson, B., Momayezi, M. and Warburton, W.K., *Nucl. Instr. Meth.* **A422** 411 (1999).
23. Grzywacz,R.K. et al, *Eur. Phys. J.*, **A** in press.
24. Grzywacz, R.K., *Nucl. Instr. Meth. in Phys. Res.* in press.
25. Batchelder, J.C. et al., *Phys. Rev.* **C57**, R1042 (1998).
26. Pfützner, M. et al., *Eur. Phys. J.* **A**, in press.
27. Królas, W., in proceedings to this conference.
28. Grodzins, L., *Phys. Lett.* **2**, 88 (1962).
29. Raman, S., Nestor, C.W. JR., and Tikkanen, P., *At. Data and Nucl. Data Tables* **78**, 1 (2001).
30. Möller, P. et al., *At. Data and Nucl. Data Tables* **59**, 185 (1995).
31. Xu, F. R., Walker, P.M., and Wyss, R., *Phys. Rev.* **C59**, 731 (1999).
32. Seweryniak, D. et al., *Phys. Rev. Lett.* **86**, 1458-1461 (2001).
33. Kruppa, A.T., Barmore, B., Nazarewicz, W. and Vertse, T., *Phys. Rev. Lett.* **84**, 4549 (2000).
34. Barmore, B., Kruppa, A.T., Nazarewicz, W. and Vertse, T., *Phys. Rev.* **C62**, 054315 (2000).
35. Kruppa, A.T. and Nazarewicz, W., private comm. (2002).
36. Hagino, K., *Phys. Rev.* **C64**, 041304R (2002).

Towards Microscopic Theory of Soft Nuclei

Vladimir Zelevinsky

Department of Physics and Astronomy and NSCL, Michigan State University, East Lansing, MI 48824-1321, USA

Abstract. Soft vibrational modes in finite quantum systems are compared with macroscopic phase transitions. I argue that in soft nuclei the quartic anharmonicity dominates the dynamics, sketch the microscopic derivation of the collective Hamiltonian and illustrate by a simple model.

INTRODUCTION

Many nuclei reveal a soft quadrupole mode; typical signatures are low 2_1^+ energies and strong E2 transitions with no clear rotational spectra. A low frequency ω implies *large amplitude motion*: the vibrational scale grows as $\omega^{-1/2}$. The mean field changes significantly in this process remaining spherical in average. Microscopic theory of soft nuclei does not exist. The RPA describes only small amplitude vibrations while the boson expansion converges too slowly. Among phenomenological models [1, 2, 3] the IBM [4] was successful in description of low-lying states and their symmetry classes (famous *Triangle* [5], a motto of this Conference). The restrictive character of the IBM forced to return to the collective quadrupole Hamiltonian [3]. Many mean field calculations predict a potential surface that is flat in the quadrupole coordinate β near spherical shape but rises steeply for large deformations. A phenomenological model [6, 7] with *quartic anharmonicity* (QA) $\sim \beta^4$ described many soft nuclei with very few parameters. Recently a model based on the "deformation box" was suggested [8]. Below I remind the QA model and sketch the contours of future microscopic theory.

WHY QUARTIC?

In soft nuclei the low-lying states are classified as emerging from a collective Hamiltonian $\mathcal{H}$. In the *harmonic limit* (I use quadrupole coordinates α_μ and momenta π_μ),

$$\mathcal{H} \Rightarrow \mathcal{H}^{(2)} = \frac{1}{2}\sum_\mu \left(B^{-1}\pi_\mu^\dagger \pi_\mu + C\alpha_\mu^\dagger \alpha_\mu \right). \tag{1}$$

The harmonic relations for spectra and transitions are badly violated in soft nuclei but it is usually possible to establish the correspondence between the actual states and those of the quadrupole vibrator. In particular, the *boson seniority* v, as a rule, is a good quantum number. The *anharmonicity* includes terms like

$$\mathcal{H}^{(3)} = \Lambda^{(30)}\alpha^3 + \Lambda^{(21)}(\alpha^2)_2\pi + \Lambda^{(12)}\alpha(\pi^2)_2 + \Lambda^{(03)}\pi^3, \tag{2}$$

CP638, *Mapping the Triangle: International Conference on Nuclear Structure*
edited by A. Aprahamian, J. A. Cizewski, S. Pittel, and N. V. Zamfir
© 2002 American Institute of Physics 0-7354-0093-8/02/$19.00

$$\mathcal{H}^{(4)} = \Lambda^{(40)}\alpha^4 + \Lambda^{(31)}(\alpha^3)_2\pi + \sum_L \Lambda_L^{(22)}(\alpha^2)_L(\pi^2)_L + \Lambda^{(13)}\alpha(\pi^3)_2 + \Lambda^{(04)}\pi^4, \quad (3)$$

and so on. With *time-even* coordinates α and *time-odd* momenta π, we can discard $\mathcal{T}$-odd terms (21), (31) and (13). The corrections to the *potential energy* (no momentum operators) dominate since $\alpha \propto \omega^{-1/2}$. The most important momentum term, $\sim [(\alpha\pi)_1]^2 \propto \mathbf{J}^2$, results from the recoupling in (22).

The signal of a soft mode is given by the *RPA instability*. The eqilibrium is restored by anharmonicities. Simple estimates [9] show that the phonon-quasiparticle coupling constant γ is of the same order as the phonon frequency $|\omega|$. Using information about *pairing* (soft phonons exist only within the superfluid gap $2\bar{E}$), we obtain the interrelation $\Omega\tau^3 \approx 1$ between the degree of collectivity Ω and the adiabaticity factor $\tau \equiv \omega/(2\bar{E})$. The effect of limiting softness works in a *finite* system only; macroscopically $\Omega \to \infty$, $\omega \to 0$. The nonlinear terms $\mathcal{H}^{(n)}$ are given by the n-fermion loops with n external phonon lines. A skeleton loop with n corners ($\sim \gamma^n \sim \omega^n$) gives

$$\mathcal{H}^{(n)} \sim \mathrm{Tr} \int dE \, \frac{\gamma^n}{E^n} \sim \Omega E\tau^n \sim \Omega\omega\tau^{n-1}. \quad (4)$$

With the above estimate, this is equivalent to $\mathcal{H}^{(n)} \sim \omega\tau^{n-4}$. The QA is singled out since $\mathcal{H}^{(4)}$ is of the order of unperturbed ω. For $n > 4$ the adiabaticity makes the anharmonic effects weaker. The odd$-n$ terms are small, as confirmed by phenomenological fits, similar to the *Furry theorem* of QED.

Since the corrections to the potential energy, $\mathcal{H}^{(40)} \sim \beta^4$, are more important (by a factor of $(\alpha/\pi)^2 \sim \tau^{-2}$) than kinetic terms $\mathcal{H}^{(22)} \sim \alpha^2\pi^2$, the basic QA Hamiltonian has a structure, in terms of d-phonon operators,

$$\mathcal{H} = \omega\left\{\sum_\mu d_\mu^\dagger d_\mu + \frac{\lambda}{4}\left(\tilde{d} + d^\dagger\right)^4\right\}, \quad (5)$$

with the only parameter $\lambda = \Lambda^{(40)}/4B^2\omega^3$ (apart from the scale). A similar Hamiltonian can be written even beyond the point $\omega^2 = 0$; the QA guarantees the stability. The next corrections include the sextic Hamiltonian $\mathcal{H}^{(60)}$ and the kinetic quartic term $\mathcal{H}^{(22)}$, both $\sim \omega\tau^2$. The $\mathcal{H}^{(22)}$ contribution contains the rotational energy $\sim ([\alpha\pi]_1)_{00}^2$. The sextic term might be important if $\mathcal{H}^{(40)}$ is numerically small, and the effective potential is more flat near the bottom and more steeply grows approaching the $E(5)$ limit [8]. The QA (5), as any γ-unstable potential, preserves the O(5) symmetry. The corresponding five-dimensional angular momentum, the boson seniority v, is a good quantum number. This brings in a degeneracy of levels with the same v and radial quantum number $\tilde{n}$, as for example 2_2^+ and 4_1^+. The degeneracy is lifted by the rotational term and by the sextic term with two possible structures, $(\beta^2)^3$ and $(\beta^3\cos(3\gamma))^2$.

PHENOMENOLOGY

The QA (5) was shown [6] to be a good zero order approach; this was recognized by Ref. [10] in relation to the new data for ^{110}Cd. With no analytic solution, a simple variational procedure exists [6] which gives quite precise results for small $\tilde{n}$; the accuracy is better for larger v. In this approximation the spectrum (in units of ω) is

$$E(\tilde{n}, v) \approx \frac{2v+5}{8}\left(3\omega_v + \frac{1}{\omega_v}\right) + 2\omega_v\tilde{n};$$ (6)

ω_v depends on seniority being an appropriate root of the cubic equation $\omega_v^3 - \omega_v = 2(2v+7)\lambda$ that goes to 1 at $\lambda \to 0$ and to $[2(2v+7)\lambda]^{1/3}$ in the quartic limit, $\lambda \to \infty$.

For pure QA the energy ratios are predicted *with no parameters*. Many nuclei are close to this limit that is precisely realized in such cases as yrast states, $\tilde{n} = 0$, $J = 2v$, in ^{100}Pd. The yrast bands of this type are intermediate between vibrational and rotational. They have the asymptotics $E_J \propto J^{4/3}$ and may stretch up far beyond the IBM limit. In many other cases the simple rotational correction gives a nice fit. The more detailed study of Xe isotopes was presented in [11], see also [12]. A lot of fits for energies, multipole moments and transition probabilities were done in [13, 14]. The quadrupole operator $\propto \alpha_\mu$ should be corrected by the *cubic* term $[\alpha^3]_{2\mu}$ as follows from the meaning of this quantity as a response of the system to the external field.

THE FIRST STEP TO MICROSCOPIC THEORY

Here the goal is to derive the collective Hamiltonian for a given number of particles and given residual interaction that should include pairing and quadrupole correlations. The convenient formalism is provided [15, 16] by the *generalized density matrix* (GDM) $R_{12} = a_2^\dagger a_1$ which is a set of operators labeled by the one-body quantum numbers [17]. The multipole moment is $Q = \mathrm{Tr}(Rq)$, where q_{12} are corresponding single-particle matrix elements. The *saturation* assumption, well satisfied for the quadrupole mode [18], means that in the exact GDM equations we can keep only the matrix elements within the collective space where the Hamiltonian is to coincide with $\mathcal{H}(\alpha, \pi)$, eq. (1), so that it should be a correspondence between the GDM and collective operators,

$$R_{12} \Rightarrow \mathcal{R}_{12} = \sum_{mn} r_{12}^{(mn)}\{\alpha^m\pi^n\},$$ (7)

with the symmetrized combinations as $\{\alpha\pi\} \equiv \frac{1}{2}(\alpha\pi + \pi\alpha)$. Thus, instead of the boson expansion or mapping of the wave functions, we map the *equations of motion* under the saturation assumption, $[R_{12}, H] \Rightarrow [\mathcal{R}_{12}, \mathcal{H}]$. The equations for the residual interaction $V_{12;34}$ include single-particle energies ε and the self-consistent field W and can be written as commutators in the combined, single-particle and full Hilbert, space,

$$[R, H] = [\varepsilon + W\{R\}, R], \quad W_{14}\{R\} = \frac{1}{2}\sum_{23} V_{12;34}R_{32},$$ (8)

Similar to (7), the field W is represented within the collective space as

$$W_{12}\{R\} \Rightarrow \mathcal{W}_{12}\{\mathcal{R}\} = \sum_{mn} W_{12}\{r^{(mn)}\}\{\alpha^m\pi^n\}.$$ (9)

Isolating the terms for each operator structure in (8) we determine the c-number coefficients $r_{12}^{(mn)}$ and construct the collective Hamiltonian in its operator form (1). The static part $r^{(00)}$ gives the mean field approximation, the linear terms $r^{(10)}$ and $r^{(01)}$ lead to the RPA and so on. The anharmonic terms $\Lambda^{(mn)}$ are not considered to be small; the method converges if the starting ideas and estimates are correct.

The operator expansion (7) within the collective space includes all *adiabatic* effects in the dynamic response of incoherent environment. The *nonadiabatic* mixing of other states, especially important for the region of the increasing level density was ignored. The corrections can be introduced by adding to the operator expansion terms connecting the collective states to the rest of the spectrum. These transitions carry no collectivity and may be thought of as "*chaotic*". The new operators are to be found from the same GDM equations but we need only their generic statistical properties known from random matrix theory.

EXAMPLE

The *Lipkin model* [19] is a standard testing ground of theoretical approaches. Here there is no problem of separating collective degrees of freedom since they are singled out by construction. The two-level Hamiltonian (energies $\pm\varepsilon/2$ with $N \leq \Omega/2$ particles (Ω is the total capacity of two levels) can be expressed in terms of the pseudomomentum $\mathbf{J}$,

$$H = \varepsilon J_z + V(J_x^2 - J_y^2), \quad J = N/2. \tag{10}$$

Near the instability point, $V^2 \approx \varepsilon^2/(4J^2)$, we have a soft RPA mode, $\omega^2 \approx \varepsilon^2 - 4V^2 J^2$. The mapping leads to the quartic Hamiltonian with $\Lambda^{(40)} \approx V(\varepsilon + 2VJ)^2$, where $V > 0$, and the collective coordinate α has a dimension of $(\text{energy})^{1/2}$. As argued above, the momentum terms are small in the soft mode region. Fig. 1 shows that the quartic anharmonicity cures the deficiency of the RPA and gives practically exact levels.

CONCLUSION

The theoretical description of the soft mode is important for a general problem of a finite many-body system in the regime of large amplitude collective vibrations. We stressed the difference with macroscopic phase transitions and showed that, for a typical situation in nuclei, the QA dominates low-lying structure, in agreement with data on a phenomenological level. We suggested the regular method of deriving the collective Hamiltonian and illustrated it with a simple model.

ACKNOWLEDGMENTS

The studies in this direction are currently in progress together with A. Volya whose active role is highly appreciated by the author. The support from the NSF via grant

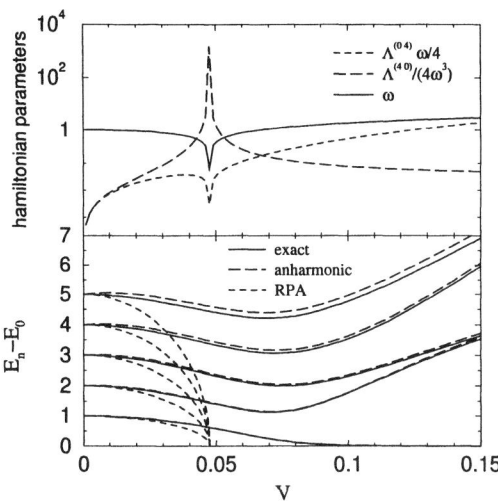

FIGURE 1. The quartic parameters (note the logarithmic scale) near the instability point, upper part. The energy spectrum in the RPA and with QA, compared to the exact solution, lower part. The lowest states are pairwise degenerate in the "Mexican hat" region; higher states above the barrier are split.

PHY-0070911 is gratefully acknowledged.

REFERENCES

1. D.M. Brink, A.F.R. de Toledo Piza, and A.K. Kerman, *Phys. Lett.* **19**, 413 (1965).
2. G. Gneuss and W. Greiner, *Nucl. Phys.* **A171**, 440 (1971).
3. A. Bohr and B. Mottelson, *Nuclear Structure*, vol. 2 (Benjamin, New York, 1975).
4. F. Iachello and A. Arima, *The Interacting Boson Model* (Camridge University, 1988).
5. R.F. Casten and D.D. Warner, *Rev. Mod. Phys.* **60**, 389 (1988).
6. O.K. Vorov and V.G. Zelevinsky, *Sov. J. Nucl. Phys.* **37**, 830 (1983).
7. O.K. Vorov and V.G. Zelevinsky, *Nucl. Phys.* **A439**, 207 (1985).
8. F. Iachello, *Phys. Rev. Lett.* **85**, 3580 (2000).
9. V. Zelevinsky, *Int. J. Mod. Phys. E* **2**, 273 (1993).
10. M.A. Caprio, R.F. Casten, and J. Jolie, *Phys. Rev. C* **65**, 034304 (2002).
11. V. Zelevinsky, *Genshikaku Kenkyu* (Tokyo) **35**, 21 (1991).
12. G.F. Bertsch, *Nucl. Phys.* **A574**, 169c (1994).
13. V.G. Zelevinsky, in: *Nuclear Structure, Reactions and Symmetries*, vol. 2 (Dubrovnik, 1986) p. 1125.
14. O.K. Vorov and V.G. Zelevinsky, in *XXI Winter School* (Leningrad, 1986) p. 195.
15. V. Zelevinsky, in *Perspectives for the Interacting Boson Model*, ed. by R.F. Casten *et al.* (World Scientific, Singapore, 1994) p. 307.
16. V. Zelevinsky and A. Volya, *in press*.
17. V.G. Zelevinsky, *Progr. Theor. Phys.* Suppl. **74-75**, 251 (1983).
18. R.V. Jolos *et al.*, *Nucl. Phys.* **A618**, 126 (1997).
19. H. Lipkin, N. Meshkov, and A. Glick, *Nucl. Phys.* **62**, 188, 199 (1965).

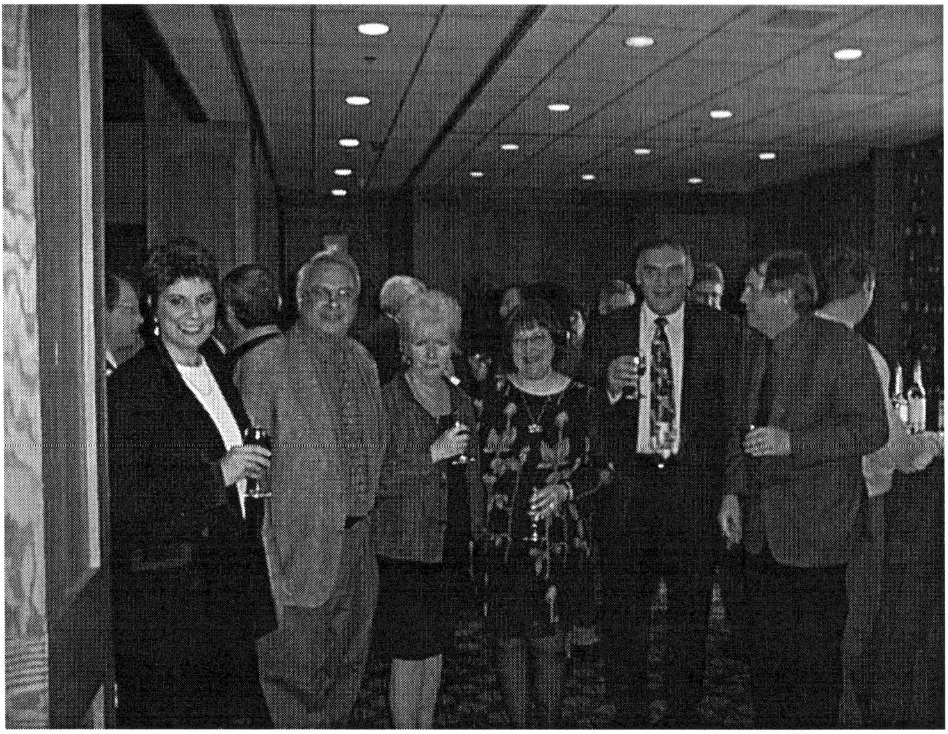

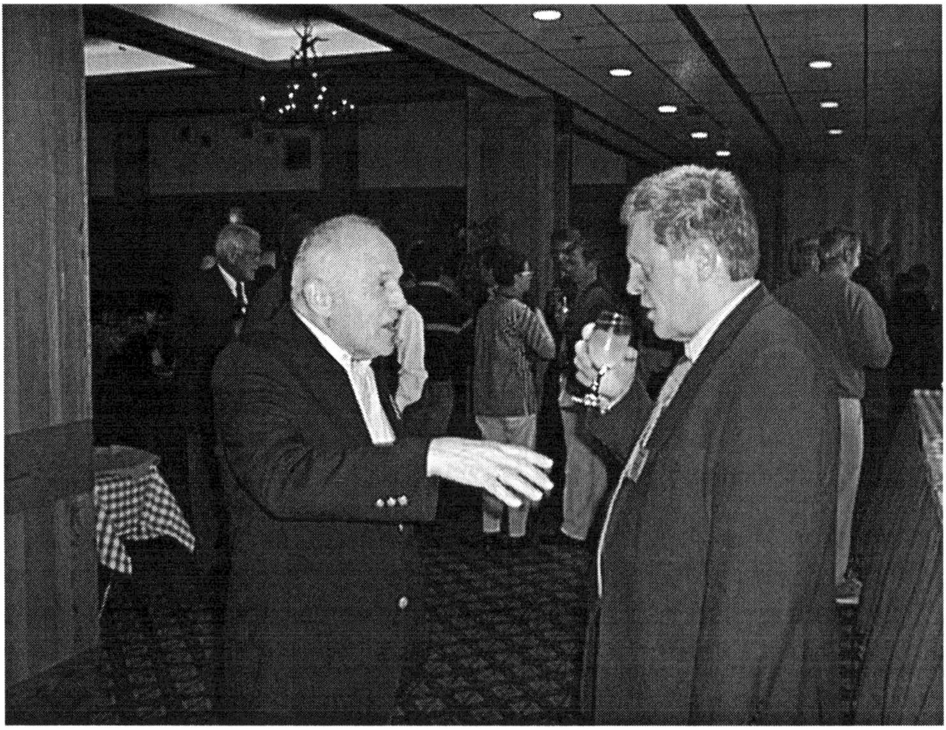

Nuclear masses and all that: an excursion with many interactions with Rick

K. Heyde*, R. Fossion* and J. E. García-Ramos†

*Department of Subatomic and Radiation Physics, Proeftuinstraat,86 B-9000 Gent, Belgium
†Depto. de Física Aplicada e Ingeniería Eléctrica, EPS La Rábida, Universidad de Huelva, Spain

Abstract. In the present contribution, we show that, although the spectroscopic properties of the monopole pairing force and a zero-range delta-function interaction are very similar, results for binding energies when filling up a major shell are radically different past mid-shell. This has significant consequences for understanding the masses and binding energies of long isotopic chains of nuclei that will be accessible with advanced exotic beam facilities. Consequently, we describe a new method to derive two-neutron separation energie S_{2n}, treating ground-state properties as well as excited-state properties in nuclei,in a consitent way. Results for the whole rare-earth mass regions are presentend. The recent high-precison results that have been obtained at ISOLDE in the neutron-deficient Pb region will be discussed and present a challenge to understand the specific mass trend.

INTRODUCTION

Let us first start with an anecdote: the way how quick a discussion with Rick can result in a paper but that it can still take a long time before it gets published.

After getting interested in nuclear masses, more in particular on S_{2n} values that had been measured with a very high precision in the neutron- deficient Pb region, we were considering various ways to look at the overall behavior of S_{2n} in a number of isotope series. The clear-cut difference in the fall-off, as a function of A (or N), when comparing experimental results with liquid-drop values gave us a thought to try to see how well simple pairing energy gain outside of closed shells might explain these systematic differences. In plotting the differences $S_{2n}^{exp} - S_{2n}^{LD}$ and comparing with pairing binding energy corrections, taking the available neutron model space as a large degenerate shell (between 50 and 82 as a j=31/2 orbital), we were getting very interesting fits, near the Sn region, but also for the heavier nuclei. Until we noticed that the sign that resulted from the fits was not the one expected. At that time, Rick was passing by for the morning and we were showing Rick these figures with some drama put into the story. Rick asked if this would remain the same when using just a zero-range δ force.

CP638, *Mapping the Triangle: International Conference on Nuclear Structure*
edited by A. Aprahamian, J. A. Cizewski, S. Pittel, and N. V. Zamfir
© 2002 American Institute of Physics 0-7354-0093-8/02/$19.00

A DIFFERENCE BETWEEN MONOPOLE PAIRING AND ZERO-RANGE EFFECTIVE INTERACTIONS FOR BINDING ENERGIES

Two well-studied interactions: the monopole pairing and the zero-range delta interaction have both been designed to reflect the short-range nature of the residual interactions and to reproduce typical spectra in nuclei near closed shells.

Although the basic properties of these interactions have long been known [1, 2, 3, 4, 5, 6], one of the key differences in their predictions is maybe not so well appreciated.

We show that these two forces are quite different in the way they affect particles filling a j-shell (in particular in their "saturation" properties), and they produce entirely different predictions for binding energies across a shell. So there appears a subtle difference between the pairing force and the zero-range force that became clear out of the discussions with Rick.

For simplicity, just consider a large single j-shell although we comment at the end on the more realistic multi-j nature of major shells. When using the monopole pairing force, and building a basis of $J^\pi = 0^+$ coupled configurations, which for seniority $v = 0$ becomes $(S_j^+)^{\frac{n}{2}}$ (with S_j^+ the pair creation operator), one can evaluate the binding energy exactly with as a result (see refs.[1, 3, 5])

$$BE^{pairing} = G(\Omega_j - N + 1)N. \tag{1}$$

Here, $N = \frac{n}{2}$ denotes the number of (valence) *pairs* with n describing the number of (valence) particles. We can easily deduce the pairing force contribution to the two-neutron separation energy, which results in the expression

$$S_{2n}^{pairing} = G\left(\Omega_j + 2 - n\right) = G(\Omega_j + 2 - 2N), \tag{2}$$

(here now, n and N refer to neutrons only, of course). These results, obtained many years ago [2] for a single-j shell, imply that with the shell fully filled with nucleons, the total binding energy is given by the diagonal expression $-G\Omega_j$. For a typical value of $G = 0.25$ MeV (using the prescription that $G \simeq 25/A$ MeV taking $A \simeq 100$) and for $j = 11/2$ ($\Omega_j = 6$), this gives a rather small value of only 1.5 MeV for the total binding energy of the closed shell. This results because the pairing force counts the number of pairs moving in the j-shell in which the Pauli effect plays an essential quenching role. This result is illustrated in figure 1 of ref.[7] (with the solid line) where binding energies for the seniority $v = 0$ 0^+ state in the $1h_{11/2}$ configuration are shown.

If we now use the zero-range $\delta(\vec{r}_1 - \vec{r}_2)$ force, and the same basis as we used when discussing the binding energy effects for the pairing force, a totally different result emerges. This may seem strange at first because one has an intuitive "feeling" that pairing and zero-range forces should give rise to very much identical results. However, the zero-range interaction does not scale with the number of *pairs* as does the pairing force but scales with all the fermions that are interacting.

If we again concentrate on the ground state and construct the $v = 0$ state, one obtains the result [2, 3]

$$\langle j^n, v = 0, JM | \sum V_{i,k} | j^n, v = 0, JM \rangle = \frac{n}{2} V_0, \tag{3}$$

which holds for $n = 0, 2, 4, \ldots \rightarrow 2j + 1$. Thus, the binding energy changes completely (see figure 1 of [7], the dashed line), giving rise to a linear behavior in n (in contrast to the "roll-over" in the binding energies for a pairing force past mid-shell). As a result, the binding energy of the filled shell for a δ-interaction amounts to $\Omega_j V_0$ where V_0 denotes the relative binding energy of the $J^\pi = 0^+$ configuration with respect to its unperturbed energy. A good estimate for this value is $V_0 = G\Omega_j$. We notice that the total binding energy contribution for the filled shell now becomes $G\Omega_j^2$.

The difference between the pairing force and zero-range delta-function force becomes important when one studies binding energies. Even though the detailed spectroscopic properties are not so different (the pairing force as well as the zero-range force both give rise to 0^+ ground states and to a n-independence in the energy spectra when filling up the shell-model orbital j with n identical nucleons), the use of the pairing force turns out to be a poor choice when studying binding energy effects. The pairing force over-saturates, whereas the zero-range force saturates. That is, the binding energy contribution for each successively added pair of nucleons decreases with n for a pairing force but is a constant for the zero-range delta interaction.

If one looks at a more realistic case, e.g., filling up nucleons within the major shell between 50 and 82 nucleons, [8, 9], one should consider the specific shell-model sequence of orbitals, and one cannot just put all particles in a large degenerate shell, with $j = 31/2$ (or $\Omega_j = 16$). In this case one fills in sequence, e.g., the $2d_{5/2}, 1g_{7/2}, 1h_{11/2}, 2d_{3/2}$ and $3s_{1/2}$ orbitals. The arguments discussed above for a δ-interaction can then be used and, as a result, one obtains a sequence of straight lines, each characterized by a slightly different slope which is fixed by the value of V_0 or by the Slater integral for that particular orbital. So we need an extra index $V_0(j)$ [2, 3, 4, 5]. As a result, one is able to reproduce rather well the trend of increasing binding energies all through a large shell [9].

Of course, attempts to treat binding energies consistently, within a shell-model context, also have to cope with the problem of modifications in the mean-field itself (varying single-particle energies ε_j) as a function of the filling of shells. This "monopole" issue is discussed in detail by Zuker et al. [10, 11].

DETERMINING NUCLEAR MASSES

Coming back to nuclear masses, improvements in the experimental methods has led to the determination of masses, even in the Pb region, with a precision of less than 100 keV [12, 13, 14]. Thereby, mass measurements have come to be critical indicators for testing nucleon-nucleon correlation effects at the level of local mass regions and the nuclear shell model. It is therefore a huge task to even try to get the overall mass surface as well as these fine details both correct. One has to remark though that both are coupled indicating the need to treat both the global mass variation and nuclear excited-state properties in a consistent and 'coupled' way.

In determining nuclear masses of an interacting many-body system, one normally derives upper bounds to the exact ground-state energy [1]. The most simple approach is just adding up the single-particle energies of all occupied orbitals i.e. $\sum_h \varepsilon_h$ (h denoting occupied orbital quantum numbers). One can do much better using Hartree-Fock theory in which the single-particle energies are determined in a self-consistent way with the interaction itself, leading to an improved estimate for the binding energy

$$\sum_h \varepsilon_h - \frac{1}{2} \sum_{h,h'} \langle h,h' \mid V \mid h,h' \rangle. \tag{4}$$

Considering the fact that nucleons may be scattered pairwise between orbitals near to the Fermi level, extra binding energy can be obtained depending on the methods used (HF+BCS or HFB binding energy values). This is what has presently been used in obtaining large-scale mass evaluations using microscopic methods [15, 16, 17] in contrast to all former methods that were using macrocopic-microscopic methods [18, 19, 20, 21]. It is still possible to improve on the binding energy by redoing calculations starting from a different intrinsic distribution of nucleons, differening from the one corresponding to the lowest energy. Thereby, a number of intrinsic configurations with corresponding higher total energies are obtained and one has to include mixing of these configurations. Attempts in that direction have recently been carried out using GCM methods [22, 23] but only in very localized situations as yet. It is clear that the conditions are very stringent in that, starting from a given interaction, one will at the same time obtain the ground-state energy and the excited state properties in a unified way. Within the context of calculating nuclear mass-tables, this is not within immediate reach. The important point though is that the binding energy of a given nucleus and the excited state properties are in no way 'independent' properties.

One may try shell-model calculations to reach that same goal, albeit from a slightly different philosophy because here, one puts major interest in the excited state properties. The drawback is that the forces used most often do not saturate well (have wrong monopole properties) which means that e.g. in the sd model space (the region spanning nuclei in between ^{16}O and ^{40}Ca), the single-particle energies $\varepsilon_{1d_{5/2}}, \varepsilon_{2s_{1/2}}$ and $\varepsilon_{1d_{3/2}}$ just beyond ^{16}O should gradually change into the observed energies for these orbitals just below ^{40}Ca [24, 25, 26, 27]. This is not the case and so one has to modify the original two-body interation until both the excited state properties and the global variations in the mean-field are well reproduced. This has, at present, been remedied by removing the monopole part of the effective forces used, and replacing this by the empirical monopole part [10, 11]. Here again, one notices the intertwining of global (binding energies) and local (excited-state energies) properties of nuclei.

AN INTERACTING BOSON MODEL DESCRIPTION OF TWO-NEUTRON S_{2N} VALUES

An interesting approach in order to try to work out the above 'program' for large mass regions is to implement the above concepts using the interacting boson model

(IBM)[28, 29]. Here one can include in a natural way the local sd boson interaction energy gain with the global variation in binding energy. So, we have set up a calculation in which ground-state binding energy and excited-state properties (energies, transition rates,..) are treated in a consistent description. The method has been desribed in a recent paper by Fossion et al. [9]and combines the global energy term

$$BE^{global} = E_0 + A.N + \frac{B}{2}.N(N-1). \tag{5}$$

where N denotes the number of nucleon pairs, counting from the nearest closed shell, with the local part contributing to the binding energy. This latter term BE^{local} then derives from diagonalizing the specific IBM Hamiltonian for a large mass region. Because we are interested in a large group of nuclei (starting near Sn,Te and ending in the Pb region), one has to be careful in counting nucleon pairs when evaluating binding energy expressions. The technical points related to this issue are presented in [9] but it comes to defining a new variable $\tilde{N}$ which is equal to N in the first half of a shell and equal to $\Omega - N$ in the second half of the shell (with Ω the shell degeneracy). Then the total binding energy reads

$$BE(\tilde{N}) = E_0 + A.\tilde{N} + \frac{B}{2}.\tilde{N}.(\tilde{N}-1) + BE^{local}(N(\tilde{N})). \tag{6}$$

It is interesting to rewrite the full expression above using the atomic mass number A.

In applying the above method to the rare-earth nuclei, it should be clear that a schematic or more realistic IBM Hamiltonian, used to describe as well as possible nuclear excited-state properties, will not per-se lead to a correct description of the nuclear ground-state binding energy. As we stressed in the section before, one needs to consider both energy parts in a 'coupled' way and so, one has to make the optimal choice in the Hamiltonian. It turned out that the extensive studies, carried out by Chou et al. [30], gives good results for both the excited states and the nuclear ground-state binding energy. Instead of displaying binding energies, we rather show two-neutron separation energies S_{2n} as most often given by the experimentalists. So, once we have fixed the local IBM Hamiltonian, it is a trivial task to deduce the linear part of the two- neutron separation energy (S_{2n} through the relation

$$S_{2n}^{global} \equiv A + B.\tilde{N} = S_{2n}^{exp.} - S_{2n}^{local}. \tag{7}$$

Results in the shell 50-82 are derived and discussed by Fossion et al. [9]. We give the resulting fit values for the A and B coefficients which, are the same along a series of isotopes (except for splitting the full interval of a shell into the first half and second half since independent fits are needed to be done). The specific behavior of these important quantities A and B need to be understood better (variations, changes when approaching closed shells,..). The general outcome is that one obtains a rather good description even in those regions where the nuclear structure characteristics are changing quite rapidly and important deviations form the overall linear dependenc of S_{2n} on nucleon number appear. We stress that we do reproduce these S_{2n} values and, at the same time, the properties of the low-lying excited states in all these nuclei. We feel that this is an interesting step forward in connecting these nuclear properties.

The above analysis points towards an intimate relation between a correct reproduction of the nuclear excited states and nuclear ground-state properties.A simultaneous description guarantees that the Hamiltonian that is used, is appropriate for the nuclei studied over a large mass region. This same point has been made before by some of us [8].

CONCLUSION

In the present contribution, we took you on a walk through the nuclear mass landscape in which Rick was more than a spectator. Some pointed questions concerning the relation of pairing and zero-range effective forces learned us something we new as some 'background' but was not really clear. It was just an element but an interesting one to tell you about in our study of nuclei and of nuclear masses.

Through the increased sensitivity on the experimental side (ISOLTRAP and MIS-TRAL at CERN-ISOLDE, GSI, GANIL, MSU,. ...), the field of nuclear masses has become fully intertwined with the more 'regular' nuclear structure studies and, as we tried to show, both aspects cannot be separated as independent fields of research. Therefore, starting from mean-field methods (using macrocopic- microscopic methods or fully microscopic ones), one needs to invoke many intrinsic configurations and their mixing in order include coupling. The shell-model and the inherent 'monopole' issue are just another example of this intertwining.

We have outlined a method to treat both the global and local energy parts in a consistent way within the IBM, thereby treating both the U(6) structure (which determines the overall binding energy variation with particle and boson number) and the internal (nuclear energy spectra themselves) structure within the various dynamical group chains of U(6) and also studying the more realistic cases. We have illustrated this method in the region of rare-earth nuclei.

This is but a first attempt but efforts are going on to address other mass regions too as well as studies to understand better the interplay between global and local contributions to the nuclear energy surface.

ACKNOWLEDGMENTS

We are grateful to the "FWO-Vlaanderen" and the "IWT" for financial support. This work has also been supported in part by the U.S.D.O.E. under grant number DE-FG02-91ER-40609. One of the authors (K.H.) would like to thank J. Äystö and the group at ISOLDE/CERN for providing the congenial atmosphere and for financial support.

REFERENCES

1.   P. Ring and P. Schuck, *The nuclear many-body problem* (Springer-Verlag, New York, Heidelberg, Berlin, 1980).
2.   A. de Shalit and I. Talmi, *Nuclear shell theory* (Academic Press, New York and London, 1963).

3. I. Talmi, *Simple models of complex nuclei* (Harwood Academic Publishers, 1993).
4. P.J. Brussaard and P.W.M. Glaudemans, *Shell-model applications in nuclear spectroscopy* (North-Holland, Amsterdam, 1977).
5. K. Heyde, *The nuclear shell-model* (Springer-Verlag, Berlin, Heidelberg, New York, 1994).
6. A. M. Lane, *Nuclear Theory - Pairing Correlations and Collective Motion* (W.A.Benjamin, Inc., New-York, Amsterdam, 1964)
7. K. Heyde, R. Fossion, J.E. García-Ramos and C. De Coster, Eur. Phys. J. **A13**, 402 (2002).
8. J.E. García-Ramos, C. De Coster, R. Fossion, and K. Heyde, Nucl. Phys. **A688**, 735 (2001).
9. R. Fossion, J.E. García-Ramos, C. De Coster, and K. Heyde, Nucl. Phys. **A697**, 703 (2002)
10. M. Dufour and A.P. Zuker, Phys. Rev. C **54**, 1641 (1996).
11. J. Duflo and A.P. Zuker, Phys. Rev. C **59**, R2347 (1999).
12. G. Bollen et al., Nucl. Instrum. Methods **A368**, 675 (1996)
13. H. Raimbault-Hartmann et al., Nucl. Instrum. Methods **B126**, 378 (1997)
14. S.Schwarz, F.Ames, G.Audi, D.Beck, G.Bollen, C.De Coster, J.Dilling, R.Fossion, J.E.Garcia-Ramos, F.Herfurth, K.Heyde, J.-J.Kluge, A.Kohl, D.Lunney, R.B.Moore, H.Raimbault-Hartmann, J.Szerypo and the ISOLDE collaboration,Nucl. Phys. **A693** , 533 (2001).
15. G.A. Lalazissis, S. Raman, and P. Ring, At. Data Nucl. Data Tables **71**,1 (1999).
16. S. Goriely, F. Tondeur, and J.M. Pearson, At. Data Nucl. Data Tables **77**, 311 (2001).
17. M. Samyn, S. Goriely, P.-H. Heenen, J.M. Pearson, and F. Tondeur, Nucl. Phys. **A700**, 142 (2002).
18. P. Möller and J.R. Nix, Nucl. Phys. A **361**,117 (1981).
19. P. Möller and J.R. Nix, At. Data Nucl. Data Tables **39**, 213 (1988).
20. P. Möller, J.R. Nix, W.D. Myers, and W.J. Swiatecki, At. Data Nucl. Data Tables **59**, 185 (1995).
21. Y. Aboussir, J.M. Pearson, A.K. Dutta, and F. Tondeur, At. Data Nucl. Data Tables **61**,127 (1995).
22. S. Peru, M. Girod, J.F. Berger, Eur. Phys. J. **A9**, 35 (2000)
23. A. Valor, P.-H. Heenen, and P. Bonche, Nucl. Phys. **A671**, 145 (2000).
24. E.K.Warburton, J.A.Becker and B.A.Brown, Phys.Rev.C **41**, 1147 (1990).
25. A.Poves and J.Retamosa, Nucl.Phys.**A571**, 221 (1994).
26. E.Caurier, F.Nowacki, A.Poves and J.Retamosa, Phys.Rev.C **58**, 2033 (1998).
27. T.Otsuka, T.Mizusaki, Y.Utsuno and M.Honma,*ENAM98, Exotic Nuclei and Atomic Masses*, eds. B.M.Sherill, D.J.Morrissey and C.N.Davids, 1998,Am.Inst.of Physics
28. F.Iachello and A.Arima, *The Interacting Boson Model*, (Cambridge University Press,1987)
29. A.Frank and P.Van Isacker, *Algebraic Methods in Molecular and Nuclear Structure Physics*, (Wiley, New-York,1994)
30. W.-T. Chou, N.V. Zamfir, and R.F. Casten, Phys.Rev.C **56**, 829 (1997).

How random are random nuclei?
Shapes, triangles and kites

R. Bijker[*] and A. Frank[*†]

[*]ICN-UNAM, AP 70-543, 04510 México D.F., México
[†]CCF-UNAM, AP 139-B, 62251 Cuernavaca, Morelos, México

Abstract. We discuss the origin of the regular features that were observed in numerical studies of the interacting boson model with random interactions.

The energy systematics of medium and heavy even-even nuclei shows very regular features, such as for example the tripartite classification of nuclear structure into seniority, anharmonic vibrator and rotor regions [1, 2]. Traditionally, this regular behavior has been interpreted as a consequence of particular interactions, such as an attractive pairing force in semimagic nuclei and an attractive neutron-proton quadrupole-quadrupole interaction for deformed nuclei.

It came as a surprise, therefore, that recent studies of even-even nuclei in the nuclear shell model [3, 4, 5] and in the interacting boson model (IBM) [6, 7, 8] with random interactions also displayed a high degree of order. Both models showed a statistical preference for $L = 0$ ground states, despite the random nature of the interactions. In addition, in the shell model evidence was found for the occurrence of pairing properties [5], and in the IBM for vibrational and rotational bands [6, 7].

These unexpected results have sparked a large number of investigations to try to understand their origin [9]. In this contribution, we discuss the phenomenom of emerging regular spectral features from the IBM with random interactions, and its relation with the underlying geometric shapes and critical points.

In order to study the geometric shapes associated with the IBM [10, 11, 12], we first consider the schematic Hamiltonian of the consistent-Q formulation (CQF) [13]

$$H = \varepsilon n_d - \kappa Q(\chi) \cdot Q(\chi) \,.$$

The parameters are restricted to the 'physically' allowed region, i.e. $\varepsilon > 0$, $\kappa > 0$ and $-\sqrt{7}/2 \leq \chi \leq \sqrt{7}/2$. The properties of the CQF Hamiltonian are investigated by taking the scaled parameters $\eta = \varepsilon/[\varepsilon + 4\kappa(N-1)]$ and $\bar{\chi} = 2\chi/\sqrt{7}$ randomly on the intervals $0 \leq \eta \leq 1$ and $-1 \leq \bar{\chi} \leq 1$. In Fig 1 we present the results in a shape phase diagram as a function of the coefficients r_1 and r_2 which were introduced in [14] as the essential control parameters to classify the equilibrium configurations of the IBM Hamiltonian. r_1 and r_2 are determined by particular combinations of the interaction parameters. In Fig. 1 we have identified each of the dynamical symmetries of the IBM: $U(5)$, $SU(3)$ with prolate/oblate symmetry and $SO(6)$. The transitions between the dynamical symmetries are indicated by the solid lines. The resulting figure is that of a kite. The socalled critical

CP638, *Mapping the Triangle: International Conference on Nuclear Structure*
edited by A. Aprahamian, J. A. Cizewski, S. Pittel, and N. V. Zamfir
© 2002 American Institute of Physics 0-7354-0093-8/02/$19.00

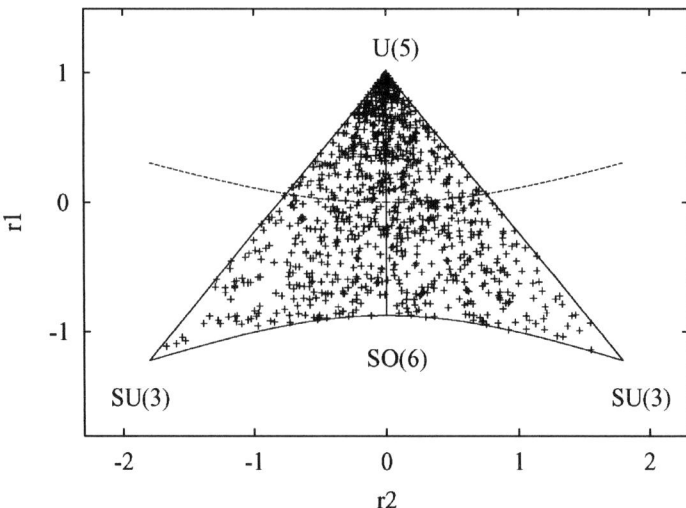

FIGURE 1. Shape phase diagram for the random CQF Hamiltonian obtained for $N = 16$ and 1000 runs. The dashed line separates the spherical from the deformed shape.

point symmetries $E(5)$ and $X(5)$ [15] are related to the points at the intersections of the solid lines and the separatrix (dashed line) which separates the spherical and deformed shapes. The prolate and oblate deformed shapes are separated by $r_2 = 0$, $r_1 < 0$. The associated critical point symmetry coincides with the $SO(6)$ limit [16].

Each CQF Hamiltonian corresponds to a point in the $r_2 r_1$ plane and is labeled by a $+$ sign in Fig. 1. The random ensemble of CQF Hamiltonians covers the interior part of the kite: 50% for the spherical shape, and 25% each for the prolate and oblate deformed shapes. For all cases, the ground state has angular momentum $L = 0$. The existence of two definite geometric shapes, a spherical and an axially symmetric deformed one, is also evident from a plot of the probability distribution $P(R)$ of the energy ratio $R = [E(4_1) - E(0_1)]/[E(2_1) - E(0_1)]$. Fig. 2 shows that for the CQF there are two characteristic peaks, one at the vibrator value $R = 2$ and one at the rotor value $R = 10/3$ (solid line).

So far, we have discussed the properties of a random ensemble of IBM Hamiltonians with minimal constraints to 'realistic' interactions. Surprisingly enough, the results for a general IBM Hamiltonian with random one- and two-body interactions chosen independently from a Gaussian distribution of random numbers with zero mean and width σ are very similar [6]. Also in this case, the probability distribution $P(R)$ exhibits peaks at $R \sim 1.9$ and $R \sim 3.3$ (dashed line). The vibrational and rotational nature of these peaks has been confirmed by a simultaneous study of the quadrupole transitions between the levels [6]. Despite the random nature of the interaction strengths both in relative size and sign, the ground state still has $L = 0$ in $\sim 63\%$ of the cases. In Fig. 3 we show the percentages of ground states with $L = 0$ and $L = 2$ as a function of the boson number N (solid line). We see a clear dominance of ground states with $L = 0$ with $\sim 60\text{-}75\%$. For

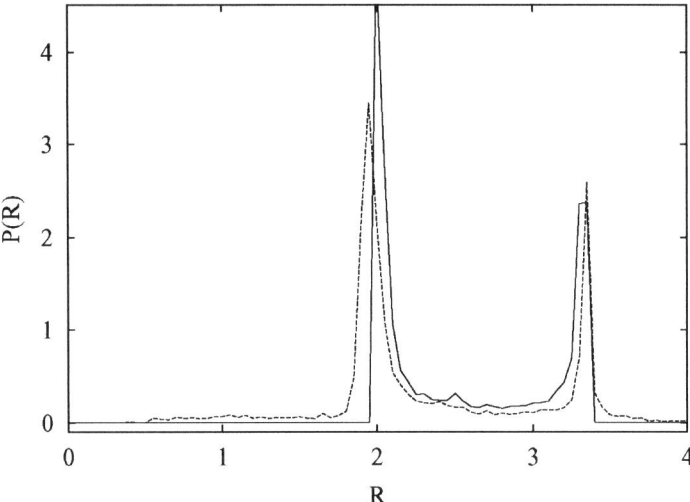

FIGURE 2. Probability distribution $P(R)$ of the energy ratio R in the IBM for the random CQF (solid line) and for the general case of random one- and two-body interactions (dashed line), obtained for $N = 16$ and 10000 runs.

$N = 3k$ (a multiple of 3) we see an enhancement for $L = 0$ and a decrease for $L = 2$. The sum of the two hardly depends on the number of bosons.

These are surprising results in the sense that, according to the conventional ideas in the field, the occurrence of $L = 0$ ground states and the existence of vibrational and rotational bands are due to very specific forms of the interactions. The basic ingredients of the numerical simulations, both for the nuclear shell model and for the IBM, are the structure of the model space, the ensemble of random Hamiltonians, the order of the interactions (one- and two-body), and the global symmetries, i.e. time-reversal, hermiticity and rotation and reflection symmetry. The latter three symmetries cannot be modified, since we are studying many-body systems whose eigenstates have real energies and good angular momentum and parity. It has been shown that the observed spectral order is a robust property that does not depend on the specific choice of the ensemble of random interactions [3, 4], the time-reversal symmetry [4], or the restriction of the Hamiltonian to one- and two-body interactions [7]. These results suggest that that an explanation of the origin of the observed regular features has to be sought in the many-body dynamics of the model space and/or the general statistical properties of random interactions.

For the IBM, the emergence of regular features from random interactions can be explained in a Hartree-Bose mean-field analysis of the random ensemble of Hamiltonians, in which different regions of the parameter space are associated with particular intrinsic states, which in turn correspond to definite geometric shapes [17]. There are three solutions: a spherical shape carried by a single state with $L = 0$, a deformed shape which corresponds to a rotational band with $L = 0, 2, \ldots, 2N$, and a condensate of quadrupole

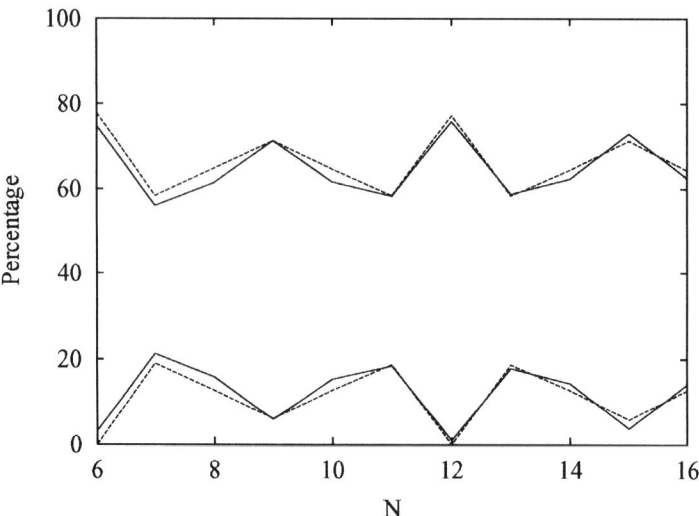

FIGURE 3. Percentages of ground states with $L = 0$ (top) and $L = 2$ (bottom) in the IBM with random one- and two-body interactions calculated exactly for 10000 runs (solid line) and in mean-field approximation (dashed line).

bosons which has a more complicated angular momentum content. We note, that the latter solution does not occur for the random ensemble of CQF Hamiltonians. The ordering of rotational energy levels depends on the sign of the corresponding moments of inertia, which have been evaluated with the Thouless-Valatin formula. In Fig. 3 we show the percentages of ground states with $L = 0$ and $L = 2$ as a function of the number of bosons N. A comparison of the results of the mean-field analysis (dashed lines) and the exact ones (solid lines) shows excellent agreement. The oscillations with N are entirely due to the contribution of the condensate of quadrupole bosons. The mean-field analysis explains both the distribution of ground state angular momenta and the occurrence of vibrational and rotational bands. The same conclusions hold for the vibron model for which a large part of the results has been obtained analytically [17].

In this contribution, we addressed the origin of the regular features obtained in numerical studies of the IBM with random interactions, in particular the dominance of $L = 0$ ground states and the occurrence of vibrational and rotational band structures. It was shown that the geometric shapes associated with IBM Hamiltonians play a crucial role in understanding these regular properties. Different regions of the parameter space are associated with definite geometric shapes, such as spherical and deformed shapes and a condensate of quadrupole bosons. For a random ensemble of CQF Hamiltonians the latter solution is absent, and the shape phase diagram assumes the simple form of a kite or a double triangle.

For the nuclear shell model the situation is less clear. Although a large number of investigations to explain and further explore the properties of random nuclei have shed light on various aspects of the original problem, i.e. the dominance of $L = 0$ ground states, in our opinion, no definite answer is yet available, and the full implications for nuclear structure physics are still to be clarified.

It is a great pleasure to dedicate this contribution to Rick Casten on the occasion of his 60th birthday in appreciation of the numerous occasions where his profound and stimulating comments have had a strong impact on our work. In particular, we gratefully acknowledge his enthousiastic support of the random ideas of his theoretician friends.

This work was supported in part by CONACyT under projects 32416-E and 32397-E, and by DPAGA-UNAM under project IN106400.

REFERENCES

1. R.F. Casten, N.V. Zamfir and D.S. Brenner, Phys. Rev. Lett. **71**, 227 (1993).
2. N.V. Zamfir, R.F. Casten and D.S. Brenner, Phys. Rev. Lett. **72**, 3480 (1994).
3. C.W. Johnson, G.F. Bertsch and D.J. Dean, Phys. Rev. Lett. **80**, 2749 (1998).
4. R. Bijker, A. Frank and S. Pittel, Phys. Rev. C **60**, 021302 (1999).
5. C.W. Johnson, G.F. Bertsch, D.J. Dean and I. Talmi, Phys. Rev. C **61**, 014311 (2000).
6. R. Bijker and A. Frank, Phys. Rev. Lett. **84**, 420 (2000).
7. R. Bijker and A. Frank, Phys. Rev. C **62**, 014303 (2000).
8. D. Kusnezov, N.V. Zamfir and R.F. Casten, Phys. Rev. Lett. **85**, 1396 (2000).
9. R. Bijker and A. Frank, Nuclear Physics News, Vol. 11, No. 4, 15 (2001).
10. A.E.L. Dieperink, O. Scholten and F. Iachello, Phys. Rev. Lett. **44**, 1747 (1980).
11. F. Iachello and A. Arima, *The interacting boson model* (Cambridge University Press, 1987).
12. F. Iachello, N.V. Zamfir and R.F. Casten, Phys. Rev. Lett. **81**, 1191 (1998).
13. D.D. Warner and R.F. Casten, Phys. Rev. Lett. **48**, 1385 (1982).
14. E. López-Moreno and O. Castaños, Phys. Rev. C **54**, 2374 (1996).
15. F. Iachello, Phys. Rev. Lett. **85**, 3580 (2000); *ibid.* **87**, 052502 (2001).
16. J. Jolie, R.F. Casten, P. von Brentano and V. Werner, Phys. Rev. Lett. **87**, 162501 (2001).
17. R. Bijker and A. Frank, Phys. Rev. C **64**, 061303 (2001); *ibid.* **65**, 044316 (2002).

174

Two neutron separation energies and phase transitions in the Interacting Boson Model

J.E. García-Ramos*, R. Fossion† and K. Heyde†

*Department of Applied Physics. EPS La Rábida, University of Huelva. 21819 Palos de la Frontera, Spain.
† Department of Subatomic and Radiation Physics, Proeftuinstraat, 86 B-9000 Gent, Belgium.

Abstract. In the framework of the interacting boson model the three transitional regions (rotational-vibrational, rotational–γ-unstable and, vibrational–γ-unstable transitions) are reanalyzed. A new kind of plot is presented for studying phase transitions in finite systems such as atomic nuclei. The importance of analyzing binding energies and not only energy spectra and electromagnetic transitions, describing transitional regions is emphasized and we discuss a new method in order to provide a consistent description of both, ground-state and excited-state properties.

INTRODUCTION

In the last few years, interest for the study of phase transitions and phase coexistence in atomic nuclei has been revived [1, 2, 3, 4, 5, 6, 7] in particular making use of the Interacting Boson Model (IBM) [8]. In the chart of nuclei, three transitional regions can be distinguished: (a) where it is observed a change from spherical to well-deformed nuclei when moving from the lighter to the heavier isotopes; (b) where one notices that the lighter isotopes are spherical while the heavier ones indicate a γ-unstable character; (c) where the lighter isotopes are well deformed while the heavier shown γ-unstable properties. Although these three transitional regions have been studied extensively in the framework of the IBM, the discussion of phase transitions has not always been treated in a proper way. In particular one of the weak points is how to define an appropriate control parameter in a system such as the atomic nucleus, where this parameter is fixed (for given N and Z) and cannot be externally controlled [9]. This problem has been recently considered in a study by Casten *et al.* [3]. In the present contribution and in reference [10] we present a different approach. We additionally show that it is convenient to explore nuclear binding energies (BE), together with excited states properties, in order to obtain a consistent description, especially when one is dealing with chains of isotopes.

CP638, *Mapping the Triangle: International Conference on Nuclear Structure*
edited by A. Aprahamian, J. A. Cizewski, S. Pittel, and N. V. Zamfir
© 2002 American Institute of Physics 0-7354-0093-8/02/$19.00

TRANSITIONAL REGIONS AND PHASE TRANSITIONS IN THE IBM: TRADITIONAL AND NEW POINT OF VIEW

In order to treat with transitional regions in the framework of the IBM, a very convenient Hamiltonian can be written as follows:

$$\hat{H} = \kappa\left(N\frac{1-\xi}{\xi}\hat{n}_d - \hat{Q}\cdot\hat{Q}\right) + \kappa'\hat{L}\cdot\hat{L}, \tag{1}$$

with $0 \leq \xi \leq 1$, where $\hat{L} = \sqrt{10}(d^\dagger \times \tilde{d})^{(1)}$ and $\hat{Q} = s^\dagger\tilde{d} + d^\dagger\tilde{s} + \chi(d^\dagger \times \tilde{d})^{(2)}$. With this parametrization one can move easily through the three legs of Casten's triangle modifying the values of χ and ξ. The three dynamical limits of the IBM correspond to the following values of (χ,ξ): $U(5)$, $(\chi,0)$; $SU(3)$, $(-\sqrt{7}/2, 1)$; and $O(6)$, $(0,1)$.

One of the most striking facts that characterize the transitional regions is the possibility of observation of phase transitions. Although phase transitions are strictly well defined in macroscopic systems and the atomic nucleus is a finite system, a number of studies have shown that the concept of a phase transition retains its validity and usefulness in small systems too [11, 12]. A useful tool in order to discuss phase transitions in finite system are the coherent states which, in the case of the IBM, are also known as intrinsic states [13, 14].

$$|c\rangle = \frac{1}{\sqrt{N!}}(\Gamma_c^\dagger)^N|0\rangle, \qquad \text{where} \qquad \Gamma_c^\dagger = \frac{1}{\sqrt{1+\beta^2}}\left(s^\dagger + \beta\, d_0^\dagger\right). \tag{2}$$

Here the parameter β will act as order parameter.

The appearance of a phase transition is denoted by a discontinuity in the first or second order derivative of the BE with respect to the control parameter and, at the same time, by an abrupt change in the value of the order parameter [9, 10]. In the case of the transitional regions $U(5)$-$SU(3)$ and $U(5)$-$O(6)$ the control parameter is ξ, giving rise to a first order phase transition in the first case, and a second order phase transition in the second one. The transitional region $O(6)$-$SU(3)$ should be described using χ as control parameter although no phase transition appear.

Though this analysis provides a very straightforward way of treating phase transitions, there exits a snag on it: the control parameter ξ (or χ) is not a genuine control parameter, because it cannot be modified externally. The control parameter only changes when moving along a chain of nuclei and also assumes a change in the number of bosons. So, a consistent treatment of phase transitions should go beyond the analysis presented above and vary, at the same time, the control parameter and the number of bosons.

The appropriate way of treating phase transitions and transitional regions is to plot in the same figure the value of the binding energy versus the control parameter for different values of N. In such a plot, a transition will develop through changes in the control parameter, but at the same time going through curves with a different number of bosons. In order to illustrate this new procedure, we discuss the transitional regions $U(5)$-$SU(3)$, $U(5)$-$O(6)$ and, $SU(3)$-$O(6)$ in figures 1a, 1b, and 1c respectively. In these figures, only the laboratory results (IBM diagonalization) are presented. As an example, trajectories for real nuclei are also plotted in each figure. The more interesting cases appear in figures

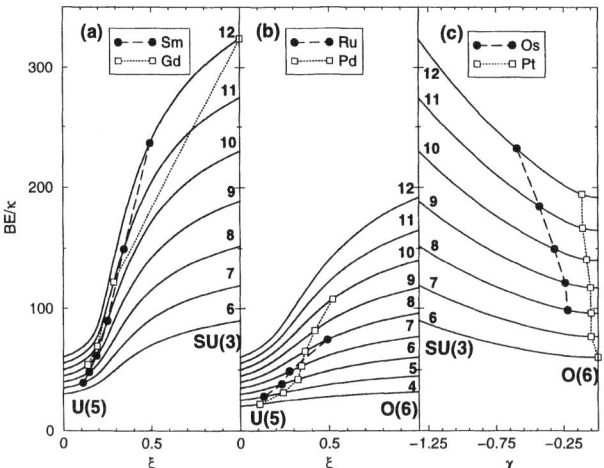

FIGURE 1. Binding energies in the laboratory system for different numbers of bosons, N, as a function of the control parameter, in the transitional regions (a) $U(5)$-$SU(3)$, (b) $U(5)$-$O(6)$, and (c) $SU(3)$-$O(6)$. Full circles and open squares correspond to the theoretical positions of different chains of isotopes.

1a and 1b because in this two cases, phase transitions indeed happen. So, inspecting these figures, one can easily see if a given nucleus is situated in the spherical region (ξ near to 0), in the deformed region (ξ around 1) or at the critical point (ξ around 0.2).

Next question will be: which observable can inform us about the existence of a phase transition? The natural answer should be BE or equivalently two-neutron separation energies (S_{2n}). However it is not clear, a priori, how S_{2n} will behave when crossing a phase transition point. Let us start with the definition of S_{2n}:

$$S_{2n}(N) = \mathscr{A} + \mathscr{B}N + BE_{IBM}(N) - BE_{IBM}(N-1), \qquad (3)$$

where $\mathscr{A}$ and $\mathscr{B}$ take into account the bulk contribution to the nuclear interaction and come from the linear and quadratic $U(6)$ Casimir invariants. The coefficients $\mathscr{A}$ and $\mathscr{B}$ remain as constants for a given major shell [15]. When the linear dependence is excluded from S_{2n}, we will refer to the remaining part as S'_{2n}. Next we will simulate a phase transitions establishing a connection between the control parameter and the number of bosons. We thereby focus on the case of the $U(5)$-$SU(3)$ transition in a chain of isotopes with 5 protons pairs, $N_\pi = 5$, and a variable number of neutrons pairs, ranging from $N_\nu = 0$ to $N_\nu = 10$. Two functional dependences will be used, firstly linear and secondly quadratic,

$$\xi = 0.099N_\nu + 0.01, \qquad (4)$$
$$\xi = 0.0099N_\nu^2 + 0.01, \qquad (5)$$

with $\chi = -\sqrt{7}/2$. In order to establish the correspondence we use the empirical observation that the system moves from the $U(5)$ to the $SU(3)$ limit for increasing number of bosons. For convenience, a linear dependence equal to $S_{2n}^{lin}/\kappa = 200 - 20N_\nu$ has been

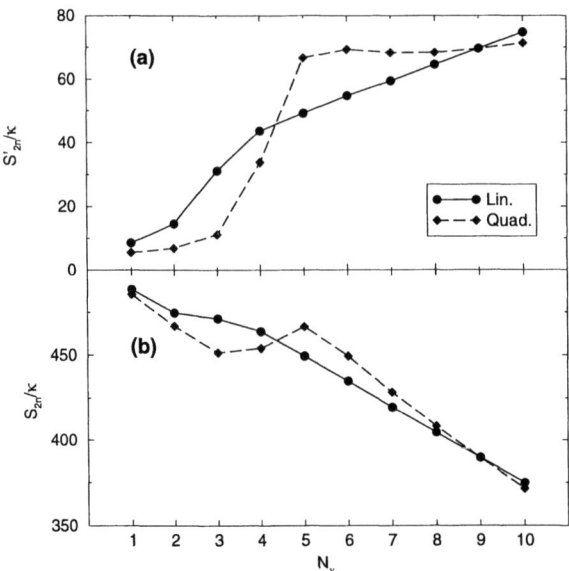

FIGURE 2. Schematic plot of (a) S'$_{2n}$ and (b) S$_{2n}$ in the $U(5)$-$SU(3)$ transitional region. Full lines corresponds to a linear variation of ξ with respect to N_v while dashed lines corresponds to a quadratic dependence.

chosen as a reference value in order to obtain a realistic behavior in S$_{2n}$. We depict the results in Fig. 2 and it can be observed an anomaly in the linear dependence of S$_{2n}$ and S'$_{2n}$, right at the place where the phase transition takes place. This anomaly is clearly observed in the Nd-Sm-Gd region [8].

REALISTIC CALCULATIONS

Within the framework of the IBM, the energy spectra and transition rates of many medium-mass and heavy nuclei have been successfully analyzed. However, in many of these studies the binding energies have been ignored. Here, we will try to calculate the BE through the knowledge of the excitation energies and transitions probabilities. At the same time we will show that the description of BE is not a trivial task in the framework of the IBM.

Let us consider the $U(5)$-$SU(3)$ transitional region and in particular the Gd isotopes which form a clear example of nuclei becoming deformed ($SU(3)$ limit). So it seems natural to take $\chi = -\sqrt{7}/2$ for the whole chain. The remaining parameters are taken from [10]. Another possible parametrization can be obtained following reference [10, 16] where a value $\chi = -0.6$ was used. Both parametrization provides reasonable descriptions of excitation energies (see reference [10]), but if we plot S$_{2n}$ we can see that one of the descriptions (the one with $\chi = -\sqrt{7}/2$) clearly fails in describing S$_{2n}$ values.

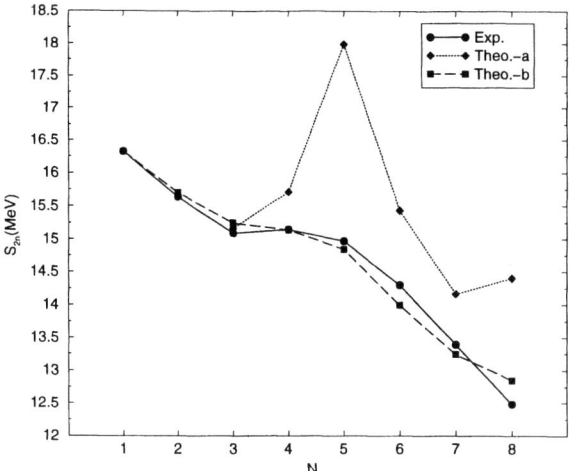

FIGURE 3. Two-neutron separation energies for Gd isotopes. Full lines correspond to experimental data, while dotted lines correspond to $\chi = -\sqrt{7}/2$ and dashed lines correspond to $\chi = -0.6$.

CONCLUSION

In this contribution we present a new approach for treating phase transitions in finite systems, taking into account that a change in the control parameter implies a change of nucleus. On the other hand we study how the crossing of a phase transition point can induce an anomalous behavior in the value of S_{2n}. Finally, we show that the study of S_{2n} for long chains of isotopes can help in fixing the parameters of the IBM Hamiltonian.

REFERENCES

1. Rowe, D.J., Bahri, C., and Wijesundera, W., Phys. Rev. Lett. **80**, 4394 (1998).
2. Iachello, F., Zamfir, N.V., and Casten, R.F. , Phys. Rev. Lett. **81**, 1191 (1998).
3. Casten, R.F., Kusnezov, D., and Zamfir, N.V., Phys. Rev. Lett. **82**, 5000 (1999).
4. Iachello, F., Phys. Rev. Lett. **85**, 3580 (2000).
5. Casten, R.F. and Zamfir, N.V., Phys. Rev. Lett. **85**, 3584 (2000).
6. Iachello, F., Phys. Rev. Lett. **87**, 052501 (2001).
7. Casten, R.F. and Zamfir, N.V., Phys. Rev. Lett. **87**, 052503 (2001).
8. Iachello, F. and Arima, A., *The interacting boson model*, Cambridge University Press, Cambridge, 1987.
9. Dieperink, A.E.L. and Scholten, O., Nucl. Phys. A **346**, 125 (1980).
10. García-Ramos, J.E., De Coster, C., Fossion, R., and Heyde, K., Nucl. Phys. A **688**, 735 (2001).
11. Gilmore, R. and Feng, D.H., Nucl. Phys. A **301**, 189 (1978).
12. Gilmore, R., *Catastrophe theory for scientists and engineers* (Wiley, New York, 1981).
13. Diepering, A.E.L., Scholten, O., and Iachello, F., Phys. Rev. Lett. **44**, 1747 (1980).
14. Ginocchio, J.N. and Kirson, M.W., Nucl. Phys. A **350**, 31 (1980).
15. Fossion, R., De Coster, C., García-Ramos, J.E., Werner, T., and Heyde, K., Nucl. Phys. A **697**, 703 (2002).
16. Chou, W.-T., Zamfir, N.V., and Casten, R.F., Phys. Rev. C **56**, 829 (1997).

Geometry of random interactions

P. Chau Huu-Tai*, A. Frank†, N. Smirnova** and P. Van Isacker*

*GANIL, BP 55027, F-14076 Caen Cedex 5, France
†Instituto de Ciencias Nucleares, UNAM, AP 70-543, 04510 México DF, Mexico
Centro de Ciencias Físicas, UNAM, AP 139-B, 62251 Cuernavaca, Mexico
**University of Leuven, Instituut voor Kern- en Stralingsfysica, Celestijnenlaan 200D, B-3001
Leuven, Belgium

Abstract. It is argued that spectral features of quantal systems with random interactions can be given a geometric interpretation. This conjecture is investigated in the context of two simple models: a system of randomly interacting d bosons and one of randomly interacting fermions in a $j = \frac{7}{2}$ shell. In both examples the probability for a specific state to become the ground state is shown to be related to a geometric property of a (hyper)polyhedron which is interaction-independent and is determined solely by particle number, shell size and symmetry character of the state. Extensions of these ideas to more general situations are discussed.

RICK'S RANDOM CONUNDRUM

Recent developments in nuclear structure physics have been a source of considerable preoccupation for Rick Casten, the person whose sixtieth birthday we are celebrating with this conference. The studies in question have unveiled a high degree of order and, in particular, a marked statistical preference was found for ground states with angular momentum $J = 0$. Surprisingly, these observations were made with entirely random interactions, either attractive or repulsive with an arbitrary strength. Whence Rick's preoccupation: Will nuclear structure physics be relegateded to the realms of randomland where all is explained by chance and an understanding can be obtained without recourse to simple, intuitive arguments, many of which are so admirably explained in Rick's book [1]?

Fortunately for us (and for Rick!) that appears not to be the case. A growing number of studies both in the nuclear shell model [2, 3, 4, 5] and the interacting boson model [6, 7] have led to an increased understanding of the nature of the regularities obtained with random interactions. It is now clear that, although some of the features observed in nuclei arise randomly, certainly not all of them do. The extent to which regularities occur randomly and the fundamental reasons why that is so, are not yet fully known. It is, therefore, of interest to study the problem from different angles and one novel approach is offered in this contribution. Specifically, we argue here that spectral features of quantal systems with random interactions can be given a geometric interpretation which allows the computation of the probability for the quantum mechanical ground state to have a given angular momentum, based on purely geometric considerations. Although these results are obtained in the context of a variety of simple models which do not cover the full complexity of random interactions, we believe them to be sufficiently general to

CP638, *Mapping the Triangle: International Conference on Nuclear Structure*
edited by A. Aprahamian, J. A. Cizewski, S. Pittel, and N. V. Zamfir
© 2002 American Institute of Physics 0-7354-0093-8/02/$19.00

conjecture the possibility of an entirely geometric analysis of the problem.

GEOMETRY OF DIAGONAL RANDOM MODELS

Consider a system consisting of n interacting particles (bosons or fermions) carrying angular momentum j, integer or half-integer. Eigenstates of a rotationally invariant Hamiltonian are characterized as $|j^n \alpha J M\rangle$ where J and M are the total angular momentum and its projection, and α is any other index needed for a complete labeling of the state. We assume here that the one-body contribution is constant for all eigenenergies and that the energy spectrum is generated by two-body interactions only. This assumption is for simplicity's sake only and spectral properties of Hamiltonians with both one- and two-body interactions can be analyzed in the same way as explained below. Under the assumption of the two-body character of the Hamiltonian, its matrix elements can be written as

$$\langle j^n \alpha J | \hat{H} | j^n \alpha' J \rangle = \frac{n(n-1)}{2} \sum_L c_{n\alpha J}^L c_{n\alpha' J}^L G_L, \tag{1}$$

where M is omitted since energies do not depend on it. The quantities G_L are two-particle matrix elements, $G_L \equiv \langle j^2 L | \hat{H} | j^2 L \rangle$, and completely specify the two-body interaction while $c_{n\alpha J}^L$ are interaction-independent coefficients. They can be expressed in terms of coefficients of fractional parentage (CFPs) [8] and, as such, are entirely determined by the symmetry character of the n-particle states.

We consider here the special case when a basis $|j^n \alpha J M\rangle$ can be found in which the expansion (1) is diagonal. In this case the Hamiltonian matrix elements reduce to the energy eigenvalues

$$E_{n\alpha J} = \frac{n(n-1)}{2} \sum_L b_{n\alpha J}^L G_L, \tag{2}$$

with $b_{n\alpha J}^L \equiv \left(c_{n\alpha J}^L \right)^2$. This is obviously a simplification of the more general problem (1) but nevertheless a wide variety of simple model situations can be accommodated by it. For example, this property is valid for any interaction between identical fermions if $j \leq \frac{7}{2}$ and remains so approximately for larger j values; it is also exactly valid for p, d, or f bosons. We shall refer to this class of problems as diagonal. For a constant interaction, $G_L = 1$, all n-particle eigenstates are degenerate with energy $\frac{1}{2} n(n-1)$ and consequently the coefficients $b_{n\alpha J}^L$ satisfy the properties $\sum_L b_{n\alpha J}^L = 1$ and $0 \leq b_{n\alpha J}^L \leq 1$. Equation (2) can thus be rewritten in terms of scaled energies as

$$e_{n\alpha J} \equiv \frac{2 E_{n\alpha J}}{n(n-1)} = G_{L'} + \sum_L b_{n\alpha J}^L (G_L - G_{L'}), \tag{3}$$

for arbitrary L'. This shows that, in the case of N interaction matrix elements G_L, the energy of an arbitrary eigenstate can, up to a constant scale and shift, be represented as a point in a vector space spanned by $m \equiv N - 1$ differences of matrix elements. Note that the position of these states is fixed by $b_{n\alpha J}^L$ and hence independent of the interaction. Furthermore, since $0 \leq b_{n\alpha J}^L \leq 1$, all states are confined to a compact region of this space with the size of one unit in each direction. For independent variables G_L with covariance matrix $\langle G_L G_{L'} \rangle = \delta_{LL'}$, states are represented in an orthogonal basis. The differences in (3) are not independent but have a covariance matrix of the form $1 + \delta_{LL'}$; this leads to a representation in a m-simplex basis (i.e., an equilateral triangle in $m = 2$ dimensions, a regular tetrahedron for $m = 3$, etc.).

The following procedure can now be proposed to determine the probability $P_{n\alpha J}$ for a specific state $n\alpha J$ to become the ground state. First, construct all points corresponding to the energies $e_{n\alpha J}$. Next, build from them the enveloping *convex* polytope (a polytope is a polyhedron in m dimensions) which is defined as the smallest possible convex polytope that contains all points. The relevance of this enveloping polytope as regards the probability of specific state to be the ground state becomes clear from the following two statements:

1. All points (i.e. states) other than the vertices of the enveloping polytope can *never* be the ground state for *whatever* choice of matrix elements G_L and thus have $P_{n\alpha J} = 0$. This eliminates all points that are located inside the polytope or on its (hyper)surface as possible ground-state candidates.
2. The probability $P_{n\alpha J}$ of a state which is a vertex of the enveloping polytope can be expressed in terms of a geometric property at that vertex.

The first statement can be proved as follows. Any point inside the enveloping polytope can be written as $E = \sum_v \beta_v E_v$, where the sum runs over all vertices of the polytope and with $\beta_v \geq 0$ and $\sum_v \beta_v = 1$. It then follows that $\sum_v \beta_v (E_v - E) = 0$. Since all β_v are positive, this equality cannot be satisfied if all differences $E_v - E$ are simultaneously positive and so E cannot be the ground state. The second statement is, at the moment, just a conjecture. Examples and proofs of specific cases and possible generalizations will be discussed in the next two sections.

TWO EXAMPLES

The general procedure outlined in the previous section can be illustrated with some examples. Consider first a system of d bosons ($j = 2$). In this case there are three interaction matrix elements, G_0, G_2, and G_4 with $G_L \equiv \langle d^2 L | \hat{H} | d^2 L \rangle$; thus, $N = 3$ and the problem can be represented in a plane. Because of the $U(5) \supset SO(5) \supset SO(3)$ algebraic structure, an analytic solution is known for n interacting d bosons with eigenenergies [9, 10]

$$
\begin{aligned}
e_{n\nu J} = \ & G_4 + \frac{n(n+3) - \nu(\nu+3)}{5n(n-1)}(G_0 - G_4) \\
& + \frac{2n(n-2) + 2\nu(\nu+3) - J(J+1)}{7n(n-1)}(G_2 - G_4),
\end{aligned} \tag{4}
$$

where ν is the seniority quantum number which counts the number of d bosons not in pairs coupled to angular momentum zero. In fact, the set $n\nu J$ is not complete but must be supplemented with another label that involves triplets of d bosons coupled to zero [11]; for a discussion of energies this label is not essential and can be omitted. Any

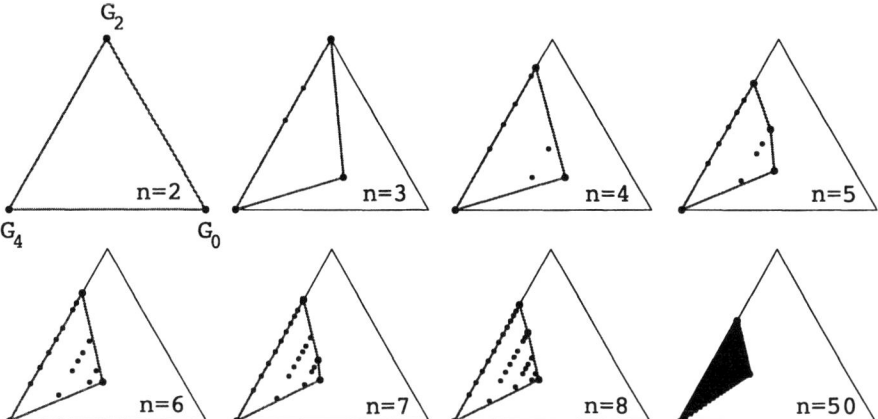

FIGURE 1. Polygons corresponding to several systems of interacting d bosons. All states are represented by a dot. The smaller dots are inside or on the edge of a convex polygon (indicated in grey, with the larger dots as vertices) and are never the ground state.

energy $e_{n\nu J}$ can be represented as a point inside an equilateral triangle with vertices G_0, G_2, and G_4 of which the position is determined by the appropriate values on the edges G_0–G_4 and G_2–G_4. Examples for several boson numbers n are shown in Fig. 1. For $n = 2$ there are three states with $J = 0, 2, 4$ and energies $e = G_0, G_2, G_4$; clearly, they have equal probability of being the ground state. As n increases, more states appear in the triangle. What emerges is a pattern of points grouped in straight lines parallel to the axis G_2–G_4 which correspond to states with a constant seniority ν. There is a single point closest to G_0 with minimum seniority ($\nu = 0$ if n is even or $\nu = 1$ if n is odd) and lines with increasing seniority as the axis G_2–G_4 is approached until this axis itself is reached for $\nu = n$. This pattern guarantees that there are (at least) three possible ground-state candidates since they always correspond to a vertex of the enveloping polygon: one minimum-seniority state [either $J(\nu) = 0(0)$ or $J(\nu) = 2(1)$, depending on whether n is even or odd] and two maximum-seniority $\nu = n$ states, one with minimal and the other with maximal angular momentum. The majority of states, shown as the smaller dots, are inside the convex polygon indicated in grey and can never be the ground state for whatever the choice of G_L. If we translate or rotate the convex polygon inside the triangle, its properties should not change since the points G_0, G_2, and G_4 are equivalent. Thus the probability for a point to be the ground state can only be related to the angle subtended at that vertex. The relation can be formally derived but also inferred from a few simple examples. If the polygon is an equilateral triangle, square, regular pentagon,...each vertex is equally probable with probability $\frac{1}{3}, \frac{1}{4}, \frac{1}{5}, \ldots$. One deduces the relation (see the $m = 2$ polygons in Table 1)

$$P_v^{(2)} = \frac{1}{2} - \frac{\theta_v}{2\pi},\qquad(5)$$

between the angle θ_v at the vertex v of the polygon and the probability $P_v^{(2)}$ for the state associated with that vertex to be the ground state. This relation has been verified [12]

TABLE 1. The angle θ_v or the angle sum $\sum_{f \ni v} \theta_{vf}$ and the vertex probability $P_v^{(m)}$ of regular polygons ($m = 2$) and polyhedra ($m = 3$).

polygon	m	θ_v	$P_v^{(2)}$	polyhedron	m	$\sum_{f \ni v} \theta_{vf}$	$P_v^{(3)}$
triangle	2	$\pi/3$	$1/3$	tetrahedron	3	π	$1/4$
square	2	$\pi/2$	$1/4$	octahedron	3	$4\pi/3$	$1/6$
pentagon	2	$3\pi/5$	$1/5$	cube	3	$3\pi/2$	$1/8$
...	...	...	...	icosahedron	3	$5\pi/3$	$1/12$
p-gon	2	$(p-2)\pi/p$	$1/p$	dodecahedron	3	$9\pi/5$	$1/20$

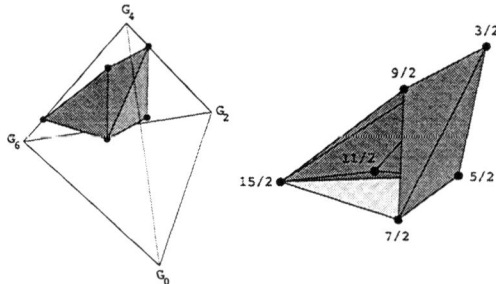

FIGURE 2. The polyhedron for a system of $n = 3$ $j = \frac{7}{2}$ fermions The left part indicates the position of the polyhedron in the tetrahedral coordinate system. The right part shows the same (enlarged) polyhedron with vertices that correspond to states with angular momentum J. For representation purposes the polyhedral face 7/2–9/2–15/2 has been removed in the right part.

to hold by comparing its results with those obtained from numerical tests for several systems of randomly interacting d bosons. These tests indicate that the relation (5) is valid independently of the nature of the random distribution, as long as it has zero mean and is the same for all parameters.

A second example concerns a system of fermions in a $j = \frac{7}{2}$ shell which was discussed recently by Zhao and Arima [4]. In this case there are four interaction matrix elements, G_0, G_2, G_4, and G_6 with $G_L \equiv \langle j^2 L | \hat{H} | j^2 L \rangle$, and this leads to a problem that can be represented in three-dimensional space. Any state in the $j = \frac{7}{2}$ shell can be labeled with particle number n, seniority ν, and total angular momentum J with analytically known expansion coefficients $b_{n\nu J}^L$ [8]. For $n = 3$ there are six different states which can be represented in three-dimensional space spanned by the three differences $G_0 - G_6$, $G_2 - G_6$, and $G_4 - G_6$ and define a convex polyhedron (see Fig. 2). The relation between the geometry of the polyhedron and the probability $P_v^{(3)}$ of each vertex state to be the ground state can again be inferred from a few simple examples. The relevant quantity in this case is $\sum_{f \ni v} \theta_{vf}$ where the sum is over all faces that contain the vertex v and θ_{vf} is the plane angle at vertex v in face f. A few examples with regular polyhedra (see the $m = 3$ polyhedra in Table 1) demonstrate that the relation is

$$P_v^{(3)} = \frac{1}{2} - \frac{1}{4\pi} \sum_{f \ni v} \theta_{vf}. \tag{6}$$

TABLE 2. Probabilities $P_{n\nu J}$ (in %) of $n = 3$ $j = \frac{7}{2}$ states obtained with the analytic formula (6) and from a numerical calculation.

	Probability $P_{n\nu J}$			Probability $P_{n\nu J}$	
$J(\nu)$	Analytic	Numerical	$J(\nu)$	Analytic	Numerical
3/2(3)	31.37	31.33	9/2(3)	3.67	3.59
5/2(3)	27.85	27.64	11/2(3)	0.07	0.07
7/2(1)	9.03	9.11	15/2(3)	28.01	28.25

Table 2 compares the probabilities for the different states to become the ground state as calculated in two approaches. One column gives the analytic results obtained from (6) while the next column lists numerical results obtained from 50000 runs with random matrix elements with a Gaussian distribution.

GENERALIZATION TO CONTINUOUS SPECTRA

The notions exemplified in the previous section can be generalized in several ways. The first is towards energies that depend on a set of continuous variables $\{t_1, \ldots, t_q\}$ as follows [compare with Eq. (2)]:

$$E(t_1, \ldots, t_q) = \sum_L b_L(t_1, \ldots, t_q) G_L. \tag{7}$$

The analogous problem now consists in the determination of the probability *density* $dP(t_1, \ldots, t_q)$ for obtaining the lowest energy at $\{t_1, \ldots, t_q\}$ with random interactions G_L. We assume by way of example that the number of variables q is one less than the number N of interactions G_L, $q = N - 1$. In that case Eq. (7) represents a q-dimensional hypersurface Σ^q embedded in a $(q+1)$-dimensional Euclidean space $\mathbf{E}^{q+1}$ with a metric that follows from the covariance matrix $\langle G_L G_{L'} \rangle = \delta_{LL'}$. Let us suppose that Σ^q is a closed orientable manifold [13]. It can then be shown that the probability density is

$$dP(t_1, \ldots, t_q) = \frac{1}{S_q} K_q(t_1, \ldots, t_q) dv, \tag{8}$$

where K_q is the *Gaussian curvature* of Σ^q, S_q is the volume of a unit hypersphere,

$$S_q = \frac{2^{(q+2)/2} \pi^{q/2}}{(q-1)!!}, \quad q \text{ even}, \qquad S_q = \frac{2\pi^{(q+1)/2}}{((q-1)/2)!}, \quad q \text{ odd}, \tag{9}$$

and dv denotes an infinitesimal element of Σ^q.

In the simplest example of one parameter $t_1 \equiv t$ and two interactions G_1 and G_2, the energy is parametrized as a curve in a plane, $E(t) = x(t)G_1 + y(t)G_2$. The condition that the energy is minimal at a certain value t_0 is expressed by $\partial E/\partial t|_{t=t_0} = 0$, and from it one deduces the probability density

$$dP(t) = \frac{1}{2\pi} \frac{x''y' - y''x'}{x'^2 + y'^2} dt = \frac{1}{2\pi} \frac{x''y' - y''x'}{(x'^2 + y'^2)^{3/2}} \sqrt{x'^2 + y'^2} dt \equiv \frac{1}{2\pi} K_1(t) ds, \tag{10}$$

185

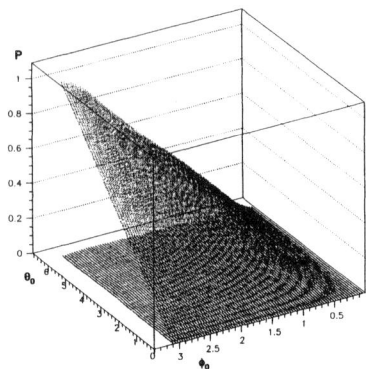

FIGURE 3. The probability $P(\theta_0, \phi_0)$ associated with a two-dimensional ellipsoidal surface in $\mathbf{E}^3$. The curved surfaces are calculated numerically and from Eq. (11) and are indistinguishable; the flat surface shows the difference between them.

in terms of the curvature $K_1(t)$ and an infinitesimal segment ds of the curve. In fact, the formula (10) is also valid for piecewise curves and precisely leads to the result (5) for a polygon.

The validity of the result (8) can be checked by comparing the probability obtained by integration of (8) over a part of Σ^q to the numerically calculated one. As an example we discuss a two-dimensional ellipsoid in $\mathbf{E}^3$ with one semi-axis c different from the others a. The ellipsoid is defined through $E(\theta, \phi) = x(\theta, \phi)G_x + y(\theta, \phi)G_y + z(\theta, \phi)G_z$ with $x(\theta, \phi) = a \cos\theta \sin\phi$, $y(\theta, \phi) = a \sin\theta \sin\phi$, and $z(\theta, \phi) = a \cos\theta$. The probability associated with the part of the surface with spherical coordinates (θ, ϕ) satisfying $0 \leq \theta \leq \theta_0$ and $0 \leq \phi \leq \phi_0$ is given by

$$P(\theta_0, \phi_0) = \frac{1}{4\pi} \int\limits_0^{\theta_0} \int\limits_0^{\phi_0} K_2(\theta, \phi)dv = \frac{\theta_0}{4\pi}\left(1 - \frac{a\cos\phi_0}{\sqrt{c^2 + (a^2 - c^2)(\cos\phi_0)^2}}\right). \tag{11}$$

Figure 3 compares this expression with the numerically calculated probability; the difference is seen to be close to zero. We have analyzed likewise the case of a three-dimensional hyperellipsoid in $\mathbf{E}^4$, showing that our approach can be generalized to higher dimensions.

These results also open up the possibility for an extension to higher-dimensional polytopes, by replacing the right-hand side of Eq. (8) with an appropriate characteristic for each vertex of the polytope. Indeed, it can be shown that the probabilities $P_v^{(m)}$ in (5) and (6) are related to the *exterior angle* at vertex v [14, 15] of either a convex polygon ($m = 2$) or a convex polyhedron ($m = 3$).

186

CONCLUDING REMARKS

We believe that, although derived for a restricted form of interaction Hamiltonians, these results suggest that generic n-body quantum systems, interacting through two-body forces, can be associated with a geometrical shape defined in terms of CFPs. Geometry thus arises as a consequence of strong correlations implicit in such systems and is independent of particular two-body interactions. Random tests can be understood in this context as sampling experiments on this geometry. In this contribution we have shown that geometric aspects of n-body systems determine some of their essential characteristics. In particular, for diagonal Hamiltonians surface curvature defines the probability to be the ground state. Other correlations could also be related to geometric features. These results generalize and put onto a firm basis previous work which hinted at a purely geometric interpretation of randomly interacting boson systems [6, 7]. Further work is required to fully explore this geometry and its consequences for our understanding of n-body dynamics.

ACKNOWLEDGMENTS

It is a great pleasure to dedicate this contribution to Rick Casten. The elder authors of this contribution have known Rick for more than 20 years and have always appreciated (and will continue to do so) his inspiring physics discussions and talks, his boundless energy, and his friendship.

AF is supported by CONACyT, Mexico, under project 32397-E. NAS thanks L. Vanhecke for a helpful discussion.

REFERENCES

1. Casten, R.F., *Nuclear Structure From a Simple Perspective* (Oxford University Press, Oxford, 2000).
2. Johnson, C.W., Bertsch, G.F., and Dean, D.J., *Phys. Rev. Letters* **80**, 2749 (1998);
   Johnson, C.W., Bertsch, G.F., Dean, D.J., and Talmi, I., *Phys. Rev. C* **61**, 014311 (2000).
3. Bijker, R., Frank, A., and Pittel, S., *Phys. Rev. C* **60**, 021302 (1999);
   Velazquez, V., and Zuker, A.P., *Phys. Rev. Letters* **88**, 072502 (2002).
4. Zhao, Y.M., and Arima, A., *Phys. Rev. C* **64**, 04131 (2001).
5. Zhao, Y.M., Arima, A., and Yoshinaga, N., nucl-th/0112075;
   Mulhall, D.M., Volya, A., and Zelevinsky, V., *Phys. Rev. Letters* **85**, 4016 (2000).
6. Bijker, R., and Frank, A., *Phys. Rev. Letters* **84**, 420 (2000); *Phys. Rev. C* **64**, 061303 (2001); *Nucl. Phys. News* Vol 11, **20** (2001).
7. Kusnezov, D., *Phys. Rev. Letters* **85**, 3773 (2000);
   Santos, L.F., Kusnezov, D., and Jacquod, Ph., nucl-th/0201049.
8. Talmi, I., *Simple Models of Complex Nuclei* (Harwood, Chur, Switzerland, 1993).
9. Brink, D.M., De Toledo Piza, A.F.R., and Kerman, A.K., *Phys. Letters* **19**, 413 (1965).
10. Arima, A., and Iachello, F., *Ann. Phys. (NY)* **99**, 253 (1976).
11. Castaños, O., Frank, F., and Moshinsky, M., *J. Math. Phys.* **19**, 1781 (1978).
12. Chau Huu-Tai, P., Frank, A., Smirnova, N., and Van Isacker, P., to be published.
13. Kobayashi, Sh., and Nomizu, K., *Foundations of Differential Geometry* (Wiley, New York, 1969).
14. Grünbaum, B., *Convex Polytopes* (Wiley, New York, 1967).
15. Banchoff, T., *J. Diff. Geom.* **1**, 245 (1967).

Generalized Partial Dynamical Symmetries in Nuclear Spectroscopy

A. Leviatan

Racah Institute of Physics, The Hebrew University, Jerusalem 91904, Israel

Abstract. Explicit forms of IBM Hamiltonians with a generalized partial dynamical O(6) symmetry are presented and compared with empirical data in ^{162}Dy.

A dynamical symmetry corresponds to a situation in which the Hamiltonian is written in terms of the Casimir operators of a chain of nested algebras

$$G_1 \supset G_2 \supset \ldots \supset G_n , \tag{1}$$

and has the following properties. (i) Solvability. (ii) Quantum numbers related to irreducible representations (irreps) of the algebras in the chain. (iii) Symmetry-dictated structure of wave functions independent of the Hamiltonian's parameters. The merits of a dynamical symmetry are self-evident, however, in most applications to realistic systems, one is compelled to break it. Partial dynamical symmetry (PDS) corresponds to a particular symmetry breaking for which some (but not all) of the above virtues of a dynamical symmetry are retained. Two types of partial symmetries were encountered so far. The first type correspond to a situation for which **part** of the states preserve **all** the dynamical symmetry. This is the case for the SU(3) PDS found in the IBM-1 [1, 2] and the Symplectic Shell Model [3, 4], and for the F-spin PDS in the IBM-2 [5]. The corresponding PDS Hamiltonians have a subset of solvable states with good symmetry while other eigenstates are mixed. A second type of partial symmetries correspond to a situation for which **all** the states preserve **part** of the dynamical symmetry. This occurs, for example, when the Hamiltonian preserves only some of the symmetries G_i in the chain (1) and only their irreps are unmixed [6, 7]. In this case there are no analytic solutions, yet selected quantum numbers (of the conserved symmetries) are retained. In the present contribution we show that it is possible to combine both types of partial symmetries, namely, to construct a Hamiltonian for which **part** of the states have **part** of the dynamical symmetry. We refer to such a structure as a generalized partial dynamical symmetry [8].

Partial symmetry of the second kind was recently considered in [7] in relation to the chain

$$U(6) \supset O(6) \supset O(5) \supset O(3) . \tag{2}$$

The Hamiltonian employed has two- and three-body interactions of the form

$$H_1 = \kappa_0 P_0^\dagger P_0 + \kappa_2 \left(\Pi^{(2)} \times \Pi^{(2)} \right)^{(2)} \cdot \Pi^{(2)} . \tag{3}$$

CP638, *Mapping the Triangle: International Conference on Nuclear Structure*
edited by A. Aprahamian, J. A. Cizewski, S. Pittel, and N. V. Zamfir
© 2002 American Institute of Physics 0-7354-0093-8/02/$19.00

The κ_0 term is the $O(6)$ pairing term defined in terms of monopole (s) and quadrupole (d) bosons, $P_0^\dagger = d^\dagger \cdot d^\dagger - (s^\dagger)^2$. It is diagonal in the dynamical symmetry basis $|[N], \sigma, \tau, L\rangle$ of Eq. (2) with eigenvalues $\kappa_0(N-\sigma)(N+\sigma+4)$. The κ_2 term is composed only of the $O(6)$ generator: $\Pi^{(2)} = d^\dagger s + s^\dagger \tilde{d}$, which is not a generator of $O(5)$. Consequently, H_1 cannot connect different $O(6)$ irreps but can induce $O(5)$ mixing. The eigenstates have good σ but not good τ quantum numbers.

To consider a generalized $O(6)$ PDS, we introduce the following IBM-1 Hamiltonian,

$$H_2 = h_0 P_0^\dagger P_0 + h_2 P_2^\dagger \cdot \tilde{P}_2 . \tag{4}$$

The h_0 term is identical to the κ_0 term of Eq. (3), and the h_2 term is defined in terms of the boson pair $P_{2,\mu}^\dagger = \sqrt{2} s^\dagger d_\mu^\dagger + \sqrt{7}(d^\dagger d^\dagger)_\mu^{(2)}$ with $\tilde{P}_{2,\mu} = (-)^\mu P_{2,-\mu}$. The latter term can induce both $O(6)$ and $O(5)$ mixing. Although H_2 is not an $O(6)$ scalar, it has an exactly solvable ground band with good $O(6)$ symmetry. This arises from the fact that the $O(6)$ intrinsic state for the ground band

$$|c; N\rangle = (N!)^{-1/2}(b_c^\dagger)^N|0\rangle , \quad b_c^\dagger = (d_0^\dagger + s^\dagger)/\sqrt{2} , \tag{5}$$

has $\sigma = N$ and is an exact zero energy eigenstate of H_2. Since H_2 is rotational invariant, states of good angular momentum L projected from $|c; N\rangle$ are also zero-energy eigenstates of H_2 with good $O(6)$ symmetry, and form the ground band of H_2. It follows that H_2 has a subset of solvable states with good $O(6)$ symmetry $(\sigma = N)$, which is not preserved by other states. All eigenstates of H_2 break the $O(5)$ symmetry but preserve the $O(3)$ symmetry. These are precisely the required features of a generalized partial dynamical symmetry as defined above for the chain of Eq. (2).

In Fig. 1 we show the experimental spectrum of ^{162}Dy and compare with the calculated spectra of H_1 and H_2. The spectra display rotational bands of an axially-deformed nucleus, in particular, a ground band $(K = 0_1)$ and excited $K = 2_1$ and $K = 0_2$ bands. An $L \cdot L$ term was added to both Hamiltonians, which contributes to the rotational splitting but has no effect on wave functions. The parameters were chosen to reproduce the excitation energies of the $2^+_{K=0_1}$, $2^+_{K=2_1}$ and $0^+_{K=0_2}$ levels. The $O(6)$ decomposition of selected bands is shown in Fig. 2. For H_2, the solvable $K = 0_1$ ground band has $\sigma = N$ and exhibits an exact $L(L+1)$ splitting. The $K = 2_1$ band is almost pure with only 0.15% admixture of $\sigma = N - 2$ into the dominant $\sigma = N$ component. The $K = 0_2$ band has components with $\sigma = N\,(85.50\%)$, $\sigma = N - 2\,(14.45\%)$, and $\sigma = N - 4\,(0.05\%)$. Higher bands exhibit stronger mixing, e.g., the $K = 2_3$ band shown in Fig. 2, has components with $\sigma = N\,(50.36\%)$, $\sigma = N - 2\,(49.25\%)$, $\sigma = N - 4\,(0.38\%)$, and $\sigma = N - 6\,(0.01\%)$. The $O(6)$ mixing in excited bands of H_2 depends critically on the ratio h_2/h_0 in Eq. (4) or equivalently on the ratio of the $K = 0_2$ and $K = 2_1$ bandhead energies. In contrast, all bands of H_1 are pure with respect to $O(6)$. Specifically, the $K = 0_1, 2_1, 2_3$ bands shown in Fig. 2 have $\sigma = N$ and the $K = 0_2$ band has $\sigma = N - 2$. In this case the diagonal κ_0 term in Eq. (3) simply shifts each band as a whole in accord with its σ assignment. All eigenstates of both H_1 and H_2 are mixed with respect to $O(5)$.

To gain more insight into the underlying band structure of H_2 we perform a band-mixing calculation by taking its matrix elements between large-N intrinsic states. The

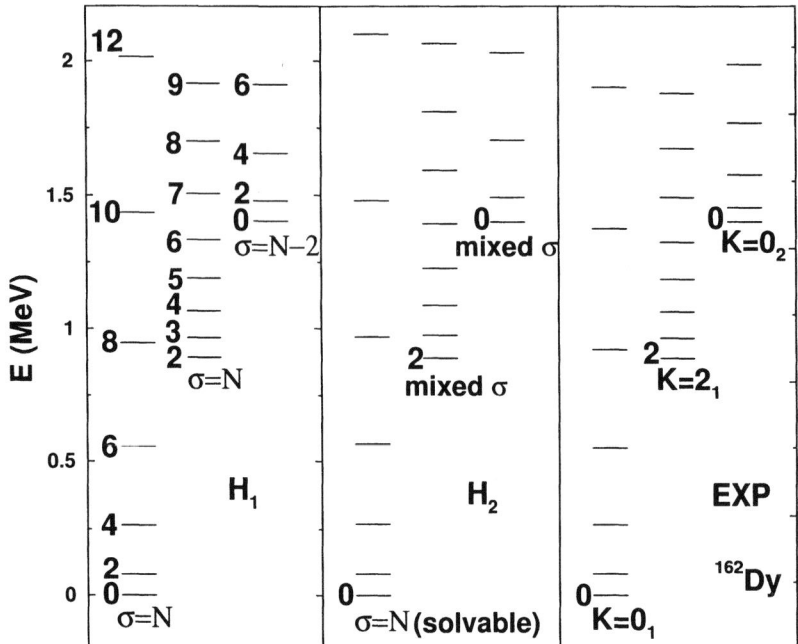

FIGURE 1. Experimental spectra (EXP) of ^{162}Dy [9,10] compared with calculated spectra of $H_1 + \lambda_1 L \cdot L$, Eq. (3), and $H_2 + \lambda_2 L \cdot L$, Eq. (4), with parameters $\kappa_0 = 8$, $\kappa_2 = 1.364$, $\lambda_1 = 8$, and $h_0 = 28.5$, $h_2 = 6.3$, $\lambda_2 = 13.45$ keV and $N = 15$.

latter are obtained in the usual way by replacing a condensate boson in $|c; N\rangle$ (5) with orthogonal bosons $b_\beta^\dagger = (d_0^\dagger - s^\dagger)/\sqrt{2}$ and $d_{\pm 2}^\dagger$ representing β and γ excitations respectively. By construction, the intrinsic state for the ground band of H_2, $|K = 0_1\rangle = |c; N\rangle$, is decoupled. For the lowest excited bands we find

$$\begin{aligned} |K = 0_2\rangle &= A_\beta |\beta\rangle + A_{\gamma^2} |\gamma_{K=0}^2\rangle + A_{\beta^2} |\beta^2\rangle , \\ |K = 2_1\rangle &= A_\gamma |\gamma\rangle + A_{\beta\gamma} |\beta\gamma\rangle . \end{aligned} \tag{6}$$

Using the parameters of H_2 relevant to ^{162}Dy (see Fig. 1) we obtain that the $K = 0_2$ band is composed of 36.29% β, 63.68% $\gamma_{K=0}^2$ and 0.03% β^2 modes, *i.e.*, it is dominantly a double-gamma phonon excitation with significant single-β phonon admixture. The $K = 2_1$ band is composed of 99.85% γ and 0.15% $\beta\gamma$ modes, *i.e.* it is an almost pure single-gamma phonon band. An $O(6)$ decomposition of the intrinsic states in Eq. (6) shows that the $K = 0_2$ intrinsic state has components with $\sigma = N$ (86.72%), $\sigma = N - 2$ (13.26%) and $\sigma = N - 4$ (0.02%). The $K = 2_1$ intrinsic state has $\sigma = N$ (99.88%) and $\sigma = N - 2$ (0.12%). These estimates are in good agreement with the exact results mentioned above in relation to Fig. 2.

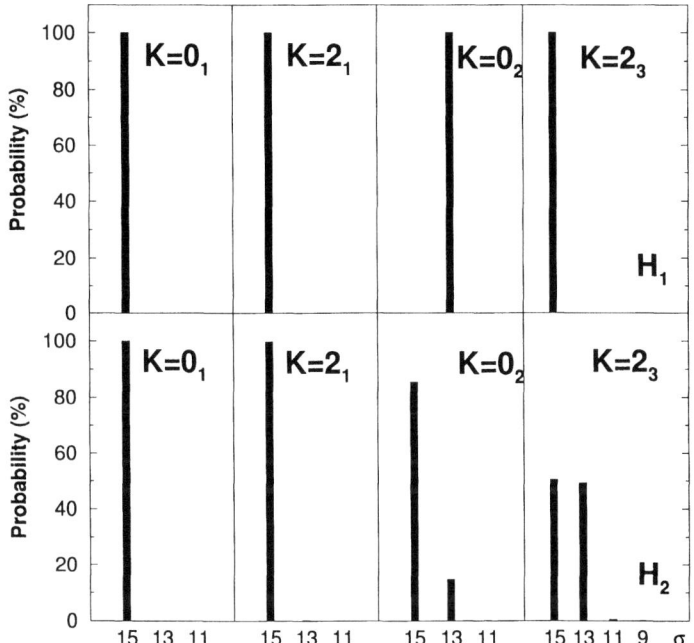

FIGURE 2. $O(6)$ decomposition of wave functions of the $K = 0_1, 2_1, 0_2, 2_3$ bands for H_1 (upper portion) and H_2 (lower portion).

In Table 1 we compare the presently known experimental $B(E2)$ values for transitions in ^{162}Dy with the values predicted by H_1 and H_2 using the $E2$ operator

$$T^{(2)} = e\left[\Pi^{(2)} + \chi\,(d^\dagger\tilde{d})^{(2)}\right]. \qquad (7)$$

The parameters e and χ in Eq. (7) were fixed for each Hamiltonian by the empirical $2^+_{K=0_1} \to 0^+_{K=0_1}$ and $2^+_{K=2_1} \to 0^+_{K=0_1}$ $E2$ rates. The $B(E2)$ values predicted by H_1 and H_2 for $K = 0_1 \to K = 0_1$ and $K = 2_1 \to K = 0_1$ transitions are very similar and agree well with the measured values. On the other hand, their predictions for interband transitions from the $K = 0_2$ band are very different. For H_1, the $K = 0_2 \to K = 0_1$ and $K = 0_2 \to K = 2_1$ transitions are comparable and weaker than $K = 2_1 \to K = 0_1$. This can be understood if we recall the $O(6)$ assignments for the bands of H_1: $K = 0_1, 2_1\,(\sigma = N)$, $K = 0_2\,(\sigma = N - 2)$, and the $E2$ selection rules of $\Pi^{(2)}\,(\Delta\sigma = 0)$ and $(d^\dagger\tilde{d})^{(2)}\,(\Delta\sigma = 0 \pm 2)$, which imply that in this case only the $(d^\dagger\tilde{d})^{(2)}$ term contributes to interband transitions from the $K = 0_1$ band. In contrast, for H_2, $K = 0_2 \to K = 2_1$ and $K = 2_1 \to K = 0_1$ transitions are comparable and stronger than $K = 0_2 \to K = 0_1$. This behaviour is due to the underlying band structure discussed above, and the fact that $\langle K = 0_2|\Pi^{(2)}_0|K = 0_1\rangle = 0$, while both terms in Eq. (7) contribute to $\Delta K = 2$ interband $E2$ intrinsic matrix elements. Recently the $B(E2)$ ratios $R_1 = \dfrac{B(E2; 0^+_{K=0_2} \to 2^+_{K=2_1})}{B(E2; 0^+_{K=0_2} \to 2^+_{K=0_1})} = 10(5)$

TABLE 1. Calculated and observed [10,11] $B(E2)$ values (e^2b^2) for ^{162}Dy. The $E2$ parameters in Eq. (7) are $e = 0.138\ (0.126)\ eb$ and $\chi = -0.22\ (-0.55)$ for $H_1\ (H_2)$.

Transition	H_1	H_2	Expt.	Transition	H_1	H_2	Expt.
$2^+_{K=0_1} \to 0^+_{K=0_1}$	1.06	1.05	1.07(2)	$2^+_{K=2_1} \to 0^+_{K=0_1}$	0.024	0.024	0.024(1)
$4^+_{K=0_1} \to 2^+_{K=0_1}$	1.50	1.49	1.51(6)	$2^+_{K=2_1} \to 2^+_{K=0_1}$	0.038	0.0395	0.042(2)
$6^+_{K=0_1} \to 4^+_{K=0_1}$	1.62	1.61	1.57(9)	$2^+_{K=2_1} \to 4^+_{K=0_1}$	0.0025	0.0026	0.0030(2)
$8^+_{K=0_1} \to 6^+_{K=0_1}$	1.65	1.65	1.82(9)	$3^+_{K=2_1} \to 2^+_{K=0_1}$	0.0428	0.0425	
$10^+_{K=0_1} \to 8^+_{K=0_1}$	1.63	1.64	1.83(12)	$3^+_{K=2_1} \to 4^+_{K=0_1}$	0.022	0.023	
$12^+_{K=0_1} \to 10^+_{K=0_1}$	1.58	1.60	1.68(21)	$4^+_{K=2_1} \to 2^+_{K=0_1}$	0.0123	0.0114	0.0091(5)
				$4^+_{K=2_1} \to 4^+_{K=0_1}$	0.046	0.047	0.044(3)
$0^+_{K=0_2} \to 2^+_{K=0_1}$	0.0014	0.0022		$4^+_{K=2_1} \to 6^+_{K=0_1}$	0.0061	0.0061	0.0063(4)
$0^+_{K=0_2} \to 2^+_{K=2_1}$	0.0012	0.1707		$5^+_{K=2_1} \to 4^+_{K=0_1}$	0.0345	0.033	0.033(2)
$2^+_{K=0_2} \to 0^+_{K=0_1}$	0.0002	0.0004		$5^+_{K=2_1} \to 6^+_{K=0_1}$	0.029	0.031	0.040(2)
$2^+_{K=0_2} \to 2^+_{K=0_1}$	0.0003	0.0005		$6^+_{K=2_1} \to 4^+_{K=0_1}$	0.0085	0.0071	0.0063(4)
$2^+_{K=0_2} \to 2^+_{K=2_1}$	0.0003	0.0365		$6^+_{K=2_1} \to 6^+_{K=0_1}$	0.046	0.047	0.050(4)

and $R_2 = \dfrac{B(E2; 2^+_{K=0_2} \to 4^+_{K=0_1})}{B(E2; 2^+_{K=0_2} \to 0^+_{K=0_1})} = 65(28)$ have been measured [9]. The corresponding predictions are $R_1 = 0.86$, $R_2 = 4.00$ for H_1 and $R_1 = 77.59$, $R_2 = 3.25$ for H_2. As noted in [9], the empirical value of R_2 deviates 'beyond reasonable expectations' from the Alaga rule value $R_2 = 2.6$. A measurement of absolute B(E2) values for these transitions is highly desirable to clarify the origin of these discrepancies.

It is a pleasure to dedicate this article to Rick Casten on the occasion of his 60th birthday, and thank him for many years of illuminating discussions. This work was done in collaboration with P. Van Isacker (GANIL) and was supported in part by the Israel Science Foundation.

REFERENCES

1. Leviatan, A., *Phys. Rev. Lett.*, **77**, 818 (1996).
2. Leviatan, A., and Sinai, I., *Phys. Rev. C*, **60**, 061301 (1999).
3. Escher, J., and Leviatan, A., *Phys. Rev. Lett.*, **84**, 1866 (2000).
4. Escher, J., and Leviatan, A., *Phys. Rev. C*, **65**, 054309 (2002).
5. Leviatan, A., and Ginocchio, J. N., *Phys. Rev. C*, **61**, 024305 (2000).
6. Talmi, I., *Phys. Lett. B*, **405**, 1 (1997).
7. Van Isacker, P., *Phys. Rev. Lett.*, **83**, 4269 (1999).
8. Leviatan, A., and Van Isacker, P., submitted for publication (2002).
9. Zamfir, N. V., *et al.*, *Phys. Rev. C*, **60**, 054319 (1999).
10. Helmer R. G., and Reich, C. W., *Nucl. Data Sheets*, **87**, 317 (1999).
11. Warner, D. D., *et. al.*, in *Proc. 6th Conf. on Capture Gamma-Ray Spectroscopy*, edited by K. Abrahams and P. Van Assche, Institute of Physics, Bristol, 1988, p. 562.

Pseudospin Symmetry: A Relativistic Symmetry in Nuclei

Joseph N. Ginocchio

MS B283, Los Alamos National Laboratory, NM 87545, USA

Abstract. We briefly review the evidence that pseudospin symmetry is an SU(2) symmetry of the Dirac Hamiltonian.

INTRODUCTION

I first met Rick in 1978 in Erice at the first "Symposium on Interacting Bosons in Nuclear Physics". I have always been impressed with his efforts to discover empirical evidence for the interacting boson model and, now, for critical points. I only wish he would do the same for pseudospin symmetry in nuclei.

The spherical shell model orbitals with non-relativistic quantum numbers (n_r, ℓ, $j = \ell + 1/2$) and ($n_r - 1, \ell + 2, j' = j + 1 = \ell + 3/2$), where n_r, ℓ, and j are the single-nucleon radial, orbital, and total angular momentum quantum numbers, respectively, were observed to be quasi-degenerate in medium weight and heavy nuclei [1, 2]. This doublet structure is expressed in terms of a "pseudo" orbital angular momentum $\tilde{\ell} = \ell + 1$, the average of the orbital angular momentum of the two states in doublet, and "pseudo" spin, $\tilde{s} = 1/2$. For example, $(n_r s_{1/2}, (n_r - 1)d_{3/2})$ will have $\tilde{\ell} = 1$, $(n_r p_{3/2}, (n_r - 1)f_{5/2})$ will have $\tilde{\ell} = 2$, etc. Since $j = \tilde{\ell} - \frac{1}{2}$ and $j' = \tilde{\ell} + \frac{1}{2}$, the energy of the two states in the doublet are then appproximately independent of the orientation of the pseudospin.

Pseudospin "symmetry" was shown to exist in deformed nuclei as well [3] and has been used to explain features of deformed nuclei, including superdeformation and identical bands [4, 5, 6, 7].

However, the origin of pseudospin symmetry remained a mystery and "no deeper understanding of the origin of these (approximate) degeneracies" existed [8]. In this paper we shall review more recent developments that show that pseudospin symmetry is a relativistic symmetry [9, 10, 11].

SYMMETRIES OF THE DIRAC HAMILTONIAN

The success of the shell model implies that nucleons move in a mean field produced by the interactions between the nucleons. Normally, it suffices to use the Schrodinger equation to describe the motion of the nucleons in this mean field. However, in order to understand the origin of pseudospin symmetry, we need to take into account the motion

CP638, *Mapping the Triangle: International Conference on Nuclear Structure*
edited by A. Aprahamian, J. A. Cizewski, S. Pittel, and N. V. Zamfir
© 2002 American Institute of Physics 0-7354-0093-8/02/$19.00

of the nucleons in a relativistic mean field and thus use the Dirac equation. The Dirac Hamiltonian, H, with an external scalar, $V_S(\vec{r})$, and vector, $V_V(\vec{r})$, potentials is given by:

$$H = \vec{\alpha} \cdot \vec{p} + \beta(M + V_S(\vec{r})) + V_V(\vec{r}), \tag{1}$$

where we have set $\hbar = c = 1$, $\vec{\alpha}$, β are the usual Dirac matrices, M is the nucleon mass, and $\vec{p}$ is the three momentum. The Dirac Hamiltonian is invariant under an SU(2) algebra for two limits: $V_S(\vec{r}) - V_V(\vec{r}) = $ constant and $V_S(\vec{r}) + V_V(\vec{r}) = $ constant [12]. The first condition leads to a spin symmetry which is relevant for mesons [12, 13] while the second leads to pseudospin symmetry [10].

The generators for the SU(2) pseudospin symmetry, $\hat{\tilde{S}}_i$, which commute with the Dirac Hamiltonian, $[H, \hat{\tilde{S}}_i] = 0$, are given by $\hat{\tilde{S}}_i = \begin{pmatrix} \hat{\tilde{s}}_i & 0 \\ 0 & \hat{s}_i \end{pmatrix}$ where $\hat{s}_i$ is the usual spin operator, $\hat{\tilde{s}}_i = U_p \, \hat{s}_i \, U_p$ and $U_p = 2\frac{\hat{s} \cdot \vec{p}}{p}$ is the unitary helicity operator [14]. These pseudospin generators have the spin operator $\hat{s}_i$ operating on the lower component of the Dirac wave function which has the consequence that the spatial wavefunctions for the two states in the pseudospin doublet are identical to within an overall phase.

This symmetry for $V_S(\vec{r}) + V_V(\vec{r}) = $ constant is general and applies to spherical, axially deformed, triaxial, gamma unstable, etc., nuclei. In the case for which the potentials are spherically symmetric, the Dirac Hamiltonian conserves the pseudo-orbital angular momentum, the generators of which are, $\hat{\tilde{L}}_i = \begin{pmatrix} \hat{\tilde{\ell}}_i & 0 \\ 0 & \hat{\ell}_i \end{pmatrix}$, where $\hat{\tilde{\ell}}_i = U_p \, \hat{\ell}_i \, U_p$, $\hat{\ell}_i = \vec{r} \times \vec{p}$. Since U_p conserves the total angular momentum but $\vec{p}$ changes the orbital angular momentum by one unit because of parity conservation, if the lower component of the Dirac wave function orbital angular momentum is $\tilde{\ell}$, the upper component also has total angular momentum j, but orbital angular momentum $\ell = \tilde{\ell} \pm 1$. If $j = \tilde{\ell} + 1/2$, then it follows that $\ell = \tilde{\ell} + 1$, whereas if $j = \tilde{\ell} - 1/2$, then $\ell = \tilde{\ell} - 1$. This agrees with the pseudospin doublets originally observed [1, 2] and discussed at the beginning of this paper. This relativistic interpretation also gives the physical significance of the pseudo-orbital angular momentum $\tilde{\ell}$ as the "orbital angular momentum" of the lower component.

For axially symmetric deformed nuclei, there is a U(1) generator corresponding to the pseudo-orbital angular momentum projection along the body-fixed symmetry axis which is conserved in addition to the pseudospin, $\hat{\tilde{\Lambda}} = \begin{pmatrix} \hat{\tilde{\Lambda}} & 0 \\ 0 & \hat{\Lambda} \end{pmatrix}$, where $\hat{\tilde{\Lambda}} = U_p \, \hat{\Lambda} \, U_p$.

The eigenfunctions of the Dirac Hamiltonian, $H\Phi_{\tau,\tilde{\mu}} = E_\tau \Phi_{\tau,\tilde{\mu}}$, are doublets ($\tilde{S} = 1/2$, $\tilde{\mu} = \pm 1/2$) with respect to the SU(2) generators $\hat{\tilde{S}}_i$

$$\hat{\tilde{S}}_z \Phi_{\tau,\tilde{\mu}} = \tilde{\mu} \Phi_{\tau,\tilde{\mu}} , \hat{\tilde{S}}_\pm \Phi_{\tau,\tilde{\mu}} = \sqrt{(1/2 \mp \tilde{\mu})(3/2 \pm \tilde{\mu})} \, \Phi_{\tau,\tilde{\mu}\pm 1} , \tag{2}$$

where $\hat{\tilde{S}}_\pm = \hat{\tilde{S}}_x \pm i\hat{\tilde{S}}_y$. The eigenvalue τ refers to the other necessary quantum numbers.

However, the exact symmetry limit can not be realized in nuclei, because, if $V_S(\vec{r}) + V_V(\vec{r}) = $ constant, there are no Dirac bound valence states and hence nuclei can not exist [15, 9, 16].

194

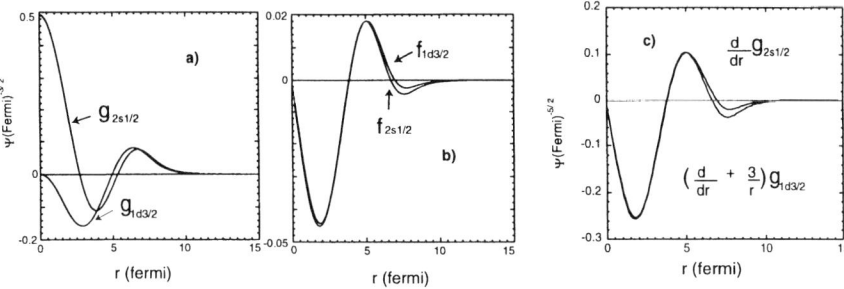

FIGURE 1. The a) upper, g, and b) lower components, f, of the $(2s_{1/2}, 1d_{3/2})$ Dirac eigenfunctions in ^{208}Pb, and c) the derivative relations between upper components, g.

REALISTIC RELATIVISTIC MEAN FIELDS AND QCD SUM RULES

A near equality in the magnitude of mean fields, $V_S(\vec{r}) \approx -V_V(\vec{r})$, is a universal feature of the relativistic mean field approximation (RMF) of relativistic field theories with interacting nucleons and mesons [17] and relativistic theories with nucleons interacting via zero range interactions [18]. Recently realistic relativistic mean fields were shown to exhibit approximate pseudospin symmetry in both the energy spectra and wave functions [19, 20, 21]. The pseudospin-orbit splittings calculated in the RMF [19]in the spherical nucleus ^{208}Pb and are larger than the measured splittings which demonstrates that the pseudospin symmetry is better conserved experimentally than mean field theory would suggest.

Applying QCD sum rules in nuclear matter [22], the ratio of the scalar and vector self-energies were determined to be $\frac{V_S(\vec{r})}{V_V(\vec{r})} \approx -\frac{\sigma_N}{8m_q}$ where σ_N is the sigma term which can be measured in pion-nucleon scattering and m_q is the average current quark mass in the nucleon. For reasonable values of σ_N and quark masses, this ratio is approximately -1.125, uncannily close to the ratio of the central potential depths of realistic mean fields. The implication of these results is that pseudospin symmetry may have a more fundamental basis rooted in QCD.

THE EIGENFUNCTIONS OF THE PSEUDOSPIN PARTNERS

Since the pseudospin generators connect the pseudospin partners (see Eq. (2)), pseudospin symmetry implies relationships between the doublet wavefunctions. From the fact that the generators, $\tilde{\vec{S}}_i = \begin{pmatrix} \hat{\tilde{s}}_i & 0 \\ 0 & \hat{s}_i \end{pmatrix}$ have the spin operator $\hat{s}_i$ operating on the lower

195

component of the Dirac eigenfunction has the consequence that the spatial wavefunctions of the lower component of the Dirac wavefunctions will be equal in shape and magnitude for the two states in the doublet within an overall phase. In Fig. 2b, for example, the lower components of the $(2s_{1/2}, 1d_{3/2})$ Dirac eigenfunctions are approximately identical whereas the upper components have a different number of nodes [19, 16] (Fig. 2a). Deformed nuclei also demonstrate quasi-similarity in the lower components [23].

Because the lower components are small, in order to test the pseudospin symmetry prediction that the lower components are almost identical we must observe transitions for which the upper components are not dominant. Magnetic dipole and Gamow-Teller transtions between the states in the doublet are forbidden non-relativistically since the obital angular momentum of the two states differ by two units, but are allowed relativistically. Pseudospin symmetry predicts that, if the magnetic moments of the two states are known, the magnetic dipole transition between the states can be predicted. Likewise if the Gamow - Teller transitions between the states with the same quantum numbers are known, the transition between the states with different quantum numbers can be predicted [24, 25].

Since the upper components are related by $\hat{\tilde{s}}_i$, spin and space are intertwined and the relationship between the upper components is more complex. For spherical nuclei the upper component are related by

$$(\frac{\partial}{\partial r} + \frac{\tilde{\ell}+2}{r})g_{\tilde{n}_r,\tilde{\ell},\tilde{\ell}+\frac{1}{2}}(r) = (\frac{\partial}{\partial r} - \frac{\tilde{\ell}-1}{r})g_{\tilde{n}_r,\tilde{\ell},\tilde{\ell}-\frac{1}{2}}(r). \tag{3}$$

In Fig. 2c we plot the each side of the Eq. (3) for $\tilde{\ell} = 1$ and we find good agreement [26, 23].

SUMMARY AND FUTURE

We have shown that pseudospin symmetry is an SU(2) symmetry of the Dirac Hamiltonian for $V_S(\vec{r}) = -V_V(\vec{r}) + constant$. One prediction of this relativistic symmetry is that the radial wavefunctions of the lower components of the two eigenstates in the doublet are identical up to a phase. Relativistic mean field calculations show these lower components to be almost identical. Furthermore, since the realistic relativistic mean field calculations overestimate the energy splitting between doublets compared to their measured values, these same calculations may overestimate the difference between the experimental eigenfunctions. The upper components of the pseudospin partners satisfy the predicted differential relationships quite well [26].

This equality of the lower component radial wavefunction leads to a prediction for the "ℓ" forbidden magnetic dipole transitions, if the magnetic moments of the states in the doublet are known. Comparison with experimental transitions shows reasonable agreement. Similar relationships for Gamow-Teller transitions between states in the doublet hold as well. These relations need to be tested also.

Although not discussed in this short survey, pseudospin symmetry has been shown to be approximately conserved in medium energy nucleon-nucleus scattering from nuclei [27, 28] but badly broken in low energy nucleon-nucleus scattering [29]. The study of pseudospin symmetry in nucleon-nucleon scattering as a function of energy needs a generalization of the generators to include non-zero space components of the vector potential [30].

Pseudospin symmetry will be investigated more thoroughly in deformed nuclei where it is most prevalent [23]. A recent study [31] suggests that pseudospin symmetry will improve for neutron rich nuclei. This needs further investigation. Finally, pseudospin symmetry has been linked via QCD sum rules to chiral symmetry breaking in nuclei suggesting a more fundamental significance which needs to be explored.

ACKNOWLEDGMENTS

This work was supported by the United States Department of Energy under contract W-7405-ENG-36. A. Leviatan, D. G. Madland, and P. von Neumann - Cosel collaborated on different aspects of the work reported in this survey.

REFERENCES

1. Arima, A.,Harvey, M., and Shimizu, K., *Phys. Lett.* B **30**, 517 (1969).
2. Hecht, K. T., and Adler, A., *Nucl. Phys.* A **137**, 129 (1969).
3. Bohr, A., Hamamoto, I., and Mottelson, B. R., *Phys. Scr.* **26**, 267 (1982).
4. Nazarewicz, W., *et. al.*, *Phys. Rev. Lett.* **64**, 1654 (1990);
5. Stephens, F. S., *et. al.*, *Phys. Rev. Lett.* **65**, 301 (1990);
6. Stephens, F. S. *et. al.*, *Phys. Rev.* C **57**, R1565 (1998);
7. Bruce, A. M., *et. al.*, *Phys. Rev.* C **56**, 1438 (1997).
8. Mottelson, B., *Nucl. Phys.* A **522**, 1 (1991).
9. Ginocchio, J. N., *Phys. Rev. Lett.* **78**, 436 (1997).
10. Ginocchio, J. N. and Leviatan, A., *Phys. Lett. B* **425**, 1 (1998).
11. Ginocchio, J. N., *Phys. Rep.* **315**, 231 (1999).
12. Bell, J. S., and Ruegg, H., *Nucl. Phys.* **B98**, 151 (1975).
13. Page, P. R., Goldman, T., and Ginocchio, J. N., *Phys. Rev. Lett.* **86**, 204 (2001).
14. Blokhin, A. L., Bahri, C., and Draayer, J. P., *Phys. Rev. Lett.* **74**, 4149 (1995).
15. Su, R. K., and Ma, Z. Q., *J. Phys.* A **19**, 1739 (1986).
16. Leviatan, A. and Ginocchio, J. N., Physics Letters **B 518** 214 (2001).
17. Serot, B. D., and Walecka, J. D. in *Advances in Nuclear Physics*, edited by J. W. Negele and E. Vogt, Vol. **16**, New York, 1998.
18. Nikolaus, B. A., Hoch, T., and Madland, D. G., *Phys. Rev.* C **46**, 1757 (1992).
19. Ginocchio, J. N., and Madland, D. G., *Phys. Rev.* C **57**, 1167 (1998).
20. Lalazissis, G. A., *et. al.*, *Phys. Rev.* C **58**, R45 (1998).
21. Meng, J., *et. al.*, *Phys. Rev.* **58**, R628 (1998).
22. Cohen, T. D., Furnstahl, R. J., Griegel, K ., and Jin, X., *Prog. in Part. and Nucl. Phys.* **35**, 221 (1995).
23. Ginocchio, J. N., and Leviatan, A., in preparation, (2002).
24. Ginocchio, J. N., *Phys. Rev.* C **59**, 2487 (1999).
25. von Neumann-Cosel, P., and Ginocchio, J. N., *Phys. Rev.* C **62**, 014308 (2000).
26. Ginocchio, J. N., and Leviatan, A., Phys. Rev. Lett. **87**, 072502 (2001).
27. Ginocchio, J. N., *Phys. Rev. Lett.* **82**, 4599 (1999).
28. Leeb, H., and Wilmsen, S., *Phys. Rev.* C **62**, 024602 (2000).
29. Bowlin, J. B., Goldhaber, A. S., and Wilkin, C., *Zeitschrift für Physik A* **331**, 83 (1988).
30. Ginocchio, J. N., Phys. Rev. **C 65**, 054002 (2002).
31. Alberto, P., Fiolhais, M., Malheiro, M., Delfino, A., and Chiapparini, M., *Phys. Rev. Lett.* **86**, 5015 (2001).

Nuclear Excitation by Electronic Transition - NEET

J. A. Becker

Lawrence Livermore National Laboratory
Livermore, California 94550

Abstract. Experiments seeking to demonstrate nuclear excitation induced by synchrotron radiation have been enabled by the development of intense synchrotron radiation. The phenomena has been demonstrated in [197]Au, while realistic upper limits for [189]Os have been established. These new experiments report probabilities for NEET which are orders of magnitude below earlier experiments. A new experiment to measure atomic-nuclear mixing in [189]Os is described. The experimental claim of NEET in isomeric [178]Hf is not credible.

INTRODUCTION

Nuclear excitation by electronic transition (NEET) is a rare decay mode for excited atomic states resulting in nuclear excitation. (Atomic states ordinarily decay via x-ray emission and Auger emission.) NEET requirements include energy degeneracy between the atomic and nuclear states, and the same transition multipolarity between the states. The development on intense beams of synchrotron radiation has enabled experiments designed to measure the probability of NEET (P_{NEET}) induced by synchrotron radiation in nuclei where the NEET conditions are met. Rare as the phenomenon is, observation of NEET induced by synchrotron radiation has been reported by Kishimoto, et al. [1]. The focus of this manuscript is on describing the process, some recent experiments, and some ideas for future experiments.

WHAT is NEET?

Nuclear excitation through Electronic Excitation (NEET) was discussed by Morita in 1973 [2]. The two dominant decay processes for excited atomic states are x-ray emission and Auger electron emission. Exchange of a virtual photon induces NEET, a second order effect. The NEET probability is small, many orders of magnitude less than atomic excitation loss by x-ray emission.

Fig. 1 isolates (schematically) NEET and gives the relevant formula in a self-evident notation. The matrix element is similar to the (inverse) internal conversion matrix element. Important conditions for NEET to occur include:

- An overlap of the approximately degenerate states
- Common multipolarity of the atomic and nuclear transitions
- $\Gamma_1 > \Gamma_2$ (Widths of the initial and final atomic-hole states, respectively).

CP638, *Mapping the Triangle: International Conference on Nuclear Structure*
edited by A. Aprahamian, J. A. Cizewski, S. Pittel, and N. V. Zamfir
© 2002 American Institute of Physics 0-7354-0093-8/02/$19.00

The last condition allows NEET to compete with real photon emission. The width of the nuclear state is so small it is treated as zero in the expression for P_{NEET}. The formula in Fig. 1 is given in [3]. Ahmad, et al., [4] independently motivate a similar formula.

Kishimoto, et al., [1] have made a marvelous experiment and find that P_{NEET} is $(5.0 \pm 0.6) \times 10^{-8}$ of the K x-ray emission rate in ^{197}Au for the K $\rightarrow$ M_1 hole transition, while Ahmad, et al., [4] report an upper limit in ^{189}Os for the probability that a K-vacancy results in nuclear excitation of ^{189}Os ($E_x = 69.5$ keV), $P_{NEET} < 3 \times 10^{-10}$. These experiments take advantage of the intense monochromatic beams available from 3rd generation synchrotrons to prepare the ionized atom to optimize conditions for the observation of NEET. Both experiments are discussed below.

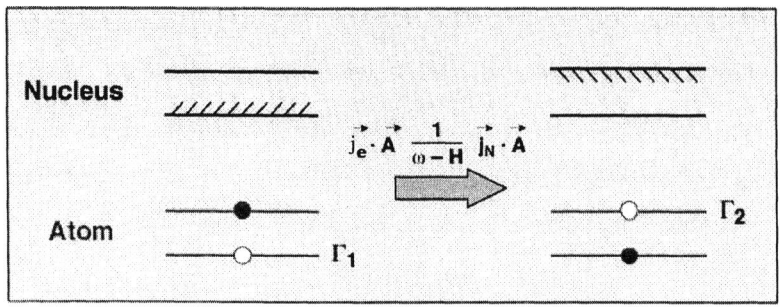

$$P_{NEET} = (1 + \frac{\Gamma_1}{\Gamma_2}) \times \frac{W^2}{\left(\delta^2 + \frac{(\Gamma_1 + \Gamma_2)^2}{4} \right)}$$

$$W^2 = 4\pi e^2 \; \frac{\omega_N^{2(L+1)}}{[(2L+1)!!]^2} \; <j_1 1/2 \, L0 | j_2 1/2> \; |M_L(\omega_N)|^2 \, B(ML)$$

FIGURE 1. Schematic of NEET. The expression for the probability of NEET is given.

RECENT NEET EXPERIMENTS

NEET in ^{197}Au

Kishimoto et al., [1] recently demonstrated the NEET effect in the nucleus ^{197}Au. ^{197}Au has an (accidental) near degeneracy between the energy of the nuclear $1/2^+$ state at $E_x = 77.351$ keV and the atomic K $\rightarrow$ M_1 hole transition ($1S_{1/2} \rightarrow 3S_{1/2}$), ($80.725 \rightarrow$

3.425) keV, ΔE_{ATOMIC} = 77.300. The energy difference in the atomic configuration is 53 eV, and the atomic and nuclear transitions can have the same multipolarity.

Therefore, two important conditions for NEET are clearly realized: a near energy degeneracy between the nuclear and atomic transitions (ΔE = 51 eV), and common multipolarity for the transitions. Kishimoto, et al., irradiated [197]Au with monochromatic photons (ΔE = 19(2) eV) at several energies. They measured photon absorption at the nuclear resonance energy 77.351 keV and also measured NEET at an incident photon energy E = 80.989 keV where they found the maximum yield. They report P_{NEET} = 5.0 (6) x 10^{-8} relative to the production of K-vacancies. Measurements of both resonance absorption and of NEET allowed Kishimoto, et al., to use ratios to remove uncertainties in their report of P_{NEET}, and in particular to avoid a determination of detection efficiency. (A calculation of the K-vacancy rate is however required.) The experimental signal of excitation of the nuclear state included measuring its decay with time, $t_{1/2}$ = 1.9 ns for the 77.35 keV state. The experimental result for NEET agrees with the recent calculations of Tkalya [5], who finds $P_{NEET} \sim 3.8 \times 10^{-8}$ and Harston [6], who finds $P_{NEET} \sim 3.6 \times 10^{-8}$.

NEET in [189]Os (I)

Ahmad, et al., [4] searched for NEET in [189]Os. They irradiated [189]Os with an intense monochromatic beam of photons delivered at the Advance Photon Source (APS), sited at Argonne National Laboratory. The incident energy was E = 98.74 keV, well above the Os K-edge at 73.9 keV. The atomic-nuclear degeneracy Ahmad, et al., attempted to take advantage of lies between the nuclear state at E_x = 69.537 keV (J^{Π} = 5/2$^-$) and the KM_I transition (70.822 keV, M1). (There are also nearby E2 atomic transitions where the energy overlap is not as good.) The NEET condition for energy degeneracy is not as well satisfied in this case as in the [197]Au case: ΔE (keV) = 70.822 - 69.537 = 1.285, while in [197]Au ΔE (keV) = 0.051. The experimental signal for NEET was not the direct decay of the 69.54 keV state, but rather the more easily observed and definitive decay of the 30.814 keV isomeric state in [189]Os [J^{π} = 9/2$^-$, $t_{1/2}$ $\sim$ 5.7(1) h], populated in the decay of the 69.54 keV state. Ahmad, et al. did not observe NEET, reporting an upper limit P_{NEET} < 9 × 10^{-10}, a limit improved in a subsequent experiment to P_{NEET} < 3 × 10^{-10} [7]. This result is consistent with the results of an independent experiment at the SPring-8 synchrotron radiation facility, where Aoki et al. [8], report P_{NEET} < 4.1 × 10^{-10}. These limits are orders of magnitude below earlier experimental efforts. (See *e.g.*, [4,8] for references to the earlier work.)

What is expected for P_{NEET} in these experiments on [189]Os? Ahmad, et al., make use of the expression they develop to calculate P_{NEET} (M1) = 1.3 × 10^{-10}, a value entirely consistent with their experimental result, and consistent with the measurement of Aoki, et al. The experimental upper limits are also consistent with two recent calculated values: P_{NEET} = 1.2 × 10^{-10} [5], and P_{NEET} = 1.1 x 10^{-10} [6].

NEET in ^{189}Os (II)

An experiment planned for late summer 2002 at LLNL takes advantage of the LLNL Electron Beam Ion Trap Facility (EBIT) [9]. Ionized ^{189}Os is prepared by bombardment with a variable energy electron beam and contained within the ion trap. The energy of the electron beam is carefully controlled and tuned so that the sum of the energies of the bombarding electron beam and the L-shell ionized ^{189}Os (a free-bound transition) adds up to the excitation of the nuclear ^{189}Os level at 216.6 keV. Trapped ions are periodically gathered up and counted. The signal is the energy and decay rate of the $J^\Pi = 9/2^-$, $E_x = 30.814$ keV, $t_{1/2} = 5.7$ h state, populated in the decay of the 216.6-keV nuclear state. Observation of this experimental signal of NEET and not direct observation of the decay of the 216.6 keV level improves confidence in any observed signal. Finally, since the energy degeneracy is accomplished by tuning the incident energy of the incident electron beam, observing an experimental signal while changing the electron-beam energy gives an opportunity for observing a resonance signal and improving confidence in the measurement.

NEET in ^{178}Hf?

Collins, et al., [10] have recently irradiated isomeric ^{178}Hf ($J^\Pi = 16^+$ $E_x = 2.4$ MeV, $t_{1/2} = 31$ y) at the SPring-8 facility. Incident monochromatic ($\Delta E = 0.5$ eV) x-ray energies were tuned between 9 and 13 keV, in steps of ~ 5 eV. Enhanced γ-ray decay of the isomer near the incident x-ray energies of 11.3, 11.7, and 9.56 keV are reported and ascribed to NEET, and a value of $P_{NEET} = 2 \times 10^{-3}$ relative to L-shell photo ionization in this energy region reported. Attribution to NEET of this signal (if real) is extremely unlikely. This large magnitude of this result is clearly orders of magnitude greater than any reasonable theoretical calculation would predict for P_{NEET} (or for that matter, cross-section estimates based on photoabsorption). There is no evidence for the required nuclear level(s) completing the atomic-nuclear degeneracy with the appropriate multipolarity is completely lacking. The cross section claimed by Collins and coworkers [10] is also in complete disagreement with the experimental results of Ahmad, et al. [11], who report upper limits to the cross section for induced decay of isomeric ^{178}Hf irradiated by synchrotron radiation as a function of incident x-ray energy. Conclusion: there is no credible evidence for observation of enhanced decay of isomeric ^{178}Hf induced by synchrotron radiation.

SUMMARY

Why twiddle a nucleus with atomic energies? There is interesting basic physics within the atomic-nuclear interaction. What are the mechanisms, what are the cross sections? There are applied physics interests, since there is the potential of energy on demand. The cross sections are apparently small, and some effects are just now being measured reliably. Measurements and understanding require a discipline oriented study in order to identify cases and circumstances where the small, second-order interactions can be turned to advantage and the effect observed. A recent discussion of nuclear transition induced by synchrotron radiation has been given by Gemmell [12], and a brief discussion of the proposed nuclear excitation of ^{189}Os using the EBIT facility has been given by Beiersdorfer, et al. [13].

ACKNOWLEDGMENTS

This work was funded in part by the U.S Department of Energy, and performed under the auspices of the U.S. Department of Energy by the University of California, Lawrence Livermore National Laboratory, Contract No. W-7405-Eng-48. This work was also supported in part by the U.S. Department of Energy Nuclear Energy Research Initiative (NERI).

REFERENCES

1. Kishimoto, S., et al,, *Phys. Rev. Lett.* **85**, 1831 (2000).
2. Morita, M., *Prog. Theor. Physics* **49**, 1574 (1973).
3. Meot, V., et al., EP 221.050.02, unpublished. The formula for NEET was developed by D. Gogny and M. S. Weiss, private communication.
4. Ahmad, I., Dunford, R.W., Esbensen, H., Gemmell, D.S., Kanter, E.P., Rütt, U., and Southworth, S.H., *Phys. Rev. C* **61**, 51304R (2000).
5. Tkalya, E.V., "Nuclear Excitation by Electronic Transition between Atomic Shells," *in X-Ray and Inner-Shell Processes*, edited by Dunford, R.W., Gemmell, D.S., Kanter, E.P., Krassig, B. and Southworth, S.H., AIP Conference Proceedings 506, Melville, New York, 2000, pp. 486 – 495. The value reported here for ^{178}Hf includes a small numerical factor. (Private communication to D. S. Gemmell.)
6. Harston, M.R., *Nucl. Phys.* **A 690**, 447 (2001).
7. Ahmad, I., Dunford, R.W., Gemmell, D.S., Lister, C.J., Siemssen, R.H., and Southworth, S.W., 2000, private communication.
8. Aoki, K., et al., *Phys. Rev. C* **64**, 044609 (2001).
9. Marrs, E., Beiersdorfer, P., and Schneider, D., *Physics Today* **47**, 27 (1994); Schneider, D., *Hyperfine Interactions* 99, **47**(1996).
10. Collins, C.B., et al., ***Europhys. Lett.*** **57**, 667 (2002).
11. Ahmad, I., Banar, J.C., Becker, J.A., Gemmell, D.S., Kraemer, A., Mashayekhi, A., McNabb, D.P., Miller, G.G., Moore, E.F., Pangault, L.N., Rundberg, R.S., Schiffer, J.P., Shastri, S.D., Wang, T-F., and Wilhelmy, J.B., *Phys. Rev. Lett.* **87**, 072503 (2001).
12. Gemmell, D., in Proceedings of " *X-Ray and Inner-Shell Processes*", Rome, 2002, to be published.
13. Beiersdorfer, P., et al., in Proceedings of " *X-Ray and Inner-Shell Processes*", Rome, 2002, to be published.

204

Octupole Effects at Super and Normal Deformation

R.V.F. Janssens

Physics Division, Argonne National Laboratory, Argonne, IL 60439, U.S.A.

Abstract. This presentation deals with recent results on the onset of octupole collectivity in su-perdeformed nuclei of the A $\sim$ 190 and A $\sim$ 150 regions as well as in actinide nuclei at normal deformation. It is shown that most of the properties of these negative parity sequences can be understood in terms of Random Phase Approximation (RPA) calculations, although the observations in some Pu isotopes continue to be a challenge to interpret.

Octupole correlations play an important, yet often subtle, role in the structure of many nuclei. These correlations usually manifest themselves through the presence at low excitation energy of states with odd spin and negative parity. The interest in octupole correlations comes, at least in part, from the fact that they are associated with the breaking of a symmetry of the nuclear Hamiltonian. In nuclei with a quadrupole deformed, axially symmetric shape, band structures of states with positive parity are the norm. They are associated with rotations where the shapes remain symmetric under space inversion. However, as soon as octupole degrees of freedom are involved, the potential becomes reflection asymmetric. Shell structure is responsible for the presence of octupole correlations as it is for many other nuclear phenomena. These correlations find their origin in long-range octupole-octupole interactions between nucleons. Microscopically, they can be described [1] as the result of the coupling between orbitals with $\Delta j = \Delta l = 3$, and they are especially strong when both protons *and* neutrons occupy orbitals near the Fermi surface fulfilling this condition.

A large fraction of Rick Casten's recent work as well as the vast majority of the presentations at this conference deal with low spin, positive parity excitations. In honor of Rick and as a salute to his many contributions to nuclear structure, it seemed appropriate, then, to broaden the scope somewhat by presenting some of the most recent results regarding octupole excitations and to cover both super and normal deformed nuclei.

In a number of nuclei, strong deformed shell effects are responsible for an excited minimum associated with a large, prolate deformation (major to minor axis ratio of about 2:1). In superdeformed (SD) nuclei of the mass A $\sim$ 150 region, there is much evidence for single particle behavior: observables such as the dynamic moments of inertia, $\mathfrak{I}^{(2)}$, the evolution of these moments with rotational frequency, and the measured quadrupole moments, Q_0, can in most cases be understood in terms of the occupation of specific single-particle orbitals [2, 3]. It was suggested in Ref. [4] that this behavior is a consequence of "extreme shell-model" behavior in which the SD nucleus is described by independent, non-interacting particles in a mean field.

In contrast, most excited bands in even-even SD nuclei of the A $\sim$ 190 region have

CP638, *Mapping the Triangle: International Conference on Nuclear Structure*
edited by A. Aprahamian, J. A. Cizewski, S. Pittel, and N. V. Zamfir
© 2002 American Institute of Physics 0-7354-0093-8/02/$19.00

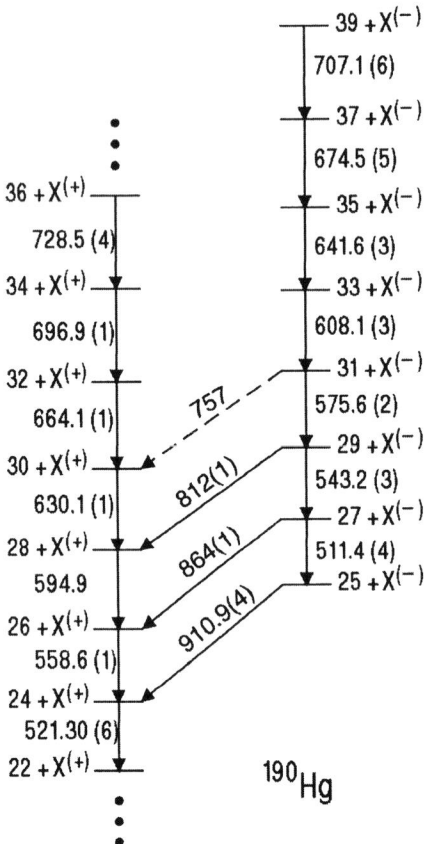

FIGURE 1. Partial level scheme of SD states in ^{190}Hg showing the E1 transitions linking SD band 2 to the yrast SD band. The uncertainties on the transition energies are given in parentheses. The X symbols reflect the fact that absolute spin values are not known at the present time.

been interpreted in terms of collective excitations. Specifically, evidence for octupole vibrations was first reported in the case of ^{190}Hg [5] where SD band 2 was found to deexcite into the yrast SD band instead of directly to levels in the normal deformed minimum, the usual decay path out of SD bands (see Figure 1). Band 2 was established to deexcite via E1 transitions [6] with rather low energy ($\sim$ 800 keV) and high transition rates (the corresponding B(E1) values are of the order of 10^{-3} $W.u.$) [7]. Similar evidence has also been reported for ^{194}Hg [8] and $^{196-198}$Pb [9]. The case of ^{194}Hg is particularly striking in that the absolute excitation energy as well as the quantum numbers of all the SD levels of interest have been established experimentally: this nucleus is one of only a handful where the transitions linking SD levels to states of normal deformation have been observed [8].

The SD minima in both the A $\sim$ 150 and A $\sim$ 190 regions are calculated [10, 11, 12] to be soft with respect to octupole deformation because of the presence of intruder orbitals ($j_{15/2}$ neutrons and $i_{13/2}$ protons) near the Fermi surface, where they are close to levels of opposite parity differing by three units ($\Delta l = 3$) in angular momentum ($g_{9/2}$ neutrons and $f_{7/2}$ protons). In fact, based on RPA calculations, Nakatsukasa *et al.* [13] proposed that most low excitations in A~190 even-even SD nuclei are associated with octupole vibrations. For example, both excited bands in ^{194}Hg have been associated with an octupole vibration over the entire frequency range where they are observed [8]; the same applies to SD bands 2 and 3 in ^{192}Hg [13], although in this case the bands are crossed by $j_{15/2}$ quasiparticle excitations at high spin. In ^{190}Hg, SD bands 2 (discussed above) and 4 are understood as being of octupole character, while band 3 is interpreted as a quasiparticle excitation [14].

Remarkably, while compelling evidence exists in the A $\sim$ 190 region, the situation is quite different near A $\sim$ 150. There is some evidence for inter-band transitions only for a single band in both ^{150}Gd and ^{152}Dy [15, 16]. In the latter nucleus, the first one where superdeformation was reported [17], five excited bands are known [16], and it is one of the weakest of those (band 6) that has been proposed to decay into the yrast SD band, i.e. the transitions of the yrast SD band were observed to be in coincidence with the γ rays of SD band 6, but the transitions linking the two bands were not observed. Nevertheless, based on this fragmentary evidence an interpretation in terms of an octupole vibration was proposed in Ref. [18]. A new Gammasphere experiment [19] has clarified the experimental situation regarding this band and has confirmed the interpretation in terms of an octupole excitation.

The large data set used recently to link the ^{152}Dy yrast SD band to the normal deformed levels [20] was also exploited to investigate the very weakly populated SD band 6. The relevant level scheme, given in Figure 2, indicates that 9 transitions with energies between 1645 and 1795 keV have been established to link SD band 6 to the yrast SD band. These transitions are of E1 character. Under the assumption that SD band 6 has the same transition quadrupole moment of 17.5 eb as the yrast SD band [21], the extracted transition rates (B(E1) $\sim$ 3-5 10^{-4} *W.u.*) are similar both to those observed among actinide nuclei exhibiting strong octupole collectivity in the normally deformed well [22] and to those of inter-band transitions in the SD wells of the A $\sim$ 190 nuclei [7] that have been interpreted in terms of an octupole vibration.

The RPA calculations by Nakatsukasa *et al.* [18] interpret SD band 6 as an octupole excitation with signature $\alpha = 1$. At zero frequency, the band is characterized by $K = 0$, but K-mixing is significant at the frequencies of interest because of the Coriolis force. In the work of Ref. [18] it was shown that the calculations reproduce the magnitude and evolution with frequency of the $\mathfrak{J}^{(2)}$ moment of inertia satisfactorily. The new data take such comparisons much further. Experiment and calculations are compared in Figure 3, where the Routhian of band 6 with respect to the yrast SD band is given as a function of the rotational frequency. The figure presents the lowest octupole excitation (dashed line), and the first 1p–1h configuration (solid line). From Figure 3, it is clear that the excitation energy and the evolution of the Routhian with frequency are well reproduced. Thus, these new results establish the first collective excitation in the SD well of an A $\sim$ 150 SD nucleus.

SD bands 2–5 are interpreted in terms of single neutron or proton excitations across the $N = 86$ and $Z = 66$ shell gaps [16, 18]. These bands are fed with intensities similar

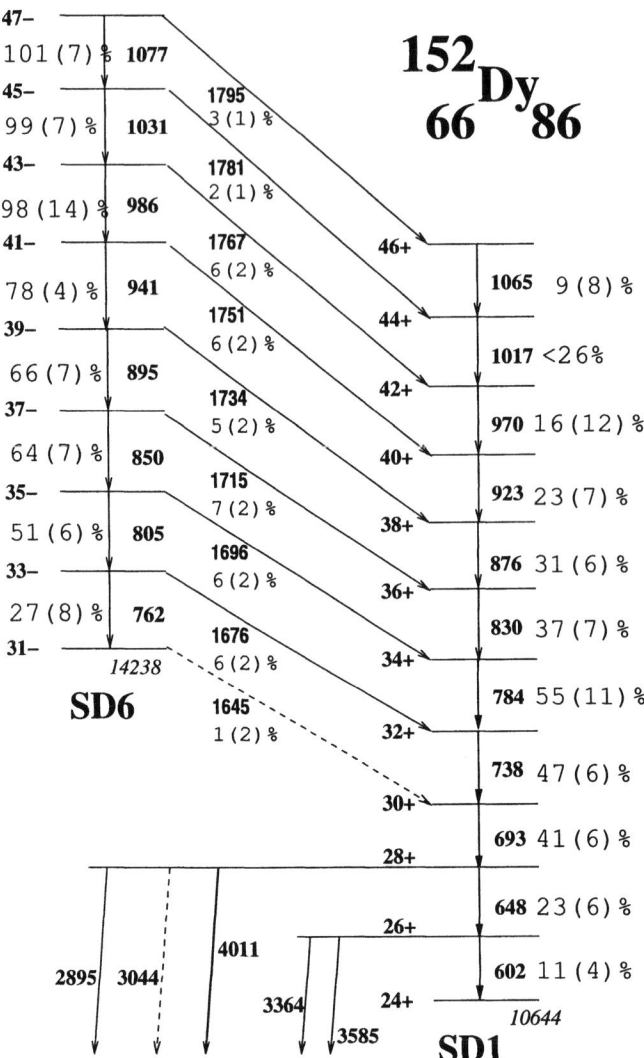

FIGURE 2. Partial level scheme of ^{152}Dy showing the lowest part of SD band 6, the lowest part of the yrast SD band 1 and the transitions that link the yrast SD band 1 to the normal states.

to that of band 6. Hence, their excitation energies with respect to the yrast SD band are likely of the same order as well in the frequency range where they are fed. Thus, the proton and neutron excitations of SD bands 2–5 are likely 1.6–1.8 MeV above the yrast SD band, and the present data also provide some measure of the SD shell gap at high frequencies.

The success of RPA calculations in reproducing the collective octupole excitations in the SD well is remarkable. However, at these large deformations, the density of states

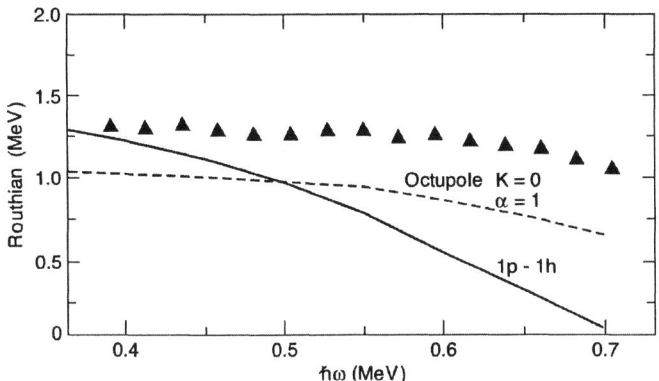

FIGURE 3. Routhians of band 6 with respect to band 1 as a function of rotational frequency in ^{152}Dy. The dashed line characterizes the lowest calculated SD excitation associated with a octupole vibration. The solid line likewise shows the lowest 1p-1h excitation which corresponds to SD band 2.

near the Fermi surface is somewhat smaller than it is at normal deformation. It is then worthwhile to investigate the predictive power of the calculations in this second regime. The nuclei of the actinide region seem particularly suited for such investigations since the orbitals located in the vicinity of their Fermi surfaces are the *same* as those found in the SD nuclei of the A~190 region. Here again, the RPA calculations have been quite successful in accounting for a good fraction of the available data. For example, the properties of negative parity states in ^{232}Th [23], ^{238}U [24], ^{244}Pu and ^{248}Cm [25] are reproduced satisfactorily. Yet, some surprises have been encountered in the lighter even-even Pu isotopes.

With Gammasphere at ATLAS, data on all Pu nuclei with mass A = 238-244, were obtained from measurements performed with Bi beams at energies $\sim$ 15% above the Coulomb barrier. This is the so-called "unsafe Coulomb Excitation" technique where the highest spin states are reached by taking advantage of both nuclear and electromagnetic processes. The measurements are described in Ref.[26]. Figure 4 compares the aligned spins i_x as a function of rotational frequency $\hbar\omega$ for the yrast sequences and the first negative parity cascades in $^{238-244}$Pu [26]. A number of interesting features clearly stand out: (i) all the Pu isotopes with mass A $\geq$ 241 exhibit a strong alignment in their respective yrast bands at $\hbar\omega$ ~0.25 MeV, (ii) this alignment is not present at all up to the highest frequencies observed in ^{239}Pu and ^{240}Pu and is delayed at least up to $\hbar\omega \geq$ 0.28 MeV in ^{238}Pu, (iii) the behavior of the alignment curve of ^{240}Pu is distinctly different from that of all the other even-even Pu isotopes, (iv) all negative parity excitations show the $\sim 3\hbar$ initial alignment, characteristic of the octupole phonon, and (v) an additional gain in alignment occurs at higher frequencies only in those isotopes where a drastic upbend occurs along the yrast line, while, in contrast, (vi) a reduction in relative alignment sets in between the two bands in ^{240}Pu. The sudden gains in i_x of 9-10$\hbar$ in the heavier Pu isotopes is consistent with the alignment under the stress of rotation of an $i_{13/2}$ proton [26]. All the available calculations (see [26] for details) indicate that the same strong proton alignment should occur around $\hbar\omega \sim 0.25$ MeV in the lighter Pu isotopes, yet it is not seen in ^{239,240}Pu and is delayed in ^{238}Pu. The effect is particularly striking in ^{240}Pu: only a small, smooth increase in alignment is observed over a range of 4-6 transitions beyond the point where the backbending occurs in the heavier Pu nuclei (Fig. 4).

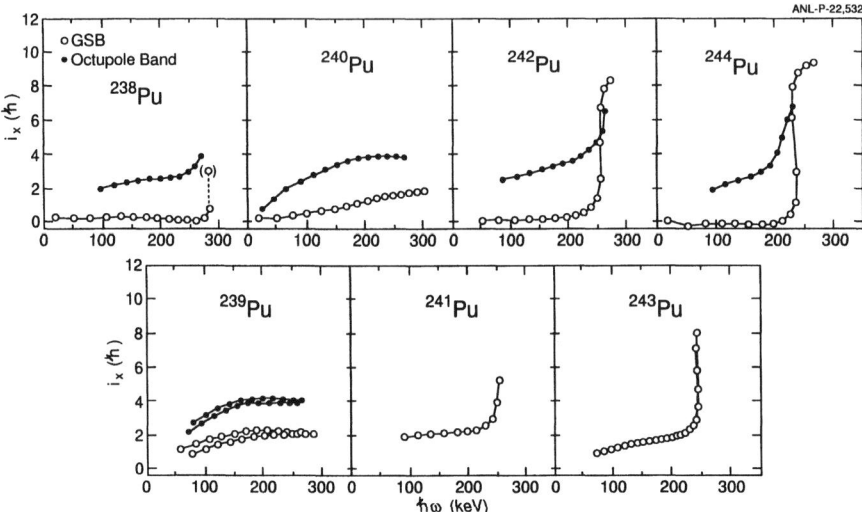

FIGURE 4. Aligned spins i_x of the yrast (open symbols) and octupole (filled symbols) rotational bands in the Pu isotopes.

In all the even-even Pu nuclei, the first excited band is of negative parity and is associated with an octupole vibration. This observation, together with the fact that octupole correlations are known to significantly alter alignment patterns seen in reflection asymmetric nuclei [27], makes it worthwhile to consider additional indications for stronger octupole correlations near A=240. In Ref. [26] it is shown that only in $^{238-240}$Pu the states of the yrast band become interleaved at high spin with the levels of negative parity to the extent that they essentially form a single band like in ^{222}Th and ^{229}Ra [28], two of the best examples of nuclei with static octupole deformation. Also, in ^{239}Pu, levels with the same spin but opposite parity are located close in energy: the $49/2^+$ and the $49/2^-$ levels are 17 keV apart, the two 53/2 states are separated by only 8 keV and the 57/2 levels are within 26 keV. Hence, these states form so-called parity doublets, as would be expected for odd nuclei with octupole deformation. Thus, it appears that, at least from the point of view of the level energies, the three lightest Pu isotopes behave like octupole deformed rotors at the highest spins. Additional evidence is provided by the measured ratios of the reduced E1 and E2 transition rates: the values of the induced intrinsic dipole moment D_0 deduced from such data become quite large at high spin only in $^{238-240}$Pu. For example, $D_0 \sim 0.2$ fm for $I \geq 21\hbar$ in ^{240}Pu [26], a value of the same order as those observed in light Th nuclei, which are among the best examples of octupole deformed nuclei. Finally, as discussed recently by Sheline and Riley [29], the hindrance factors for alpha decay in the light Pu isotopes are smaller than those seen in all neighboring nuclei and are of the same order as the values measured in the octupole deformed Ra, Rn and Th nuclei. In addition, in ^{240}Pu strong E1 transitions linking members of the second-excited 0^+ band to the negative parity band have also been observed [30]: the measured B(E1)/B(E2) ratios are of the same order as those measured for the octupole band itself. The nature of this second-excited 0^+ band is presently not understood, but the presence of strong E1 transitions can again be viewed as illustrating the importance of octupole effects in ^{240}Pu.

The experimental evidence suggests that a transition from an octupole vibration to stable octupole deformation may have occurred. Such an evolution with angular momentum has been predicted in Ref. [31]. Detailed microscopic calculations are needed to fully account for the enhanced importance of octupole correlations near ^{240}Pu. In particular, the role of the octupole-driving orbitals needs to be fully explored. In this context, it is striking that in the immediate odd-even neighbors of ^{240}Pu the $\Delta l = 3$, $\Delta\Omega = 0$ particle-hole configurations $\pi\{3/2^-[521]3/2^+[651]^{-1}\}$ and $\nu\{7/2^+[624]7/2^-[743]^{-1}\}$ come close in energy to the Fermi surface (within 0.5 MeV) and are expected to play a role in the ground state configuration.

To summarize, this presentation has focused on recent studies of octupole effects in normal deformed and superdeformed nuclei. In particular, the first case of an octupole excitation in the A $\sim$ 150 SD region has been discovered in ^{152}Dy. The absolute excitation energy of the lowest level in the octupole SD band has been measured to be 14238 keV. In other words, the octupole excitation is located at an $\sim$1.3 MeV excitation energy above the SD ground state. Octupole collectivity in the SD nuclei appears to be well described by RPA calculations. While such calculations are also quite successful in actinide nuclei, the strength of the octupole correlations near ^{240}Pu appears to be such that, at the highest spins, $^{238-240}$Pu exhibit properties associated with stable octupole deformation, suggesting that a transition with spin from a vibration to stable deformation may have occurred.

The author presents Rick Casten his most heartfelt wishes for the continuation of his stellar scientific career. May his life also remain one joyous adventure. He thanks the organizers of this conference for the opportunity to present these data. Thanks are also due to his many colleagues who have contributed to the results presented in this contribution. They are too numerous to be all listed here. Special recognition is due to his colleagues at ANL, K. Abu-Saleem, I. Ahmad, M.P. Carpenter, T. L. Khoo, F.G. Kondev, T. Lauritsen, C. J. Lister, and I. Wiedenhoever (presently at Florida State University), without whom none of this work would have been possible.

This work is supported by the U.S. Department of Energy, Nuclear Physics Division, under Contract No. W-31-109-ENG-38. The author is indebted for the use of Pu isotopes to the office of Basic Energy Sciences, U. S. Department of Energy, through the transplutonium element production facilities at the Oak Ridge National Laboratory.

REFERENCES

1. W. Nazarewicz *et al.*, Nucl. Phys. **A429**, 269 (1984).
2. R.V.F. Janssens and T.L. Khoo, Annu. Rev. Nucl. Part. Sci. **41**, 321 (1991).
3. S.T. Clark *et al.*, Phys. Rev. Lett. **87**, 172503 (2001).
4. W. Satula *et al.*, Phys. Rev. Lett. **77**, 5182 (1996).
5. B. Crowell *et al.*, Phys. Rev. **C51**, R1599 (1995).
6. A. Korichi *et al.*, Phys. Rev. Lett. **86**, 2746 (2001).
7. H. Amro *et al.*, Phys. Lett **B413**, 15 (1997).
8. G. Hackman *et al.*, Phys. Rev. Lett. **79**, 4100 (1997).
9. A. Prevost *et al.*, Eur. J. Phys. **A 10**, 13 (2001) and references therein.
10. P. Bonche *et al.*, Phys. Rev. Lett. **66**, 876 (1991)
11. T. Nakatsukasa *et al.*, Prog. Theor. Phys., **87**, 607 (1992).
12. J. Meyer *et al.*, Nucl. Phys. **A588**, 597 (1995).
13. T. Nakatsukasa *et al.*, Phys. Rev. **C53**, 2213 (1996).
14. A.N. Wilson *et al.*, Phys. Rev. **C54**, 559 (1996).
15. P. Fallon *et al.*, Phys. Rev. Lett. **73**, 782 (1994).
16. P.J. Dagnall *et al.*, Phys. Let. **B335**, 313 (1994).
17. P. J. Twin *et al.*, Phys. Rev. Lett. **57**, 811 (1986).
18. T. Nakatsukasa *et al.*, Phys. Lett. **B343**, 19 (1995).
19. T. Lauritsen *et al.*, to be published.
20. T. Lauritsen *et al.*, Phys. Rev. Lett. **88**, 042501 (2002).
21. D. Nisius *et al.*, Phys. Lett. **B392**, 18 (1997).
22. I. Ahmad and P. A. Butler, Annu. Rev. Nucl. Part. Sci., **43**, 71 (1993).
23. K. Abu Saleem *et al.*, to be published.
24. D. Ward *et al.*, Nucl.Phys. **A600**, 88 (1996).
25. G. Hackman *et al.*, Phys.Rev. **C57**, R1059 (1998).
26. I. Wiedenhöver *et al.*, Phys. Rev. Lett. **83**, 2143 (1999).
27. S. Frauendorf and V.V. Paskevich, Phys. Lett. **B141**, 23 (1984).
28. J.F. Smith *et al.*, Phys. Rev. Lett. **75**, 1050 (1995).
29. R. Sheline and M.A. Riley, Phys. Rev. **C61**, 057301 (2000).
30. I. Wiedenhöver *et al.*, to be published.
31. R.V. Jolos and P. von Brentano, Phys. Rev. **C49**, R2301 (1994) and Nucl. Phys. **A587**, 377 (1995).

New Interpretation of the O(6) Limit of the Interacting Boson Model.

J. Jolie, S. Heinze, P. von Brentano, V. Werner* and P. Cejnar†

*Institute of Nuclear Physics, University of Cologne, Zülpicherstrasse 77, 50937 Cologne, Germany.
†Institute for Particle and Nuclear Physics, Charles University, V Holešovičkách 2, 180000 Prague, Czech Republik.

Abstract.
We show that O(6) limit is simultaneously a dynamical symmetry of the U(6) group of the Interacting Boson Model and a critical point of a prolate-oblate phase transition. This observation leads to a new form of the Casten triangle of which the phase transitional behavior is studied. Empirical evidence for such a transition in the Pt region is discussed.

INTRODUCTION

The discovery in 1978 by R.F. Casten and J.A. Cizewski [1] that ^{196}Pt represents an excellent example of the O(6) limit of the Interacting Boson Model (IBM) [2] has had a major impact on the use of this model. First it gave a solid basis to the introduction of the s boson in the interacting boson approximation, which was needed to describe both vibrational (U(5)-limit) and rotational (SU(3)-limit) nuclei and gave rise to the new O(6) limit. Secondly, it was quickly demonstrated by the same authors that with the introduction of small symmetry breakings the IBM allowed a description of the complex transitional region in between γ-unstable and well deformed prolate rotors [3]. Yielding a third benchmark to which emperical nuclei can be compared, the O(6)-limit gave also rise to the very succesful introduction of the Casten triangle, which allows to classify many atomic nuclei [4]. In the last years,the study of the Casten triangle revived in three different (but related) fields: the study of quantal chaos [5, 6, 7, 8], the study of parameter symmetries [9, 6, 10, 11] and the study of quantum phase transitional behavior of atomic nuclei [12, 13]. Here we will consider some of these recent developments and especially focus on the role of the O(6) symmetry in those.

THE O(6) LIMIT

The O(6) limit of the IBM is special in several respects. Its generic features are best seen when considering the following parameterisation of the standard IBM hamiltonian:

$$\hat{H} = \varepsilon \hat{C}_1[U(5)] + \alpha \hat{C}_2[U(5)] + \beta \hat{C}_2[O(6)] + \delta \hat{C}_2[SU(3)] + \xi \hat{C}_2[O(5)] + \gamma \hat{C}_2[O(3)]. \quad (1)$$

CP638, *Mapping the Triangle: International Conference on Nuclear Structure*
edited by A. Aprahamian, J. A. Cizewski, S. Pittel, and N. V. Zamfir
© 2002 American Institute of Physics 0-7354-0093-8/02/$19.00

with $\hat{C}_n(G)$ the nth order Casimir operator of the group G. In (1) all effects connected with binding energies are removed since no U(6) Casimir operators are present. The advantage of this parametrisation in comparison to the standard multipole expansion is that it makes the group structure of the general IBM Hamiltonian clear. The O(6) symmetry is obtained when the parameters $\varepsilon, \alpha, \delta$ are zero. In this case a basis $|N, \sigma, \tau, L>$ consisting out of the irreps of the U(6), O(6), O(5) and O(3) groups classifies the energies, which are described by the analytical expression:

$$E = \beta\sigma(\sigma+4) + \xi\tau(\tau+3) + \gamma I(I+1). \tag{2}$$

There are a number of interesting features associated with this energy expression. From structural point of view the lowest states are formed by those states with the maximal value $\sigma = N$. The next class of states have $\sigma = N-2, N-4, \ldots 1$ or 0. Within a given σ the τ values can take the values $\tau = 0, 1, 2, \ldots \sigma$.

Because the O(5) quantum number τ will turn out to play a prominent role in the following, we note that all hamiltonians (1) with $\delta = 0$ conserve this quantum number [14, 15]. When also β equals 0 one obtains the U(5) limit with:

$$E = \varepsilon\hat{n}_d + \alpha\hat{n}_d(\hat{n}_d+5) + \xi\tau(\tau+3) + \gamma I(I+1), \tag{3}$$

where we deliberately use the τ quantum number instead of the more common v.

The O(5) symmetry leads to a peculiar behavior of the wavefunctions: they contain always either an even or an odd number of d bosons. This is trivially the case in the U(5) limit where the number of d bosons is a quantum number. This general structure of the wavefunctions, which can be analytically expressed in the O(6) limit, holds also for all numerical solutions when $\delta = 0$. Because the O(5) structures are similar in the O(6) and U(5) limits, detailed information on the structure of the lowest $\sigma = N - 2$ states is needed to firmly establish an O(6) nucleus. To obtain this we need to consider the electric quadrupole operator of the IBM:

$$\hat{Q} = (s^\dagger\tilde{d} + d^\dagger s)^{(2)} + \chi(d^\dagger\tilde{d})^{(2)}, \tag{4}$$

which contains one extra parameter χ which is normally chosen such that the value is consistent with the quadrupole operator used in the hamiltonian (consistent Q-formalism [16]). Atomic nuclei representing the O(6) structure have $\chi = 0$ and therefore are characterised by vanishing quadrupole moments, the $|\Delta\tau| = 1$ selection rule and the vanishing transitions between states having different σ quantum numbers. The latter is clearly established for ^{196}Pt [17] which, however, has a non-vanishing quadrupole moment for its first excited state of $Q(2_1^+) = +0.66(12)eb$. We will find out below why this is the case.

COMPLETENESS OF THE CASTEN TRIANGLE

The Casten triangle was from the beginning a tremendouos success in its application to classify actual nuclei. While it underwent many redefinitions over the years, the essential parts have not changed till recently [13]. Essentially, it concerns a parametrisation of the hamiltonian (1) in a very simple hamiltonian:

$$\hat{H} = \eta \hat{n}_d + \frac{(\eta - 1)}{N} \hat{Q}_\chi \cdot \hat{Q}_\chi \tag{5}$$

which contains two independent parameters η, describing the transition between spherical and deformed, and χ, describing the transition between prolate and γ-soft deformation. In the expression (5) we have introduced a N dependence needed for scaling later. The success of this simple parametrisation lies in the fact that most nuclei can then be located on one of the legs of the triangle (see for instance ref [18]) and that one can visualise the hamiltonian (1) in two dimensions.

It is instructive to rewrite the hamiltonian (5) in the form of hamiltonian (1)yielding:

$$\hat{H} = (\eta + \frac{2(1-\eta)}{7N}(\chi^2 + \frac{\chi\sqrt{7}}{2}))\hat{C}_1[U(5)] + \frac{2(1-\eta)}{7N}(\chi^2 + \frac{\chi\sqrt{7}}{2})\hat{C}_2[U(5)] \tag{6}$$
$$+ \frac{(1-\eta)}{N}(1 + \frac{2\chi}{\sqrt{7}})\hat{C}_2[O(6)] + \frac{(\eta-1)\chi}{N\sqrt{7}}\hat{C}_2[SU(3)]$$
$$+ \frac{(\eta-1)}{N}(1 + \frac{3\chi}{\sqrt{7}} + \frac{2\chi^2}{7})\hat{C}_2[O(5)] + \frac{(1-\eta)}{14N}(\chi^2 + 2\sqrt{7}\chi)\hat{C}_2[O(3)].$$

One notices two important features: while the number of parameters is reduced to two, the hamiltonian contains the whole rich structure of the hamiltonian (1) and the Casimir operators associated with U(6) do not appear, such it has no effects the binding energies. Being a success in comparison to experiment, a two parameter description has its limitations, which led to several alternative proposals. The first was a tetrahadron proposed to incorporate the neutron-proton degree of freedom by Dieperink [19], another one used a rectangle to incorporate that in the U(5) limit all χ values are possible [20, 21], although multiplied with zero. Recently the restrictions of the Casten triangle became more important as the IBM became an excellent model in which to study quantal chaos [5, 6, 7, 9]. In the context of the study of chaos in nuclei a pentagon was proposed[22] to incorporsate two additional limits that generally are not considered. They concern the $\overline{O(6)}$ and the $\overline{SU(3)}$ limit. While the first can be seen as a mathematical curiosity, leading to no new physics, the second one has a clear physical interpretation as it corresponds to an oblate rotor, if one consistently uses positive effective charges in the electromagnetic transition operator. We will return in detail to this question below.

PHASE TRANSITIONS AND THE O(6) LIMIT

Phase transitions involving nuclear shapes became again to the forefront of nuclear structure physics when F. Iachello developed new critical symmetries which can describe atomic nuclei at the phase transition [23, 24]. The new symmetries, called X(5) and E(5), are obtained within the framework of the collective model [25] under some simplifying approximations. This new developments were first triggered by studies of ^{152}Sm by Casten and the Cologne group [26]. Remarkably, the parameter free predictions provided

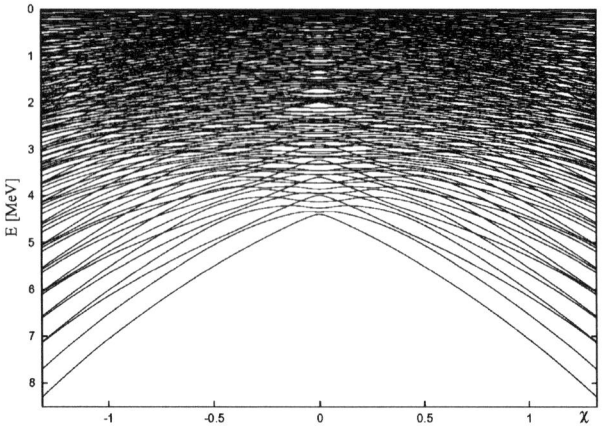

FIGURE 1. All 0^+ states as a function of χ for 40 bosons

by these symmetries are closely realised in some atomic nuclei[27, 28]. The phase transitions considered in these references are those of the ground state deformation. These quantum phase transitions take place at zero temperature and depend on the number of nucleons. The two new critical point symmetries focus on the β degree of freedom and are based on a separation of the β degree of freedom from the γ degree of freedom. An infinite square well potential in the β degree of freedom is imposed. Since these descriptions are embedded in the framework of the collective model, they do not incorporate finite N effects reflecting the finite number of nucleons in actual atomic nuclei.

However, finite N effects are important, for at least three reasons. First, classically, phase transitions only occur for systems having an infinite number of constituents and therefore an extension of the concept to systems having a limited number of constituents is of importance. Secondly, since atomic nuclei contain only an integer number of nucleons, nature does not allow us to vary the control parameters continuously in the region where the phase transition occurs. While the experimental limitation to integer nucleon number leads to discrete changes in the properties of atomic nuclei around the critical point, theoretical models do allow one to continuously vary the appropriate control parameter. Finally, the limited number of N tends to wash out the phase transitional behavior making the search for phase transitional behavior more difficult.

The way we adopted to systematically study the finite-N phase transitions relies on using IBM calculations with a large number of bosons (up to 40) [8]. By studying such systems it will be possible to: single-out observables and effects related to the phase transition, study the N dependence of these effects and last but not least apply these observables in the low N limit to study empirical nuclei. Here we consider the very simple hamiltonian (6) with $\eta = 0$. This hamiltonian contains one free parameter χ which we vary from $\chi = -\sqrt{7}/2$ (SU(3) value) over $\chi = 0$ (O(6) value) to $\chi = +\sqrt{7}/2$ ($\overline{SU(3)}$ value). Thus we extend the calculation outside the normal Casten triangle to

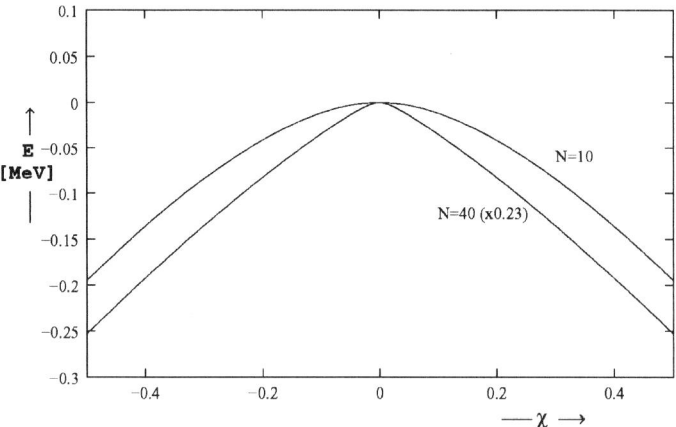

FIGURE 2. The groundstate energy as a function of χ for 10 and 40 bosons

include oblate nuclei. For $\chi < 0$ ($\chi > 0$) the potential has a minimum at $\gamma = 0°$ ($\gamma = 60°$), corresponding to a prolate (oblate) shape. For $\chi = 0$ the potential is flat in γ. In Figure 1 we show the absolute energy of all 0^+ states for N= 40. At the O(6) limit the energy of the groundstate makes a kink of which the sharpness increases with N (see Figure 2). Only at the three limits levels can cross without interaction. Avoided level crossing are sharp at N=40 when χ in not equal to zero. Despite the symmetry around $\chi = 0$ it is clear that the groundstate mixes with the first excited state and one can prolongate the states through several state-mixings. A useful quantity to prove this is the SU(3) wavefunction entropy [6, 7] which measures the overlap with the SU(3) basis vectors. As shown in Figure 3, the entropy changes rapidly at the critical point meaning that the wavefunctions are different above and below $\chi = 0$. What one observes is a first order phase transition between prolate and oblate deformation in which the O(6) limit corresponds to the critical point [13]. The origin of the phase transition lies in the form of the potential in the γ degree of freedom. This is logical because the potential has a minimum at an axially symmetric shape in SU(3) but becomes totally flat in γ at O(6). Figure 3 illustrates also the location of the minimum, showing the discontinuous behavior typical for phase transitions. Note that this result is valid for all N. So the control parameter for changes in $V(\gamma)$ is likewise χ. Here comes in a mooth point. Normally the control parameter should be the one governing a linear dependence between two hamiltonians. This is not the case for χ, but near O(6) one obtains:

$$\hat{H} = -\frac{1}{N}(s^\dagger \tilde{d} + d^\dagger s)^{(2)} \cdot (s^\dagger \tilde{d} + d^\dagger s)^{(2)} - \frac{2\chi}{N}(s^\dagger \tilde{d} + d^\dagger s)^{(2)} \cdot (d^\dagger \tilde{d})^{(2)} + O(\chi^2). \quad (7)$$

and exactly at the phase transition, the hamiltonian linearises (also for N going to infinity). The fact that χ occurs in a linear way as part of a quadratic term insures that we have no simple *mirror* reflection around $\chi = 0$. There is one clear physical observable that exemplifies this: the quadrupole moment of the first excited state changes sign at the phase transition.

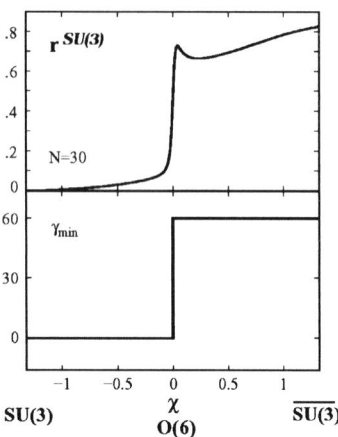

FIGURE 3. SU(3) wavefunction entropy ratio as defined in [7] (top) and location of the groundstate deformation γ as a function of χ (bottom).

So while it seems clear that we are dealing with a phase transition, the nature of this particular one is quite special, because it is a phase transition and a dynamical symmetry. That here we have a unique situation is illustrated at O(6) in Figure (1). One observes clear bunching of the 0^+ states at the phase transition and degeneracies of some excited states. They are due to the fact that τ, the O(5) quantum number, is valid at the transition, which in our case holds the higher symmetry.

A NEW EXTENDED CASTEN TRIANGLE

In the construction of the Casten triangle, solutions with positive χ have only occasionally been considered [22], because they correspond to oblate ground state deformations which are seldom found in actual atomic nuclei. However we saw that the IBM model allows solutions with positive χ as well, corresponding to an oblate deformation. By extending the symmetry triangle [13] from $\chi = -\sqrt{7}/2$ to $\chi = +\sqrt{7}/2$ we obtain a variation of the Casten triangle, which is shown in Fig 4. Here, to the usual Casten triangle, U(5)-SU(3)-O(6), we add a corresponding triangle for U(5)-$\overline{\text{SU}(3)}$-O(6). The new triangle maps also the phase transitions which occurs inside the triangle. They are the first order transition occuring on the O(6)-U(5) axis to the point where the nucleus changes from spherical to deformed. From thereon the first order transition goes outwards to the legs. The central point itselfs represent the second order transition and can be seen as the triple point of nuclear deformation where spherical, prolate and oblate deformation coexist. The phase transitions in the extended Casten triangle are exemplified in Figure 4 which shows the energy of the groundstate using (5) for N=40.

Up to now we discussed the results at large N values. In the remaining part the experimental evidence, hence at low N values, will be discussed. The question is whether there exist atomic nuclei whom exhibit the characteristics of the oblate-prolate phase

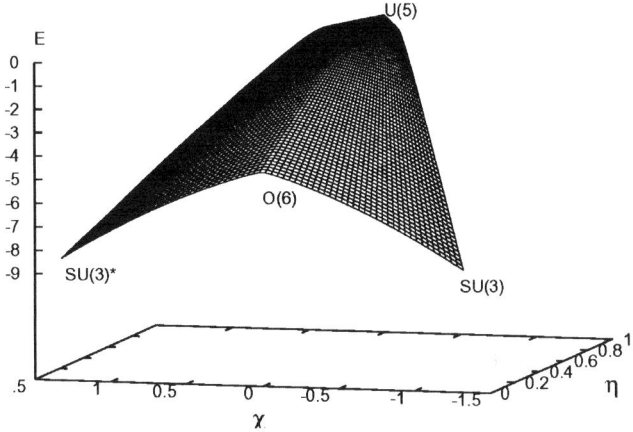

FIGURE 4. The groundstate energy as a function of η and χ for N=40.

transition. Although oblate nuclei or transitions from γ soft to oblate shapes are rare they do occur in the platinum and mercury isotopes close to the well known O(6) nucleus ^{196}Pt. In [13] three different signatures were indicated: $R(4_1^+/2_1^+)$, $Q(2_1^+)$ and $B(E2;2_2^+ \rightarrow 2_1^+)$. In an attempt to span a large part of the extended Casten triangle we plot these quantities in Figure 5 for nuclei ranging from the well deformed prolate rotor ^{180}Hf to ^{200}Hg. Clearly all experimental observables indicating the prolate-oblate phase transition [13] are observed. The very rapid change of the quadrupole moment might explain why ^{196}Pt has a non vanishing quadrupole moment while still being a good O(6) nucleus. Surprising is the structure in mercury isotopes which does not resemble a vibrational or shell model like structures whom have $R(4_1^+/2_1^+)$ around 2.

CONCLUSION

In conclusion, we have reinterpreted γ-soft nuclei as exhibiting a first-order quantum phase transitional behavior. The new shape phase transitional behavior at O(6) corresponds to jump from prolate to oblate shapes at the critical point, $\chi = 0$. Secondly, because the O(6) limit can be exactly solved one does not need any approximation to describe the behavior at the critical point analytically when $\eta = 0$ [13]. The O(6) limit is therefore unique in being, simultaneously, both a dynamical symmetry of the U(6) group of the IBA and a critical point of a prolate-oblate phase transition. Finally and most importantly, the exact solution is obtained for any N and not only for $N = \infty$. We also demonstrated that there is at least one emperical example of such a phase transition found around the Pt isotopes

The authors want to thank R.F. Casten for illuminating discussions and for inventing his triangle. The authors acknowledge the support of the DFG under Project number Br 799/10-1. One of the authors (P.C.) thanks the University of Cologne for the hospitality and financial support provided by the partnership with Charles University.

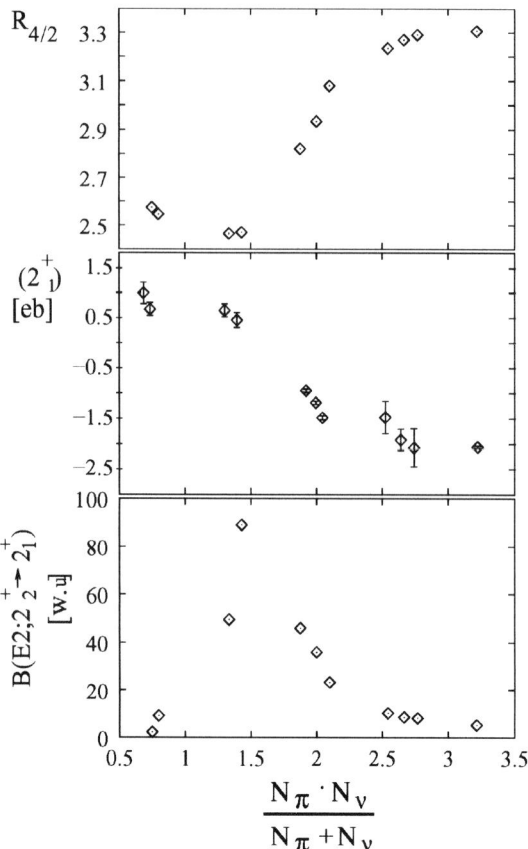

FIGURE 5. The experimental observables $R(4_1^+/2_1^+)$, $Q(2_1^+)$ and $B(E2;2_2^+ \to 2_1^+)$ for the nuclei: ^{180}Hf, $^{182,184,186}W$, $^{188,190,192}Os$, $^{194,196}Pt$ and $^{198,200}Hg$.

REFERENCES

1. R.F. Casten and J. A. Cizewski, Nucl. Phys. **A309**, 477 (1978).
2. F. Iachello and A. Arima, The Interacting Boson Model (Cambridge University Press, Cambridge, 1987).
3. R.F. Casten and J. A. Cizewski, Phys. Lett. **79B**, 5 (1978).
4. R.F. Casten in *Interacting Bose-Fermi Systems in Nuclei*, ed. F. Iachello Plenum, p 1 (1981).
5. N. Whelan, and Y. Alhassid, Nucl. Phys. **A556**, 42 (1993)
6. P. Cejnar and J. Jolie. Phys. Lett **B420**, 241 (1998).
7. P. Cejnar and J. Jolie, Phys. Rev. **E58**, 387 (1998).
8. P. Cejnar and J. Jolie, Phys. Rev. **E61**, 6237 (2000).
9. D. Kusnezov, Phys. Rev. Lett. **79**, 537 (1997).
10. A.M. Shirokov, N.A. Smirnova, Yu.F. Smirnov, Phys. Lett. **B434**, 237 (1998).
11. P. Cejnar, and H.B. Geyer, Phys. Rev. **C64** 034307 (2001).
12. F. Iachello, N.V. Zamfir, R.F. Casten Phys. Rev. Lett. **81**, 1191 (1998).
13. J. Jolie, R.F. Casten, P. von Brentano, V. Werner Phys. Rev. Lett. **87**, 162501 (2001).

14. A. Leviatan, A. Novoselsky and I. Talmi, Phys. Lett B **172** 144 (1986).
15. J. Jolie and H. Lehmann, Phys. Lett B **342** 1 (1995).
16. D.D. Warner and R.F. Casten Phys. Rev. Lett. SSbf 48 1385 (1982).
17. H.G. Boerner, J. Jolie, S.J. Robinson, R.F. Casten, J.A. Cizewski, Phys. Rev. C **42** R2271 (1990).
18. R.F. Casten and D.H. Feng, Physics Today Vol 37 No 11 p26 (1984).
19. A.E.L. Dieperink, Progress in Nucl. and Part. Phys. Vol 9 (1983).
20. V. Werner et al., Phys. Rev. **C61**, 021301 (2000).
21. D. Bucurescu et al. Nucl. Phys. A **672** 21 (2000).
22. J. Jolie and P. Cejnar, J. Phys. G **25**, 843 (1999).
23. F. Iachello, Phys. Rev. Lett. **85**, 3580 (2000).
24. F. Iachello, Phys. Rev. Lett. **87** 052502 (2001).
25. A. Bohr and B.R. Mottelson, Nuclear Structure (Benjamin, New York, 1975).
26. R.F. Casten, M. Wilhelm, E. Radermacher, N.V. Zamfir, and P. von Brentano, Phys. Rev. C **57**, R1553 (1998).
27. R.F. Casten and N.V. Zamfir, Phys. Rev. Lett. **85**, 3584 (2000).
28. R.F. Casten and N.V. Zamfir, Phys. Rev. Lett. **87** 052503 (2001).

222

Are There X(5) Nuclei In The A~80 and A~100 Regions?

Daeg S. Brenner

Clark University, Worcester, MA 01610, USA

Abstract. Iachello has introduced a new class of symmetries based on solutions to differential equations to model phase transition and critical point behavior in nuclei. For the shape transition region between a spherical vibrator and an axial rotor the dynamical symmetry is denoted X(5). A search of existing data for the A~80 and A~100 regions provides tantalizing hints of nuclei with X(5) character in both regions.

INTRODUCTION

The evolution of nuclear structure and shape between spherical closed shells and deformed spheroids has often been characterized as a phase transition. However, the application of a concept normally applied to macroscopic systems consisting of huge numbers of entities to nuclei containing of the order of a hundred quantum particles has lacked a firm footing in theory. Recently, Iachello [1] introduced a framework, based on symmetries, for understanding nuclear phase/shape transitions near or at the critical point for the transition region between a spherical vibrator and an axial rotor. In Iachello's theory the nucleus is composed of an infinite number of bosons and the potential is separable into $V(\beta)$ and $V(\gamma)$ components with the former represented by a square well and the latter by a harmonic oscillator potential. He was able to show that there are analytical solutions to this model at the critical point that in this instance is a first-order phase transition. The critical point is designated by a special dynamical symmetry, X(5), and levels are assigned quantum numbers, s, that determine their energies and transition rates. Signatures of X(5) nuclei include the energy ratios $E(4_1^+)/E(2_1^+)$ for a given s sequence of levels, $E(0_2^+)/E(2_1^+)$ between the $s = 1$ and s = 2 sequences, $E(J^+)/E(2_1^+)$ as a function of J for the $s = 1$ sequence, and intra- and inter-sequence B(E2) values. Examples of nuclei that fit the X(5) description have been found in the N = 90 region by Casten and Zamfir [2] and Caprio [3].

THE A~80 REGION

The beauty of Iachello's approach is that specific predictions arise naturally from the solutions of differential equations. For example, the X(5) symmetry is characterized by level energy ratios, $E(4_1^+)/E(2_1^+) = 2.91$ and $E(0_2^+)/E(2_1^+) = 5.67$,

CP638, *Mapping the Triangle: International Conference on Nuclear Structure*
edited by A. Aprahamian, J. A. Cizewski, S. Pittel, and N. V. Zamfir
© 2002 American Institute of Physics 0-7354-0093-8/02/$19.00

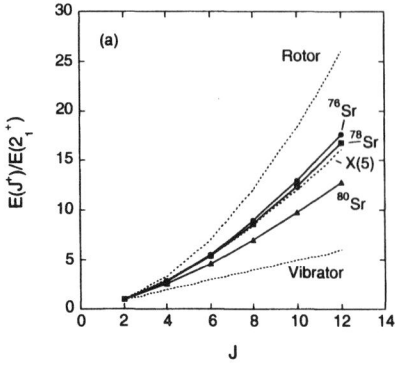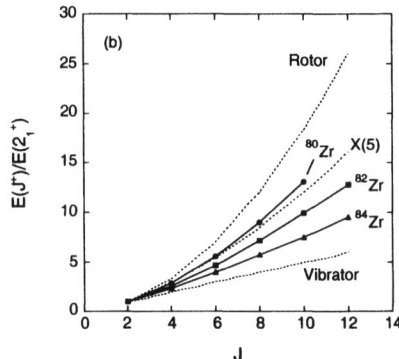

FIGURE 1. Yrast bands for (a) ^{76,78,80}Sr and (b) ^{80,82,84}Zr, the energy levels for the $s = 1$ sequence in X(5), and the harmonic vibrator and symmetric rotor limits.

respectively. In addition, inter- and intra-band transition probabilities are predicted relative to the B(E2; $2_1^+ \rightarrow 0_1^+$) and the energy sequences of levels as a function of J for given s values are defined by the ($2_1^+ - 0_1^+$) energy difference. Thus, for a given X(5) nucleus, the low-lying structure is defined by a simple normalization to one energy difference and one transition probability.

A search of the A~80 region for nuclei with E(4_1^+)/E(2_1^+) = 2.91±0.10 finds three that meet this criteria; ^{76}Sr, ^{78}Sr, and ^{80}Zr. The yrast band energy levels of relevant Sr and Zr and their neighboring isotopes are shown in Fig. 1. It is clear from these plots that ^{76}Sr, ^{78}Sr, and ^{80}Zr yrast sequences closely follow the X(5) prediction.

Unfortunately, there are only limited data available for comparison with X(5) because these nuclei lie near the N = Z line, far from stability. The 0_2^+ level has not been identified in any of these nuclides and level lifetimes, from which B(E2) values can be calculated, are known only for the 2_1^+ and 4_1^+ levels in ^{78}Sr. However, as noted above, all that needs to be known in order to predict the energy levels and transition probabilities for any nucleus are the 2_1^+ energy and B(E2; $2_1^+ \rightarrow 0_1^+$). Thus the low energy structure of ^{78}Sr, assuming it to be an X(5) nucleus, can be predicted as shown in Fig. 2. The lifetime of the 4_1^+ level then provides the only additional new piece of information concerning the character of ^{78}Sr. The excellent agreement between the data, 171(17) units relative to the B(E2; $2_1^+ \rightarrow 0_1^+$), and the X(5) prediction, 169 units, is encouraging but clearly insufficient evidence to confirm X(5) structure.

THE A~100 REGION

The apparent rapid onset of deformation that occurs in Sr and Zr nuclei around neutron number N = 60 has been a subject of interest and controversy for more than

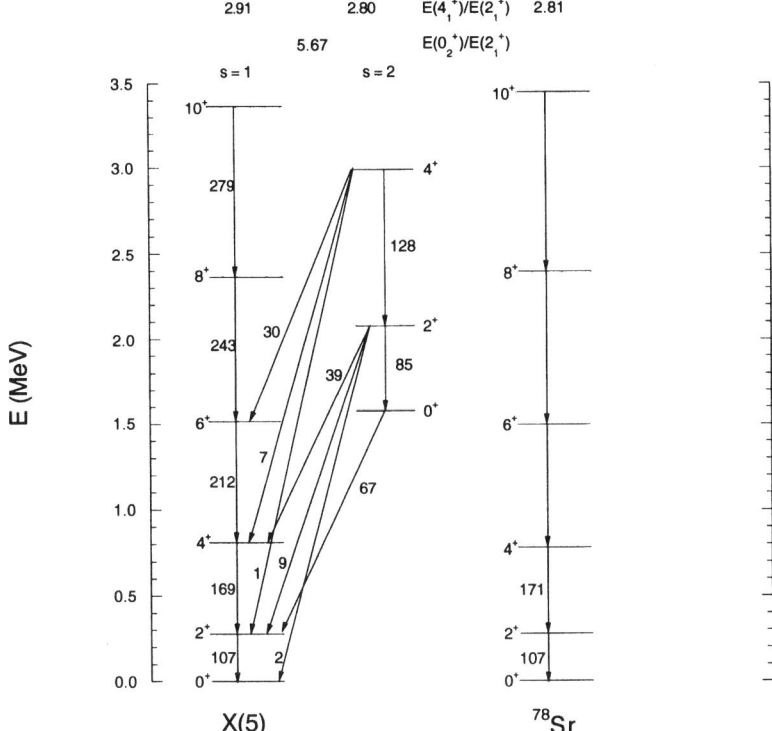

FIGURE 2. Comparison of X(5) with data for ^{78}Sr [4]. The E(2_1^+) value and the B(E2; $2_1^+ \rightarrow 0_1^+$) value are normalized to the data. The E(4_1^+)/E(2_1^+) values refer to relative energies within a given s sequence of levels.

thirty years. Experimentalists and theorists alike have delighted in the richness and complexity of structural evolution and shape coexistence in this region, a region that is often subject to diverse and conflicting interpretations. Urban *et al.* [5] have published a thorough review of the current status of the A~100 transition region and concluded that the onset of deformation is more gradual than previously thought, taking place from N = 56 where there is a subshell closure to N = 62-64. Given this more gradual evolution of shape, a search for an X(5) candidate seemed worth pursuing.

The procedure used for the A~80 region was followed for A~100. Fig. 3 shows yrast band data for Zr and Mo isotopes that suggest that ^{100}Zr and ^{104}Mo are reasonable candidates for X(5). However, a more detailed examination of ^{100}Zr discounts it as X(5) because the yrast band is much less collective than expected and there are two low lying 0^+ states at 331 and 829 keV but none near 1200 keV, the X(5) prediction.

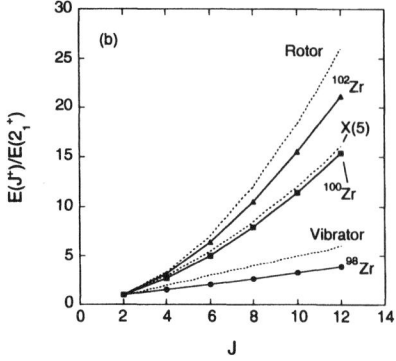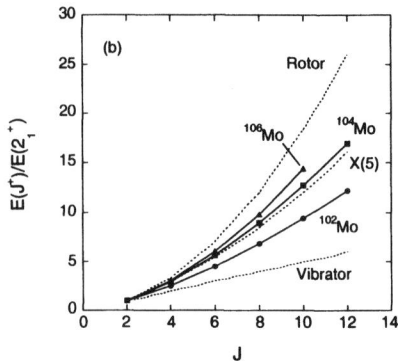

FIGURE 3. Yrast bands for (a) ^{98,100,102}Zr and (b) ^{102,104,106}Mo, the energy levels for the $s = 1$ sequence in X(5), and the harmonic vibrator and symmetric rotor limits.

More detail on ^{104}Mo is shown in Fig. 4. Here we see that the energy sequence of the experimental yrast band agrees well with X(5). In addition, the 0_2^+ level at 886 keV is not unreasonably far from the X(5) prediction of 1089 keV. Clearly, more data, especially lifetime measurements, are needed before any definitive conclusion can be reached.

CONCLUSION

There are several nuclides in the N = 90 region (^{150}Nd, ^{152}Sm, ^{154}Gd, and ^{156}Dy) that have yrast band energies and B(E2) values in close agreement with expectations for the X(5) critical point symmetry. In addition, the intra-sequence B(E2) values between the $s = 2$ and $s = 1$ levels agree qualitatively with the pattern predicted for X(5) [2,3]. Normally phase transition and critical point phenomena are associated with systems containing large numbers of particles. The good agreement obtained for nuclei in the N = 90 region raises the question whether examples of X(5) might be found in lighter mass regions.

Data for the A~80 and A~100 shape transition regions provide tantalizing hints of the X(5) dynamical symmetry in ^{76}Sr, ^{78}Sr, ^{80}Zr and ^{104}Mo. Confirmation will require much additional information, especially B(E2) values. Since the nuclei involved are very far from stability, such measurements will benefit from accelerated beams of radioactive nuclei that will be available at a second-generation radioactive beam facility such as the proposed Rare Isotope Accelerator (RIA).

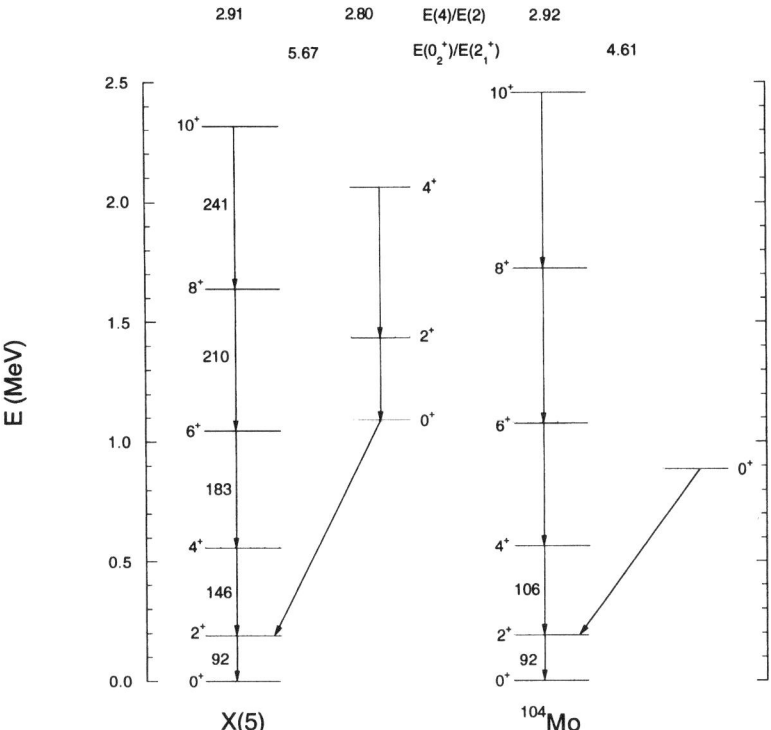

FIGURE 4. Comparison of X(5) with data for ^{104}Mo [4]. The $E(2_1^+)$ value and the $B(E2;\ 2_1^+\rightarrow0_1^+)$ value are normalized to the data. The $E(4_1^+)/E(2_1^+)$ values refer to relative energies within a given s sequence of levels.

ACKNOWLEDGMENTS

I wish to thank Victor Zamfir and Mark Caprio for helpful discussions. Support for this research was provided by the USDOE under Grant DE-FG02-88ER40417.

REFERENCES

1. Iachello, F., *Phys. Rev. Lett.* **87**, 052502 (2001) and these proceedings.
2. Casten, R.F., and Zamfir, N. V., *Phys. Rev. Lett.* **87**, 052503 (2001).
3. Caprio, M.A., these proceedings.
4. Data retrieved from the National Nuclear Data Center: http://www.nndc.bnl.gov/.
5. Urban, W., Durell, J.L. , Smith, A.G., Phillips, W.R., Jones, M.A., Varley, B.J., Rzaca-Urban, T., Ahmad, I., Morss, L.R., Bentaleb, M., Schulz, N., *Nucl. Phys.* **A689**, 605 (2001).

Triaxiality and The Wobbling Mode In ^{167}Lu

H. Amro*, W.C. Ma[†], G. Hagemann**, B. Herskind**, J.A. Winger[†], Y. Li[†],
J. Thompson[†], G. Sletten**, J.N. Wilson**, D.R. Jensen**, P. Fallon[‡], D.
Ward[‡], R.M. Diamond[‡], A. Görgen[‡], A. Machiavelli[‡], H. Hübel[§],
J. Domscheit[§] and I. Wiedenhöver[¶]

*Mississippi State University, Mississippi State, MS 39762, Wright Nuclear Structure Laboratory,
Yale University, New Haven, CT, USA
[†]Mississippi State University, Mississippi State, MS 39762
**The Niels Bohr Institute, Blegdamsvej 17, DK-2100 Copenhagen, Denmark
[‡]Lawrence Berkeley National Laboratory, Berkeley, CA 94720
[§]ISKP, University of Bonn, Nussallee 14-16, D-53115 Bonn, Germany
[¶]Florida State University, Tallahassee, Fl 32306-4350

Abstract. High spin states in ^{167}Lu nucleus were populated through the ^{123}Sb(^{48}Ca,xn) reaction at
203 MeV. Five presumably triaxial superdeformed (TSD) bands have been found. The electromagnetic properties of several connecting transitions between the yrast (TSD1) and the excited (TSD2)
bands have been established. Evidence for the assignment of TSD2 band as a wobbling mode built
on the yrast TSD band is presented. These bands coexist with bands built on quasiparticle excitations
in normal deformed (ND) minimum for which new data are also presented.

The wobbling mode is a direct consequence of rotational motion of a triaxial body.
The wobbling degree of freedom introduces sequenses of bands with increasing number
of wobbling quanta [1], and a characteristic $\Delta I=1$ decay pattern between the bands in
competition with the in-band decay. A favorable candidate for establishing this exotic
excitation mode is found in earlier in ^{163}Lu [2]. Several, presumably triaxial, superdeformed bands (TSD's) involving an aligned $i_{13/2}$ proton, which would strongly influence
the decays from a wobbling excitation [3], have been reported in recent years [4]. Here
we report on the experimental evidence for the wobbling mode in ^{167}Lu.

High spin states in ^{167}Lu were populated through the ^{123}Sb(^{48}Ca,4n) reaction at 203
MeV from the 88 inch Cyclotron at LBNL. Five TSD bands were found in ^{167}Lu. The
two strongest populated, TSD1 and TSD3, bands have been firmly linked to normal
deformed (ND) structures. Most importantly, several transitions connecting TSD2 to
TSD1 were also identified. From angular distribution and angular correlation analysis,
spins and parities for TSD1, TSD2, and TSD3 have been determind. In addition, the
mixing and branching ratios for the linking transition where experimantally determind.
From these values the $B(E2)_{out}/B(E2)_{in}$ were found to be ~ 0.22. Considering the
low excitation energy for TSD2 (~ 720 keV), these values are much larger than those
expected from the signiture partner. Furthermore, no three-quasiparticle excitation of
the correct spin and parity is expected with similar excitation energies for TSD2 from
cranking calculations. The $B(E2)_{out}/B(E2)_{in}$ ratios in ^{167}Lu are comparable to the

CP638, *Mapping the Triangle: International Conference on Nuclear Structure*
edited by A. Aprahamian, J. A. Cizewski, S. Pittel, and N. V. Zamfir
© 2002 American Institute of Physics 0-7354-0093-8/02/$19.00

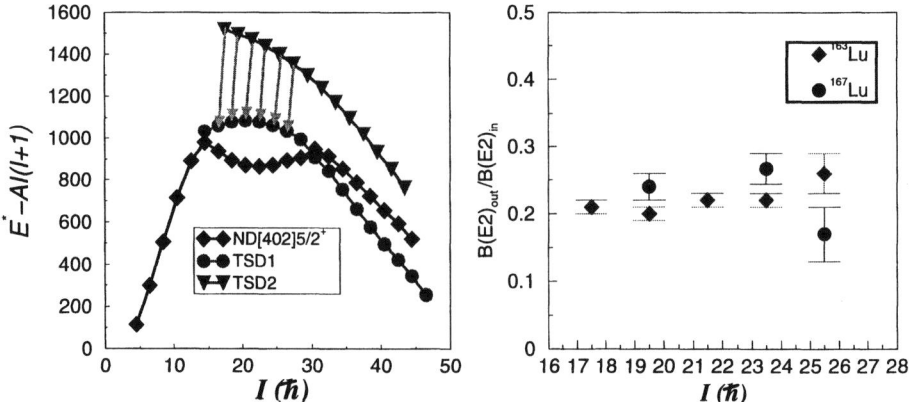

FIGURE 1. Left: Excitation energies for selected TSD and ND bands ^{167}Lu. Transitionslinking the wobbler candidate in ^{167}Lu and the yrast TSD band are indicated with arrows. Right: Properties of the TSD2→Tsd1 $\Delta I = 1$ transitions in ^{167}Lu compared to corresponding experimental values [2] for ^{163}Lu.

values found in ^{163}Lu [2] as can be seen Fig. 1, right. Similarities in the behaviour of the wobbling mode recently established in ^{163}Lu [2] and that been reported in ^{167}Lu is illustrated in Fig. 1, left. The present evidence for the wobbling mode in ^{167}Lu, four neutrons away from ^{163}Lu, is an important step to support the wobbling mode as a general phenomenon, and it establishes stable triaxiality in a broader region.

This work is supported by US DOE grants DE-FG02-95ER40939 (MSU) and DE-FG02-91ER-40609 (Yale) , the Danish Science Foundation and the German BMBF (contract No. 06 BN 907).

REFERENCES

1. A. Bohr and B. Mottelson, *NuclearStructure*, Vol. II,(Benjamin, New York,1975)
2. S.W. Ødegård et al., Phys. Rev. Lett. **86** (2001) 5866.
3. I. Hamamoto, Phys. Lett. B193 (1987) 399
4. G.B. Hagemann, Phys. Scripta T88 (2000) 77 and refs. therein.

Multiphonon Excitations in ^{124}Sn

D. Bandyopadhyay*, C. Fransen*, N. Boukharouba*, V. Werner†, J. L.
Weil*, S. W. Yates* and M. T. McEllistrem*

*University of Kentucky,Lexington, KY 40506-0055
†Institut für Kernphysik, Universität zu Köln, 50937 Köln, Germany

Abstract. Low spin states in ^{124}Sn have been studied with the (n,n'γ) reaction to explore collective
excitations and search for a multiphonon scenario in this isotope.

Since Bohr and Mottleson [1] explored their existence, collective excitations such
as quadrupole and octupole vibrations in the low energy levels of even-even spherical
nuclei has been a field of considerable interest. The semimagic, even-even tin isotopes
exhibit vibrational band structures, and hence, have been considered as potential candidates to test the degree of separation of collective excitations from other configurations
[2]. The nucleus ^{124}Sn, being a closed proton shell nucleus, was considered a suitable
candidate for the study of single particle states. However, in 1968, Clement and Baranger
[3] performed shell model calculations, which indicated collective behaviour of the 2_1^+
and 3_1^- states of tin nuclei. Since then, different theoretical approaches have been used
to explain the low lying states of different tin isotopes[4]. But still, the scenario is not
completely clear because of a lack of experimental data.

To explore the multiphonon structure in ^{124}Sn, measurements have been performed at
the University of Kentucky 7 MV Van de Graaff accelerator through the (n,n'γ) reaction
[5]. Levels in ^{124}Sn have been studied through excitation functions of γ rays with neutron
energies from 2.2 to 4.5 MeV in 100 keV steps and angular distributions of γ rays at
neutron energy of 3.8 MeV. Clear definition of decay patterns was further aided by $\gamma\gamma$
coincidence measurements.

The figure shows the proposed level scheme for the multiphonon structure in ^{124}Sn
as established in these experiments. Beside the 4_1^+ (2101.7 keV), 2_2^+ (2129.7 keV) and
0_2^+ (2192.4 keV) levels of the 2-phonon triplet, another 4^+ (2221.4 keV) level has been
observed with strong collective strength. Either this or the 2101.7 keV level may be an
intruder state. Several closely spaced levels near 3 MeV excitation energy have been
considered to belong to the 3-phonon multiplet. Spin assignments for these levels have
been made from the angular distribution data. Comparisons of excitation function data
with statistical model calculations further support the spin assignments for these levels.
Once new level (2819.5 keV, 6_1^+) is included as a member of the multiplet. Reduced E2
transition probabilities have been extracted from doppler-shift attenuation measurements
of lifetimes. It has been observed that the decays have adequate collective strength to
belong to the 3-phonon group. Reduced transition probabilities from three-phonon states
to the members of the two-phonon triplet are an order of magnitude stronger than the

CP638, *Mapping the Triangle: International Conference on Nuclear Structure*
edited by A. Aprahamian, J. A. Cizewski, S. Pittel, and N. V. Zamfir
© 2002 American Institute of Physics 0-7354-0093-8/02/$19.00

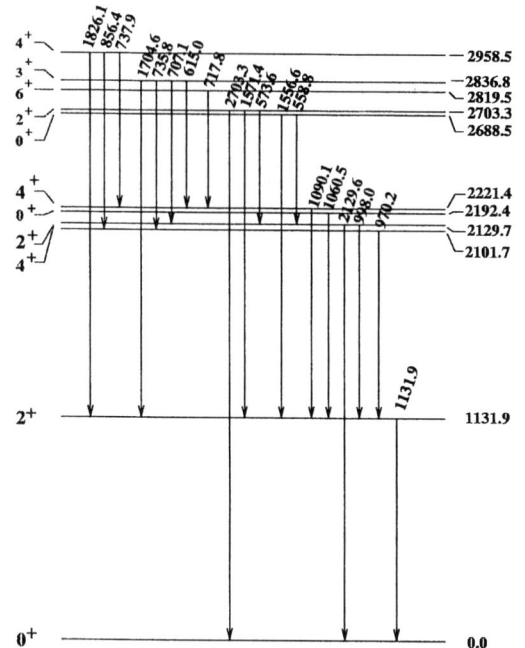

FIGURE 1. Proposed level scheme for 3-phonon multiplet

forbidden crossover transitions from the three-phonon to the one phonon level or to the ground state, thus supporting the collective nature of these states. IBM calculations have been performed including 4 neutron bosons and an inert ^{132}Sn core. The energies of the levels and the collective strengths of the transitions, on average, agree well with IBM calculations.

This work was supported by the U. S. National Science Foundation under grant No. PHY-0098813.

REFERENCES

1. A. Bohr & B. R. Mottleson, Nuclear structure, **Vol II**, 325 (Pub. Advanced Book Program, W. A. Benjamin Inc, 1975)
2. Der-San Chuu *et al.* , Prog. of Theor. Phys. **93**, 727 (1995).
3. D. M. Clement, *et al.*, Nucl. Phys. **A120**, 25 (1968).
4. I. Morrison *et al.*, Nucl. Phys. **A350**, 89 (1980).
5. P. E. Garrett *et.al*, J. Res. Nat. Inst. Standards Tech., **105**, 101 2000

g-Factor of the 2^+_1 State of ^{164}Yb

Z. Berant[1,2,3], A. Wolf[1,2,3], N. V. Zamfir[1,2], M. A. Caprio[1], D. S. Brenner[2],
N. Pietralla[1], R. L. Gill[4], R. F. Casten[1], C. W. Beausang[1], R. Kruecken[1],
C. J. Barton[1], J. R. Cooper[1], A. A. Hecht[1], D. M. Johnson[1], J. R. Novak[1],
H. Cheng[1], B. F. Albanna[1], and G. Gürdal[1,5]

[1]*Wright Nuclear Structure Laboratory, Yale University, New Haven, CT, USA*
[2]*Chemistry Department, Clark University, Worcester, MA, USA*
[3]*Nuclear Research Center Negev, Beer-Sheva, Israel*
[4]*Brookhaven National Laboratory, Upton, NY, USA*
[5]*University of Istanbul, Turkey*

Abstract. The g-factor of the first excited state of ^{164}Yb was measured by perturbed gamma-gamma angular correlation in an external static magnetic field of 5.55 Tesla. The result, $g(2^+_1)=0.32(5)$, extends the systematics of $g(2^+_1)$ for Yb isotopes down to N=94, and is discussed in the framework of different models.

An experimental setup for g-factor measurements of excited states in proton-rich nuclei was recently implemented on one of the beamlines of the tandem accelerator of the Wright Nuclear Structure Laboratory at Yale University. The setup consists of an aluminized Kapton moving tape collector system connected on one side to a reaction scattering chamber and on the other side to a superconducting coil, which provides static magnetic fields of up to 6 Tesla. Reaction products are deposited on the tape and transported to the center of the coil in a cyclic movement. This system was used to measure perturbed gamma-gamma angular correlations in nuclei produced following beta-decay of the reaction products deposited on tape. An 88 MeV ^{14}N beam with an intensity of about 15 pnA was used to produce the parent ^{164}Lu activity via the reaction ^{155}Gd(^{14}N,5n)^{164}Lu. A multi-detector angular correlation system consisting of eight HPGe detectors with efficiencies of 20%-25% was set around the center of the coil. The angles between the detectors were chosen so that 12 of the total of 28 pairs of detectors were at 35 degrees or 145 degrees. At these angles the perturbation effect due to the interaction between the nuclear magnetic moment and the static external magnetic field for $0^+{\to}2^+{\to}0^+$ gamma-gamma angular correlations is close to its maximum. An external field of 5.55 Tesla was used. About 2.7×10^8 coincidence events with field up and 2×10^8 events with field down were accumulated.

From the coincidence data we calculated the double ratio:

$$R(\theta, B) = \sqrt{\frac{I(\theta, B)}{I(\theta, -B)} : \frac{I(-\theta, B)}{I(-\theta, -B)}}$$

CP638, *Mapping the Triangle: International Conference on Nuclear Structure*
edited by A. Aprahamian, J. A. Cizewski, S. Pittel, and N. V. Zamfir
© 2002 American Institute of Physics 0-7354-0093-8/02/$19.00

$I(\pm\theta, \pm B)$ is the coincidence intensity for a particular cascade and all detector pairs at angle $\pm\theta$ and field $\pm B$. The signs of the angles were defined using the standard convention with respect to the direction of the field. For the 852 keV – 123 keV, $0^+_2 \rightarrow 2^+_1 \rightarrow 0^+_1$ cascade in ^{164}Yb we obtained at 145 degrees: $R(145, 5.55T)=1.505(60)$. The g-factor of the 2^+_1 state was extracted from this ratio: $g(2^+_1)=0.32(5)$. The double ratio of the same cascade at 180 degrees was R $(180, 5.55T)= 1.03(6)$. Since the expected value of R for any cascade at 180 degrees is 1.00, we have here a consistency check that no significant experimental asymmetries are present in our setup.

We now consider the present result in view of the systematics of $g(2^+_1)$ data for even-even Yb isotopes. In Figure 1 we present the known data and our result, compared to the simple Z/A prediction of the hydrodynamical model and to the prediction of the proton-neutron version of the interacting boson approximation, IBA-2. Although our result does not have the high accuracy of the existing data for the other isotopes, we clearly see that the g-factors are almost constant in the range N=94 to N=106, and certainly do not show the rather strong N-dependence predicted by IBA-2. A similar constancy of $g(2^+_1)$ vs. N was observed[1] for the platinum isotopes from ^{184}Pt to ^{198}Pt. This behavior indicates that most nucleons take part in the collective motion. All other elements from Gd to Pt show a stronger N-dependence, in better agreement with IBA-2, thus indicating that only the valence nucleons contribute to the magnetic moment. The difference between the two behaviors in the Gd-Pt region is not completely understood and warrants further investigation.

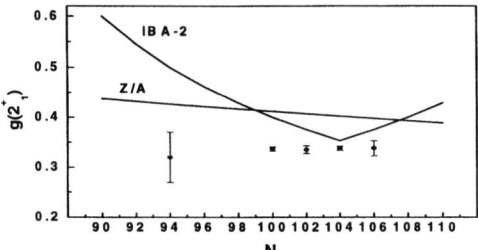

FIGURE 1. $g(2^+_1)$ vs. N for Yb isotopes compared to Z/A and IBA-2 predictions.

This work was supported by the US DOE under grants DE-FG02-91ER-40609, DE-FG02-88ER-40417, DE-AC02-98-CH10886 and by the DFG under grant Pi393/1-1.

1. A. Stuchbery et al., *Phys. Rev. Letters* **76**, 2246-2249 (1996).

Calculations of γ and γγ band B(E2) transition probabilities within Triaxial Projected Shell Model

P. Boutachkov, A. Aprahamian

Department of Physics, University of Notre Dame, Notre Dame, IN 46556, USA

Abstract. The Triaxial Projected Shell Model (TPSM) provides a unified microscopic description of rotational and multi-phonon vibrational states. We applied the TPSM model with its new capability to calculate B(E2) [1] values in the rare-earth region. The energies of the double-phonon states are reproduced. The calculated B(E2) values agree within an order of magnitude with the experimental ones. The calculations point to the identification of the new K=4$^+$ band in ^{162}Dy as a possible γγ-band.

Recently two-phonon γγ-vibrational bands have been identified in a number of nuclei [2]. While a microscopic description of the energies and transition probabilities of two-phonon vibrational excitations remains a challenge to nuclear models. The Triaxial Projected Shell Model (TPSM) has been shown to successfuly describe the energies of multi-phonon vibrational bands in the Er isotopes [3]. Within the TPSM, one calculates the γ-vibrational states by building a shell-model space truncated in a triaxially deformed basis. This is done by an exact three-dimensional angular-momentum projection of the γ-deformed Nilsson+BCS basis. Once the Hamiltonian is diagonalized in the TPSM basis, the eigenfunctions are used to calculate the electric quadrupole transition probabilities [1].

The calculated energies of the γγ-bands for different isotopes in the rare-earth region are shown in Fig 1. It can be seen from Fig 1. that the energies of ^{166,168,170}Er and ^{162,164}Dy are well-reproduced by this relatively new code. This points to the newly observed K=4$^+$ band in ^{162}Dy [4] as a candidate for a γγ-vibrational band.

The calculated B(E2) ratios for the studied nuclei are shown in Fig. 2. The predicted B(E2) values from the γγ-band of ^{166,168}Er and ^{162}Dy are much larger than the experimental values. We suspect that this is due to the fact that single-particle excitations are not included in the model space. The results for ^{170}Er show good agreement even without the inclusion of single-particle excitations.

In conclusion, the TPSM was applied to the rare-earth region. The energies of the established double γ-bands have been reproduced and a new candidate ^{162}Dy has been pointed out. The B(E2) values obtained by the calculation are bigger than the observed ones because of the lack of single-particle excitations in the TPSM basis. One can argue that the 4$^+_{K=4}$ wave function of ^{166}Er and especially of ^{170}Er is more collective than the ^{168}Er 4$^+_{K=4}$ state because the calculated B(E2) are closer to the experimental values. Therefore the model calculations suggest that the level at 2452 keV [5] in ^{170}Er is a double-phonon excitation.

CP638, *Mapping the Triangle: International Conference on Nuclear Structure*
edited by A. Aprahamian, J. A. Cizewski, S. Pittel, and N. V. Zamfir
© 2002 American Institute of Physics 0-7354-0093-8/02/$19.00

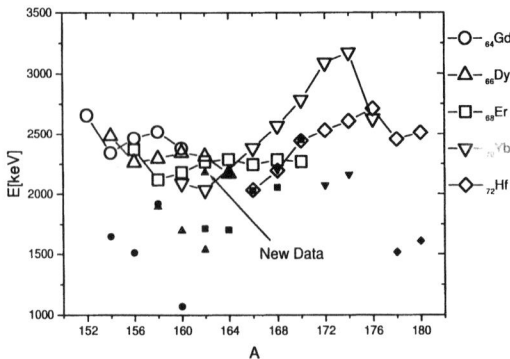

FIGURE 1. Energies of the K=4$^+$ band heads for various isotopes in the rare-earth nuclei. The experimental data are shown by solid symbols. The TPSM calculations are shown with open symbols. The square symbols represent the energies of the γ–band while the circles show the K=4+ bands.

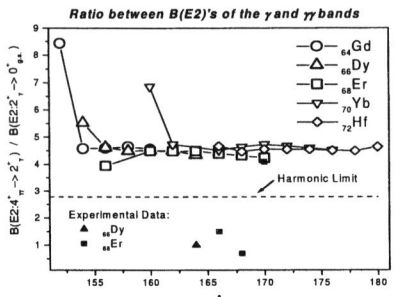

FIGURE 2. B(E2) ratios for different isotopes. The experimental data is presented with solid symbols. The TPSM calculations are shown with open symbols.

ACKNOWLEDGMENTS

The U.S. National Science Foundation under contract 99-01133 supported this research.

REFERENCES

1. P.Boutachkov, A.Aprahamian, Y.Sun, J.A.Sheikh, S.Frauendorf, submitted Eur. Phys. J (2002).
2. Y.Sun, et al., Phys. Rev. C 61, 064323 (2000).
3. X.Wu, et al., Phys. Rev. C 49, 1837 (1994); H.Borner, et al., Phys. Rev. Lett. 66, 691 (1991); T.Hartlein et al., Eur. Phys. J. A2, 253 (1998); P.E.Garrett, et al., Phys. Rev. Lett. 78, 4545 (1997); C.Fahlander, et al., Phys. Lett. B 388, 475 (1996).
4. C.Y.Wu, D.Cline, M.W.Simon, G.A.Davis, R.Teng, Phys. Rev. C 64, 064317 (2001).
5. W.Younes, et al., Nuclear Structure 98, p. 464, ORNL, AIP Conference Proceedings 481, Woodburt, NY

High-spin states in neutron-rich Rh isotopes

N. Fotiades*, J. A. Cizewski†, R. Krücken**, R. M. Clark‡, P. Fallon‡, I. Y. Lee‡, A. O. Macchiavelli‡, J. A. Becker§, D. P. McNabb§, W. Younes§ and L. A. Bernstein§

*Los Alamos National Laboratory, Los Alamos, New Mexico 87545
†Department of Physics and Astronomy, Rutgers University, New Brunswick, New Jersey 08903
**A. W. Wright NSL, Physics Department, Yale University, New Haven, Connecticut 06520
‡Nuclear Science Division, Lawrence Berkeley National Laboratory, Berkeley, California 94720
§Lawrence Livermore National Laboratory, Livermore, California 94550

Abstract. High-spin states in the neutron-rich ^{106,108,110,111,112}Rh isotopes have been investigated in the fission of the compound system formed in three heavy-ion induced reactions. Four bands were assigned to ^{106,108,110,112}Rh, respectively. Comparison with the lighter odd-odd Rh isotopes supports the assignment of the bands to the $\pi g_{9/2} \otimes \nu h_{11/2}$ configuration. In ^{111}Rh the level scheme consists of two rotational bands. In the ground-state band the odd-proton occupies the $\pi g_{9/2}$ orbital.

The study of high-spin states in neutron-rich Rh nuclei is very interesting because these isotopes lie in a transitional mass region between spherical and deformed nuclei. The present work is the first spectroscopic study of high-spin states in ^{106,108,110,111,112}Rh. The existing shortage of information on high-spin states for all five isotopes is due to the difficulty to study them as evaporation residues in heavy-ion fusion reactions. An alternative way to study these isotopes is prompt γ-ray spectroscopy of fission fragments following fusion reactions.

The Gammasphere array at LBNL was used to detect γ-ray coincidences in three experiments involving the following reactions: i) 134.5-MeV ^{24}Mg on ^{173}Yb ii) 129-MeV ^{23}Na on ^{176}Yb and iii) 91-MeV ^{18}O on ^{208}Pb. The ^{173}Yb and ^{176}Yb were gold-backed targets. The complementary fission fragment technique was used to assign transitions to ^{106,108,110,111,112}Rh. The level schemes obtained are shown in figure 1. Details on the experiments and the level schemes will be discussed elsewhere [1].

The similarities between the bands assigned to ^{106,108,110,112}Rh in figure 1 is striking suggesting common interpretation. Comparison with the lighter odd-odd Rh isotopes supports the interpretation that the bands are built on the $\pi g_{9/2} \otimes \nu h_{11/2}$ configuration. Spin and parity of the levels were assigned based on this comparison and remain tentative. The excitation energies of the bands remain uncertain. The B(M1)/B(E2) ratios [1] deduced with the adopted spin-parity values also support this configuration assignment.

In the double-midshell nucleus of ^{111}Rh 28 transitions were observed in coincidence with the previously known 211 keV, $(9/2^+) \rightarrow (7/2^+)$ transition. The proposed level scheme consists of two rotational bands. In band 1 the odd-proton occupies the $\pi g_{9/2}$ orbital. The interpretation of band 2 remains uncertain. The level scheme of ^{111}Rh is

CP638, *Mapping the Triangle: International Conference on Nuclear Structure*
edited by A. Aprahamian, J. A. Cizewski, S. Pittel, and N. V. Zamfir
© 2002 American Institute of Physics 0-7354-0093-8/02/$19.00

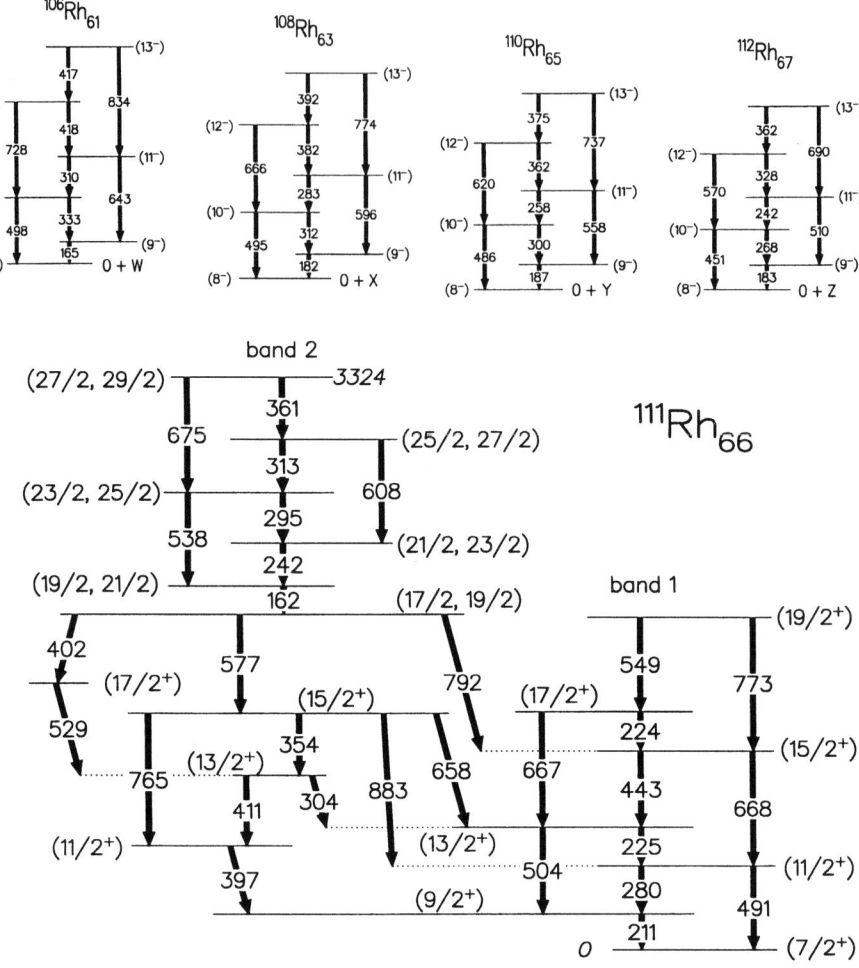

FIGURE 1. Level schemes assigned to [106,108,110,111,112]Rh in the present work.

very similar to that of [109]Rh [2].

This work has been supported by the U.S. Department of Energy under Contracts No. W-7405-ENG-36 (LANL), FG02-91ER-40609 (Yale), W-7405-ENG-48 (LLNL) and AC03-76SF00098 (LBNL) and by the National Science Foundation (Rutgers).

REFERENCES

[1] N. Fotiades *et al.*, to be published.
[2] Ts. Venkova *et al.*, Eur. Phys. J. A **6**, 405 (1999).

Investigating Low-spin States in ^{92}Zr With The (n,n′γ) Reaction

C. Fransen*, D. Bandyopadhyay*, N. Boukharouba*, P. von Brentano†,
S.R. Lesher*, M.T. McEllistrem*, N. Pietralla†, V. Werner† and S.W. Yates*

*University of Kentucky, Lexington, KY 40506-0055, USA
†Institut für Kernphysik, Universität zu Köln, 50937 Köln, Germany

Abstract. To investigate the evolution of collectivity in a transition from nuclei with a collective behavior to nearly closed shell nuclei, the low-spin level scheme of ^{92}Zr has been investigated with the (n,n′γ) reaction.

Collective excitations of heavy nuclei are among the most interesting aspects of nuclear structure. In particular, the properties of collective excitations which are not fully symmetric with respect to the proton-neutron (pn) degree of freedom, so called mixed-symmetry (MS) states [1] predicted in the pn version of the interacting boson model (IBM–2) [2], are sensitive to the isovector part of the residual nucleon-nucleon interaction. Recently, investigation of the fundamental one-phonon 2^+ MS state and searches for members of an expected multiplet of two-phonon MS states with the structure $(2_1^+ \otimes 2_{ms}^+)$ in nuclei of the A=100 region have been of special interest. In measurements on ^{94}Mo, the one-phonon 2^+ MS $(2_{1,ms}^+)$ state and the 1^+, 2^+, and 3^+ two-phonon MS states were clearly identified [3, 4, 5]. Candidates for these states were also found in neighboring ^{96}Ru [6].

To investigate the evolution of these states in a transition from nuclei with somewhat collective behavior like ^{94}Mo to nearly closed shell nuclei, measurements on ^{92}Zr were performed with the (n,n′γ) reaction at the Van de Graaff accelerator of the University of Kentucky. Gamma-ray angular distributions were measured with neutron energies of 2.4 and 3.9 MeV, and an excitation function with neutron energies from 2.4 to 4.0 MeV, in 100 keV steps, was performed. We obtained detailed information about the low-spin level scheme of ^{92}Zr, including spins, γ-ray branching ratios, γ-ray multipolarities, and level lifetimes. We were able to observe a total of 37 excited states and 92 γ-ray transitions in ^{92}Zr. Seven excited states were identified for the first time. From these data, absolute transition strengths were obtained, thus allowing the interpretation of the excited states.

^{92}Zr exhibits a much lower collectivity than neighboring ^{94}Mo, as is clear from the weak $2_1^+ \to 0_1^+$ transition with $B(E2; 2_1^+ \to 0_1^+) = 6.4_{-0.4}^{+0.5}$ W.u. and the low-lying 0_2^+ and 4_1^+ states at about 1.5 times the excitation energy of the 2_1^+ state.

The 2_2^+ state decays to the 2_1^+ state with an enhanced $M1$ transition strength of $B(M1; 2_2^+ \to 2_1^+) = 0.37(4)\,\mu_N^2$ and shows a weakly collective $E2$ ground state decay

CP638, *Mapping the Triangle: International Conference on Nuclear Structure*
edited by A. Aprahamian, J. A. Cizewski, S. Pittel, and N. V. Zamfir
© 2002 American Institute of Physics 0-7354-0093-8/02/$19.00

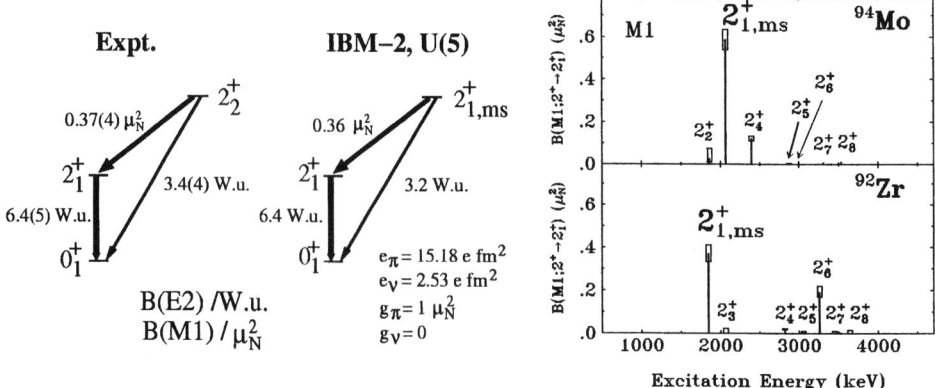

FIGURE 1. The left part depicts the identification of the $2^+_{1,ms}$ state in ^{92}Zr from a comparison of the $E2$ and $M1$ strengths with predictions in the IBM-2. The right part shows a comparison of the $M1$ strengths of all observed $2^+ \rightarrow 2^+_1$ decays for ^{94}Mo (from [7]) and ^{92}Zr giving evidence for the interpretation of the 2^+_2 state in ^{92}Zr as the $2^+_{1,ms}$ state due to the strong $2^+_2 \rightarrow 2^+_1$ $M1$ transition.

of $B(E2; 2^+_2 \rightarrow 0^+_1) = 3.4(4)$ W.u. in agreement with predictions in the IBM–2 for a one-phonon MS state both in the dynamic O(6) and U(5) limits with the ^{88}Sr core. The left part of Fig. 1 shows a comparison with predictions in the U(5) limit of the IBM–2. In the right part of Fig. 1, all $2^+ \rightarrow 2^+_1$ $M1$ transition strengths in ^{92}Zr are compared with those in ^{94}Mo [7], in which the 2^+_3 state was identified as the $2^+_{1,ms}$ state. Therefore, we interpret the 2^+_2 state as a candidate for a one-phonon MS state.

No states with clear signatures of two-phonon MS states were found. This result is not surprising in view of the low collectivity of ^{92}Zr. Only the 3^+_2 state at 3275.9 keV and a $1^{(+)}$ state at 3471.9 keV exhibit some properties of two-phonon MS states. Preliminary calculations in the shell model [8] reproduce the characteristics of ^{92}Zr better than IBM–2 calculations and affirm that ^{92}Zr is transitional between nuclei with a more collective behavior, like ^{94}Mo, and nuclei with evident shell model structure.

This work was supported by the U.S. National Science Foundation under Grant No. PHY-0098813 and the Deutsche Forschungsgemeinschaft under Grant No. Pi 393/1-2.

REFERENCES

1. F. Iachello, Lecture Notes on Theor. Phys., Groningen, 1976; T. Otsuka, Ph.D. thesis, Univ. of Tokyo, 1978.
2. F. Iachello, Phys. Rev. Lett. **53**, 1427 (1984).
3. N. Pietralla, *et al.*, Phys. Rev. Lett. **83**, 1303 (1999).
4. N. Pietralla, *et al.*, Phys. Rev. Lett. **84**, 3775 (2000).
5. C. Fransen, *et al.*, Phys. Lett. **B 508**, 219 (2001).
6. H. Klein, *et al.*, Phys. Rev. C **65**, 044315 (2002).
7. C. Fransen, *et al.*, to be published.
8. V. Werner, *et al.*, to be published.

Mixed-Mode Shell-Model Calculations

V. G. Gueorguiev* and J. P. Draayer*

*Department of Physics and Astronomy, Louisiana State University, Baton Rouge, Louisiana 70803

Abstract. A one-dimensional harmonic oscillator in a box is used to introduce the oblique-basis concept. The method is extended to the nuclear shell model by combining traditional spherical states, which yield a diagonal representation of the usual single-particle interaction, with collective configurations that track deformation. An application to ^{24}Mg, using the realistic two-body interaction of Wildenthal, is used to explore the validity of this mixed-mode shell-model scheme. Specifically, the correct binding energy (within 2% of the full-space result) as well as low-energy configurations that have greater than 90% overlap with full-space results are obtained in a space that spans less than 10% of the full-space. The theory is also applied to lower pf-shell nuclei, $^{44-48}Ti$ and ^{48}Cr, using the Kuo-Brown-3 interaction. These nuclei show strong SU(3) symmetry breaking due mainly to the single-particle spin-orbit splitting. Nevertheless, the results also show that yrast band B(E2) values are insensitive to fragmentation of the SU(3) symmetry. Specifically, the quadrupole collectivity as measured by B(E2) strengths remains high even though the SU(3) symmetry is rather badly broken. The IBM and broken-pair models are considered as alternative basis sets for future oblique-basis shell-model calculations.

Typically, two competing modes characterize the structure of a nuclear system. The spherical shell model is the theory of choice when single-particle behavior dominates. When deformation dominates, it is the Elliott SU(3) model. This manifests itself in two dominant elements in the nuclear Hamiltonian: the single-particle field, $H_0 = \sum_i \varepsilon_i n_i$, and a collective quadrupole-quadrupole interaction, $H_{QQ} = Q \cdot Q$. It follows that a simplified Hamiltonian $H = \sum_i \varepsilon_i n_i - \chi Q \cdot Q$ has two solvable limits associated with these modes.

To probe the nature of such a system, we first consider the simpler problem of a one-dimensional harmonic oscillator in a box of size $2L$. As for real nuclei, this system has a finite volume and a restoring harmonic potential $\omega^2 x^2 / 2$. Depending on the value of $E_c = \omega^2 L^2 / 2$, which plays the role of a critical energy, there are three spectral types: (1) for $\omega \to 0$ the spectrum is simply that of a particle in a box; (2) at some value of ω, the spectrum begins with E_c followed by the spectrum of a particle in a box perturbed by the harmonic potential; and (3), for sufficiently large ω there is a harmonic oscillator spectrum below E_c followed by the perturbed spectrum of a particle in a box. The last scenario is the most interesting one since it provides an example of a two-mode system. For this case, the use of two sets of basis vectors, one representing each of the two limits, has physical appeal. One basis set consists of the harmonic oscillator states; the other set consists of basis states of a particle in a box. Even thought a mixed spectrum is expected around E_c, a numerical study that includes up to 50 harmonic oscillator states below E_c shows that first order perturbation theory works well after the breakdown of the harmonic spectrum. Although the spectrum seems to be well described in this way, the wave functions near E_c have an interesting coherent structure.

An application of the theory to ^{24}Mg, using the realistic two-body interaction of

CP638, *Mapping the Triangle: International Conference on Nuclear Structure*
edited by A. Aprahamian, J. A. Cizewski, S. Pittel, and N. V. Zamfir
© 2002 American Institute of Physics 0-7354-0093-8/02/$19.00

Wildenthal, demonstrates the validity of the mixed-mode shell-model scheme. In this case the oblique-basis consists of the traditional spherical states, which yield a diagonal representation of the single-particle interaction, together with collective SU(3) configurations, which yield a diagonal quadrupole-quadrupole interaction. The results obtained in a space that spans less than 10% of the full-space reproduce the correct binding energy (within 2% of the full-space result) as well as the low-energy spectrum and structure of the states that have greater than 90% overlap with the full-space results. In contrast, for a m-scheme spherical shell-model calculation one needs about 60% of the full space to obtain results comparable with the oblique basis results.

Studies of the lower pf-shell nuclei, $^{44-48}Ti$ and ^{48}Cr, using the realistic Kuo-Brown-3 (KB3) interaction show strong SU(3) symmetry breaking due mainly to the single-particle spin-orbit splitting. Thus the KB3 Hamiltonian could also be considered a two-mode system. This has been further supported by the behavior of the yrast band B(E2) values that seems to be insensitive to fragmentation of the SU(3) symmetry. Specifically, the quadrupole collectivity as measured by the B(E2) strengths remains high even though the SU(3) symmetry is rather badly broken. This has been attributed to a quasi-SU(3) symmetry where the observables behave like a pure SU(3) symmetry while the true eigenvectors exhibit a strong coherent structure with respect to each of the two bases. This provides the opportunity for further study of the implications of two-mode calculations.

Future research may extend to multi-mode oblique calculations. An immediate extension of the current scheme might use the eigenvectors of the pairing interaction within the Sp(4) algebraic approach to the nuclear structure, together with the collective SU(3) states and spherical shell model states. Hamiltonian driven basis sets can also be considered. In particular, the method may use eigenstates of the very-near closed shell nuclei obtained from a full shell model calculation to form Hamiltonian driven J-pair states for mid-shell nuclei. This type of extension would mimic the Interacting Boson Model (IBM) and the so-called broken-pair theory. In particular, the three exact limits of the IBM can be considered to comprise a three-mode system. Nonetheless, the real benefit of this approach is expected when the spaces encountered are too large to allow for exact calculations.

ACKNOWLEDGMENTS

We acknowledge support from the U.S. National Science Foundation under Grant No. PHY-9970769 and Cooperative Agreement No. EPS-9720652 that includes matching from the Louisiana Board of Regents Support Fund. V. G. Gueorguiev is grateful to the Louisiana State University Graduate School for awarding him a dissertation fellowship and to the U. S. National Science Foundation for the support for young scientists to attend the International Nuclear Structure Conference on "Mapping the Triangle" held May 22-25, 2002 in Grand Teton National Park, Wyoming, and in so doing allowing him to present the main results of his Ph.D. dissertation.

Experiments on Chiral Symmetry Breaking in the Mass 130 Region

A.A. Hecht*, C.W. Beausang*, H. Amro*, Z. Berant*, C.J. Barton*, M.A. Caprio*, R.F. Casten*, J.R. Cooper*, G. Gürdal*, R. Krücken*, D. Meyer*, H. Newman*, J.R. Novak*, N. Pietralla*, J. Ressler*, M. Sciacchitano*, N.V. Zamfir* and Jing-ye Zhang[†]

*Wright Nuclear Structure Laboratory, Yale University, New Haven, CT, USA
[†]University of Tennessee, Knoxville, TN 37996, USA

Abstract. High spin states were studied in ^{140}Eu in a search for evidence of chiral structure.

Static chiral symmetries are common in nature. Well known examples range from the macroscopic spirals of snail shells to the microscopic handedness of certain molecules. Chiral symmetry is also well known in particle physics, where it is of a dynamic nature distinguishing between the two possible orientations of the intrinsic spin with respect to the momentum of the particle. Recently, the possibility that a type of chiral symmetry might exist in some doubly-odd, triaxial nuclei was suggested. For a triaxial nucleus, possessing a short, intermediate, and long axis, the collective angular momentum vector R tends to align along the intermediate axis, as this axis possesses the largest moment of inertia. For a suitable choice of particle numbers and deformation, the valence proton and neutron Fermi surfaces lie near the bottom and top of a high-j shell, respectively. In this case, their single particle angular momentum vectors will tend to align along the short and long nuclear axes, respectively, perpendicular to each other and to the collective angular momentum. Such a favorable situation is encountered in the mass $A \sim 130$ region, with $Z \sim 60$, $N \sim 75$. Here, the proton Fermi surface lies low in the $h_{11/2}$ shell (high-j, low-Ω) while the neutron Fermi surface lies high in the same $h_{11/2}$ shell (high-j, high-Ω). In addition, these nuclei are predicted to be soft with respect to γ-deformation, and this deformation may be stabilized by the shape driving effects of the valence particles.

In this case, the total angular momentum vector I may not lie along a principal nuclear axis or even in a principal plane. The three nuclear axes, looked at from the point of view of the total angular momentum vector, can form either a left or right-handed coordinate system which have opposite handedness or chirality. The experimental signature of such a chiral symmetry would be degenerate pairs of $\Delta I=1$ bands of the same parity. With mixing between the bands, one band would be lowered in energy and be the yrast band, the other would become the less populated chiral partner band.

In search of these chiral partner bands, high spin states in the doubly odd $N=75$ nuclei ^{136}Pm and ^{138}Eu, and the $N=77$ nucleus ^{140}Eu were investigated following the ^{116}Sn(^{24}Mg, p3n), ^{106}Cd(^{35}Cl, 2pn), and ^{92}Mo(^{51}V, 2pn) reactions, respectively. New bands were found for the nuclei. Polarization and angular correlation measurements have been performed to establish the relative spin and parity assignments for the bands.

CP638, *Mapping the Triangle: International Conference on Nuclear Structure*
edited by A. Aprahamian, J. A. Cizewski, S. Pittel, and N. V. Zamfir
© 2002 American Institute of Physics 0-7354-0093-8/02/$19.00

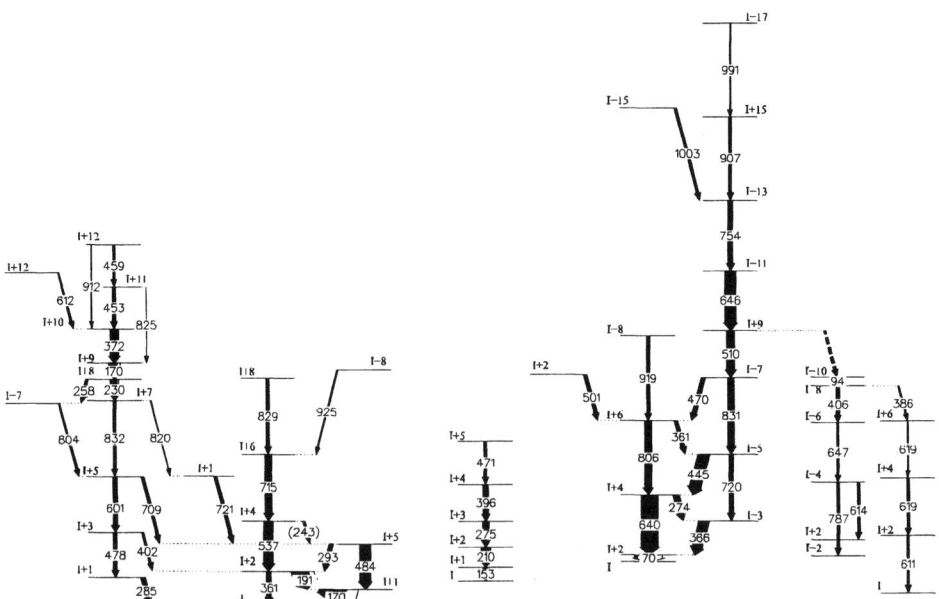

FIGURE 1. New levels seen in ^{140}Eu. Absolute spins have not been assigned, spins are given relative to the band heads.

Candidate chiral band partners have been identified in ^{136}Pm and ^{138}Eu [1, 2]. Previously known levels for ^{140}Eu are from low spin decay studies, and have been identified up to a 5^- isomer. High spin levels in ^{140}Eu have been identified for the first time in the current work and are presented in figure 1.

The various bands presented have been assigned to ^{140}Eu through a combination of excitation functions and Kα x-ray correlations. The analysis has been done on in-beam spectra, so there is no cross isomer information. At present, the three separate band groups cannot be connected. Analysis and theoretical interpretations are ongoing for ^{140}Eu.

This work is supported by the U.S. DOE under grant numbers DE-FG02-91ER-40609, DE-FG05-96ER-40983, and DE-FG02-88ER-40417.

REFERENCES

1. A.A. Hecht *et al.*, Phys. Rev C **63**, 051302(R) (2001).
2. K. Starosta *et al.*, Phys. Rev. Lett. **86**, 971 (2001).

Quasicontinuous Decay of Superdeformed ^{195}Pb

M.S. Johnson*, J.A. Cizewski*, K.Y. Ding*, N. Fotiades*, M.B. Smith*,
J.S. Thomas*, K. Hauschild†, D.P. McNabb†, J.A. Becker†,
L.A. Bernstein†, R.M. Clark**, M.A. Deleplanque**, R.M. Diamond**,
P. Fallon**, I.Y. Lee**, A.O. Macchiavelli** and F.S. Stephens**

*Physics and Astronomy, Rutgers University, New Brunswick, NJ 08903 USA
†Lawrence Livermore National Laboratory, Livermore, CA 94550 USA
**Lawrence Berkeley National Laboratory, Berkeley, CA 94720 USA

Abstract. The quasicontinuous decay spectrum of superdeformed excitations in ^{195}Pb has been extracted. The slow increase of intensity as E_γ decreases is consistent with a level density of normal deformed excitations at $\sim 1\hbar$ with no gap parameter.

Superdeformed (SD) rotational bands have been observed in many A$\sim$190 nuclei and provide evidence for a second minimum in the potential energy at large deformation. The SD states decay to normal deformed (ND) states via mixing between cold SD states and hot ND excitations. Decays to ND yrast states occur via two pathways: discrete one-step transitions, which link SD states and yrast ND excitations, and multistep decays via intermediate states above the ND yrast excitations.

The shape of the quasicontinuous (QC) decay spectrum [1] depends on the level density of the hot ND states, as well as the γ-ray strength function. In particular, the shape of the QC spectrum depends on the gap in ND level density at finite spin, a gap which arises from pairing correlations. For a finite gap, G, the QC spectrum has low intensity at energies $E_\gamma \sim U$ (energy above the yrast), then rises dramatically for energies below $E_\gamma < U - G$. For no gap in the level density, there is a gradual increase in γ-ray intensity as E_γ decreases, which starts at $E_\gamma \sim U$. The quasicontinuous decays of SD states have been previously studied [2-4] in even-even Hg and Pb isotopes.

To probe the decay of SD excitations in ^{195}Pb, the ^{174}Yb(^{26}Mg,5n) reaction was used with E(^{26}Mg) = 132 MeV, gold-backed targets and the Gammasphere array at LBNL. Four SD bands have been previously identified [5] in ^{195}Pb. The analysis focused on the two lowest-lying, signature partner, SD bands. Since the crosstalk between these two bands suggests that the QC decay spectrum gated on either band would have contributions from both bands, the QC spectra were added together.

Background-subtracted SD spectra for each ring of Gammasphere were obtained by double-gating on SD transitions and correcting for neutron events, summing effects, and escape events. Discrete transitions were removed and intensities normalized to one cascade. Angular distributions were fitted to $W(\theta) = A_0 + A_2 P_2(cos\theta)$. The sum of the A_0 spectra for SD bands 1 and 2 is displayed in Fig. 1. The shape of the statistical E1 spectrum was determined by fitting the high-energy region of the ^{195}Pb ND spectrum to $E_\gamma^3 e^{-E\gamma/T}$, and subtracted. The A_2 coefficients were used to isolate the stretched E2 and

CP638, *Mapping the Triangle: International Conference on Nuclear Structure*
edited by A. Aprahamian, J. A. Cizewski, S. Pittel, and N. V. Zamfir
© 2002 American Institute of Physics 0-7354-0093-8/02/$19.00

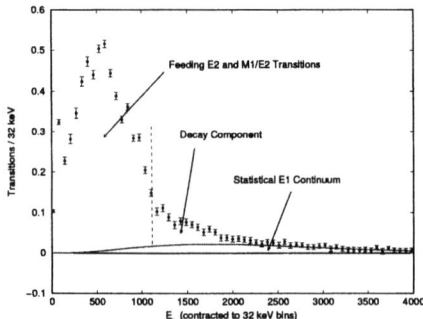

FIGURE 1. Components of non-discrete spectrum for SD band 1 and band 2 of [195]Pb.

TABLE 1. Results of QC Decay Analysis

	$\bar{m}$	$\bar{E}_\gamma$	$\bar{m} \otimes \bar{E}_\gamma$	$\bar{m} \otimes \Delta\mathbf{I}$	U
[195]Pb - 1+2	2.03(45)	1.40(9)	2.8(7)	0.6(4)	2.6(3)
[194]Pb	1.93(16)	1.44(3)	2.7(2)	0.8(3)	2.74

mixed M1/E2 transitions; the remaining spectrum is the QC decay of the SD bands.

The key results for the QC decay of SD bands 1 and 2 in [195]Pb are summarized in Table I and compared with previous results [3] for [194]Pb. The integral of the QC spectrum yields the average multiplicity $\bar{m}$, and the centroid gives the average γ-ray energy $\bar{E}_\gamma$. The energy carried off by the QC decay is $\bar{m} \otimes \bar{E}_\gamma$. The angular momentum carried off by the QC decay is $\bar{m} \otimes \Delta I$, where $\Delta I = 0.3\hbar$ is the average angular momentum of a QC transition.

The intensity of the QC decay spectrum of [195]Pb at high energy increases slowly as E_γ decreases. The shape and intensity of this spectrum is consistent with a two-step decay process and no gap in the level density of ND excitations at $\sim 1\hbar$. This shape is in contrast to those observed [3] for the QC decay of SD excitations in [192,194]Pb, which were consistent with finite gaps, 0.3 MeV and 0.9 MeV, in level density at $10\hbar$ and $6\hbar$, respectively. Since the shape of the [195]Pb QC decay spectrum is consistent with no gap in level density, U could be deduced by determining the highest lower limit at which the integral of the high-energy region of the QC spectrum is non-zero. This deduced value of U is consistent with the value determined by $\bar{m} \otimes \bar{E}_\gamma$ and also confirms that a good separation of E2 feeding transitions and QC decay transitions has been obtained. For [195]Pb $U \sim 2.6$ MeV, which is similar to the value obtained for [194]Pb from the analyses of discrete and quasicontinuous decays of the SD band [3,6].

Work supported by the National Science Foundation and US Department of Energy.

[1] T. Døssing, et al., Phys. Rev. Lett. **75**, 1276 (1995).
[2] R.G. Henry et al., Phys. Rev. Lett. **73**, 777 (1994).
[3] D.P. McNabb et al., Phys Rev C **61**, 031304(R) (2000).
[4] T. Lauritsen et al., Phys. Rev. C **62**, 044136 (2000).
[5] L.P. Farris et al., Phys Rev C **51**, R2288 (1995).
[6] K. Hauschild et al., Phys. Rev. C **55**, 2819 (1997) and references therein.

Excited states in drip line nucleus ^{140}Dy

W. Królas[1,2,3], R. Grzywacz[4,5], K. P. Rykaczewski[4,5], J. C. Batchelder[6],
C. R. Bingham[4,7], C. J. Gross[4,8], D. Fong[2], J. H. Hamilton[2], D. J. Hartley[7],
J. K. Hwang[2], Y. Larochelle[7], T. A. Lewis[4], K. H. Maier[1,4],
J. W. McConnell[1], A. Piechaczek[9], A. V. Ramayya[2], K. Rykaczewski[10],
D. Shapira[4], M. N. Tantawy[7], J. A. Winger[11], C. -H. Yu[4], E. F. Zganjar[9],
A. T. Kruppa[1,12], W. Nazarewicz[4,7,13], and T. Vertse[1,12]

[1]*Joint Institute for Heavy Ion Research, Oak Ridge, TN 37831*
[2]*Dept. of Physics and Astronomy, Vanderbilt University, Nashville, TN 37235*
[3] *H.Niewodniczański Institute of Nuclear Physics, PL-31342 Kraków, Poland*
[4]*Physics Division, Oak Ridge National Laboratory, Oak Ridge, TN 37831*
[5]*Institute of Experimental Physics, Warsaw University, PL-00681 Warsaw, Poland*
[6]*UNIRIB Oak Ridge Associated Universities,Oak Ridge, TN 37831*
[7]*Dept. of Physics and Astronomy, University of Tennessee, Knoxville, TN 37996*
[8]*Oak Ridge Institute for Science and Education, Oak Ridge, TN 37831*
[9]*Dept. of Physics and Astronomy, Louisiana State University, Baton Rouge, LA 70803*
[10]*Oak Ridge High School, Oak Ridge, TN 37830*
[11]*Dept. of Physics and Astronomy, Mississippi State University, Mississippi State, MS 39762*
[12]*Institute of Nuclear Research, Hungarian Academy of Sciences, H-4001 Debrecen, Hungary*
[13]*Institute of Theoretical Physics, Warsaw University, PL-00681 Warszawa, Poland*

Abstract. A new 7 μs isomer in the drip line nucleus ^{140}Dy was selected from the products of the ^{54}Fe (315 MeV) + ^{92}Mo reaction by a recoil mass spectrometer and studied with recoil – delayed γ-γ coincidences. Five cascading γ transitions were interpreted as the decay of an $I^\pi = 8^-$ $\{\nu 9/2^-[514] \otimes \nu 7/2^+[404]\}$ K isomer via the ground state band.

In an experiment performed at the Holifield Radioactive Ion Beam Facility at Oak Ridge we observed for the first time a 7 μs isomer in the drip line nucleus ^{140}Dy [1]. The ^{140}Dy ions were selected from the products of ^{54}Fe (315 MeV) + ^{92}Mo reaction by the Recoil Mass Spectrometer and implanted in a passive catcher placed in the center of the Clover Germanium Detector Array for Recoil Decay Spectroscopy. A Multichannel Plate Detector provided a recoil implantation reference time and enabled a recoil – delayed γ-γ coincidence study.

We have identified a new cascade of γ-rays at 202, 364, 476, 550 and 574 keV with a half-life of 7.0 ± 0.5 μs correlated with the implantation of the selected $A = 140$ recoils (see Fig. 1, left panel). The five γ lines are in coincidence with each other, and with Dy K X-rays, which places them in one cascade in ^{140}Dy. Basing on the intensities and energies of the transitions a level scheme resembling a rotational band in a deformed nucleus fed by the isomeric level, can be constructed (see Fig. 1, right panel). Also, a comparison to the decay patterns of $I^\pi = 8^-$ K isomers in the less exotic $N = 74$ even isotones of ^{134}Nd, ^{136}Sm and ^{138}Gd displayed in Fig. 1 shows striking similarity. This leads us to the interpretation of the isomeric level at 2166 keV

CP638, *Mapping the Triangle: International Conference on Nuclear Structure*
edited by A. Aprahamian, J. A. Cizewski, S. Pittel, and N. V. Zamfir
© 2002 American Institute of Physics 0-7354-0093-8/02/$19.00

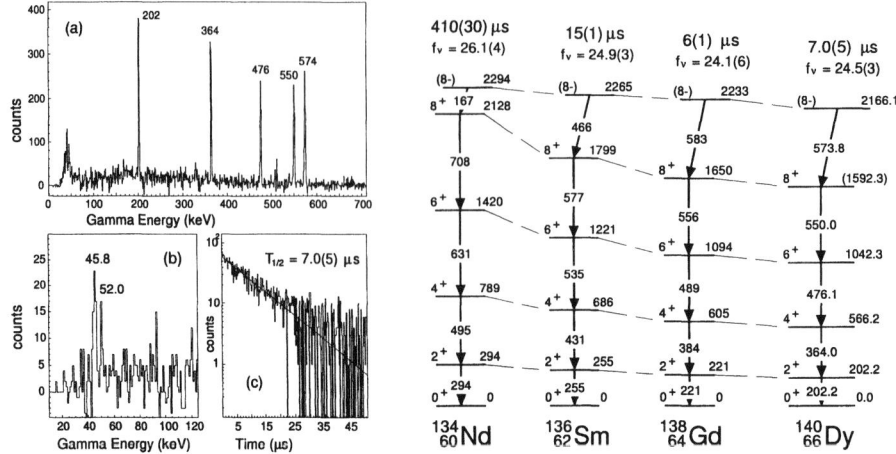

FIGURE 1. ^{140}Dy γ lines from the decay of the $I^\pi = (8^-)$ isomer obtained from double γ coincidence data by adding five spectra gated on the labeled transitions - panel (a). Dysprosium K_α and K_β X rays in coincidence with the sequence of five new γ lines from triple γ coincidence data - panel (b). Decay pattern produced by double-gating on five transitions - panel (c). Systematics of the decay properties of $I^\pi = 8^-$ isomers in $Z \geq 60$, $N = 74$ isotones and the proposed ^{140}Dy level scheme - right panel.

as an $I^\pi = (8^-)$ $\{v9/2^-[514] \otimes v7/2^+[404]\}$ K isomer decaying via the $8^+ \rightarrow 6^+ \rightarrow 4^+ \rightarrow 2^+ \rightarrow 0^+$ cascade belonging to the ground-state band in ^{140}Dy. Our findings were recently confirmed in an independent experiment [2].

The isomeric level is placed at an excitation energy close to the 2150 keV predicted in [3] for this two-quasineutron configuration. A hindrance per degree of K-forbiddenness f_v was determined to be 24.5(3), a value very close to the ones found for less exotic $N = 74$ isotones displayed in Fig. 1. This suggests that the configuration and properties of this K isomer in ^{140}Dy are not affected by the proximity of the proton drip line.

Following the global systematics of the ground to the first-excited 2^+ state transitions in even-even nuclei of Raman et al. [4] the observed excitation energy of the first 2^+ state gives a deformation parameter $\beta_2 \approx 0.234$ for ^{140}Dy. This is a somewhat smaller quadrupole deformation than the previously anticipated values (see e.g. [3, 5]) but it is close to the $\beta_2 = 0.25$ derived from the observed level schemes of ^{141gs,m}Ho [6].

REFERENCES

1. Królas, W. et al., *Phys. Rev.* **C65**, 031303 (2002).
2. Cullen, D.M. et al., *Phys. Lett.* **B529**, 42-49 (2002).
3. Xu, F. R., Walker, P.M., and Wyss, R., *Phys. Rev.* **C59**, 731-734 (1999).
4. Raman, S., Nestor, C.W. JR., and Tikkanen, P., *At. Data and Nucl. Data Tables* **78**, 1-128 (2001).
5. Rykaczewski, K. et al, *Phys. Rev.* **C60**, 011301 (1999).
6. Seweryniak, D. et al., *Phys. Rev. Lett.* **86**, 1458-1461 (2001).

Measuring angular distributions of gamma-rays emitted from fast exotic beams

Heather Olliver*, Thomas Glasmacher* and Andrew Stuchbery†

*National Superconducting Cyclotron Laboratory (NSCL) and Department of Physics and
Astronomy at Michigan State University, East Lansing, MI 48824 USA
†Australian National University, Canberra, ACT 0200, Australia

Abstract. We propose to apply the techniques of γ-γ angular correlations and angular distributions
for in-beam gamma-ray spectroscopy with fast exotic beams. Presented are angular distribution
plots which include Doppler shifts and different degrees of alignment of the m-substates. The plots
indicate that angular distribution measurements of gamma rays produced by nuclear fragmentation
reactions at relativistic velocities are feasible.

Measuring gamma-ray angular distributions is a well established experimental technique
used with low-energy beams to determine multipolarities of gamma-ray transitions.
However, this technique is not well established for fast beams and nuclear fragmentation reactions. The Coupled Cyclotron Facility at the NSCL produces exotic beams in
the energy range of 100-150 MeV/nucleon. A fragment separator separates and selects
desired nuclei by physical means after the primary beam has collided with a production target at the exit of the second cyclotron. The speed of a typical fast exotic beam is
approximately $0.4c$. To make in-beam γ-γ angular correlation and distribution measurements of fast exotic beams, an array of eighteen 32-fold segmented germanium detectors
will be used [1].

The angular distribution function, $W(\theta)$, specifies the relative probability to observe
a gamma-ray transition from spin $I_i \rightarrow I_f$ at an angle θ with respect to the beam axis and
has the form,

$$W(\theta) = 1 + A_2 B_2 P_2(\cos(\theta)) + A_4 B_4 P_4(\cos(\theta)) + ... \qquad (1)$$

in the center of mass. P_λ are the Legendre polynomials, A_λ are angular distribution
coefficients and B_λ are orientation parameters which are proportional to the statistical
tensor. The B_λ coefficients have a form directly related to the population parameters
which specify the m-state distribution of an inital state with spin I_i. The degree of alignment of a state which is dependent on the m-state distribution is not well known for a fast
fragmentation reaction. Assuming a Gaussian form with oblate alignment for the m-state
distribution, the degree of alignment can be characterized by the parameter σ/I_i where
σ is the width of the distribution. Work by F. Azaiez [2] gives preliminary evidence that
DCO ratios can distinguish dipole and quadrupole transitions in fragmentation reactions.

In Figure 1, $W(\theta)$ is shown in the lab frame, with θ measured with respect to the
beam direction, for various values of β and σ/I_i for a quadrupole and dipole transition.
The multipole mixing ratio is zero in the calculation for the 3→2 transition. $W(\theta)$

CP638, *Mapping the Triangle: International Conference on Nuclear Structure*
edited by A. Aprahamian, J. A. Cizewski, S. Pittel, and N. V. Zamfir
© 2002 American Institute of Physics 0-7354-0093-8/02/$19.00

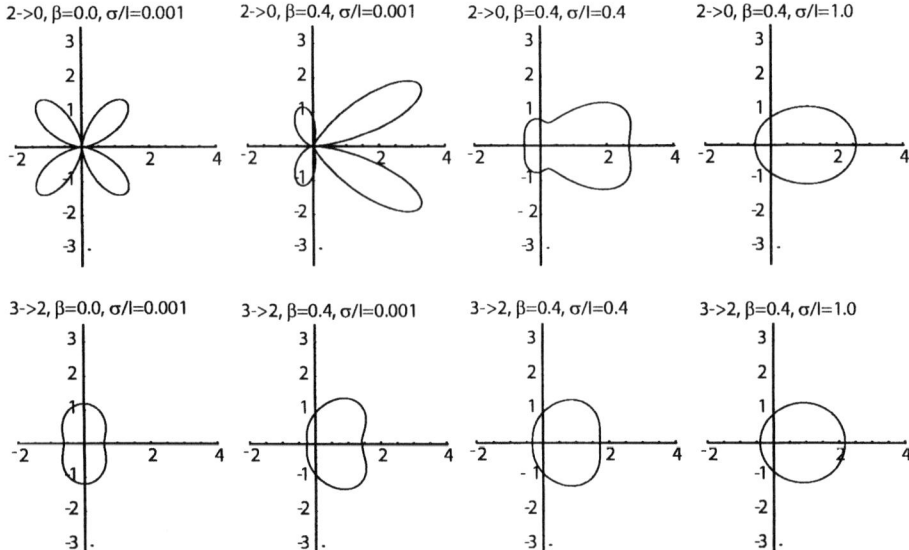

FIGURE 1. The first row contains polar plots of the angular distribution function, $W(\theta)$, for a $2{\to}0$ transition. The second row is for a $3{\to}2$ transition. The values of the beam velocity, β, and the width of the Gaussian m-state distribution, σ/I_i, are given for each plot. The beam direction is from left to right ($+0°$) with θ measured counterclockwise.

is normalized to 4π. The numerical values for $W(\theta)$, in the center of mass ($\beta{=}0$), agree with the tabulated values of Yamazaki [3] and Mateosian and Sunyar [4]. If a lower degree of alignment of 20%, corresponding to $\sigma/I_i{=}1.0$, is assumed, the angular distributions for the stretched quadrupole ($2{\to}0$) and pure dipole ($3{\to}2$) transitions are fairly similar when $\beta{=}0.4c$. However, the difference between the two can still be distinguished. If the alignment is greater than 20%, for $\beta{=}0.4c$, the distinction between a quadrupole transition and dipole transition becomes more pronounced.

In conclusion, it appears possible to distinguish a quadrupole transition from a dipole transition for a nuclear fragmentation reaction. However, the alignment present in a fragmentation reaction needs to be explored further.

This material is based on work supported by the National Science Foundation under Grants No. PHY-9875122 and PHY-0110253.

REFERENCES

1. W. F. Mueller et al., Nucl. Instrum. Methods **A** 466, 492 (2001).
2. F. Azaiez, in *Proceedings of the International Symposium on Exotic Nuclear Structures, Debrecen, Hungary 2000*, edited by Zs. Dombrádi and A. Krasznahorkay (MTA ATOMKI, Debrecen, 2000) p.149.
3. T. Yamazaki, Nucl. Data, Sect. A 3, 1 (1967).
4. E. Der Mateosian and A. W. Sunyar, At. Data Nucl. Data Tables **13**, 391 (1974).

Which dynamical symmetries does the Dirac equation have?

P.R. Page* [1]

*Theoretical Division, MS-B283, Los Alamos National Laboratory,
Los Alamos, NM 87545, USA*

Abstract. It is known that the Dirac equation has two dynamical symmetries, spin and pseudospin symmetry. Both are approximately realised in nature: spin symmetry in heavy–light mesons and pseudospin symmetry in nuclei. We prove that the spin and pseudospin symmetries are the only symmetries of their type that is possible for a parity conserving Dirac equation.

The Dirac Hamiltonian H' describing the relativistic motion of a single particle in a vector (with only a time component) and scalar potential is

$$H' = c\,\vec{\alpha}\cdot\vec{p} + \beta\left(mc^2 + V_S(\vec{r})\right) + V_V^0(\vec{r}) \quad \vec{\alpha} \equiv \begin{pmatrix} 0 & \vec{\sigma} \\ -\vec{\sigma} & 0 \end{pmatrix} \quad \beta \equiv \begin{pmatrix} 1 & 0 \\ 0 & -1 \end{pmatrix} \quad (1)$$

where the usual Dirac matrices $\vec{\alpha}$ and β are written in terms of the Pauli matrices $\vec{\sigma}$, the momentum operator is $\vec{p}$, and m is the mass of the particle. If the vector potential $V_V^0(\vec{r})$ is equal to the scalar potential $V_S(\vec{r})$ plus a constant potential U, which is independent of the spatial location $\vec{r}$ of the particle from the origin, $V_V^0(\vec{r}) = V_S(\vec{r}) + U$, then the Hamiltonian is invariant under a spin symmetry [1], i.e. there exists operators $\hat{S}_i$, $i \in \{1,2,3\}$, such that H' commutes with $\hat{S}_i$, i.e. $[H', \hat{S}_i] = 0$. If the vector potential is equal to minus the scalar potential plus a constant potential U, $V_V^0(\vec{r}) = -V_S(\vec{r}) + U$, then the Hamiltonian is invariant under a pseudospin symmetry [1]. The symmetries are called *dynamical* because they only obtain under certain assumptions about the behaviour of the potentials.

The spin symmetry is approximately realised in nature in heavy–light quark mesons systems, particularly the charm–light quark and strange–light quark systems, to the extent that the strange and charm quarks can be regarded as very heavy, so that these systems can be taken to be one–body systems [2]. The pseudospin symmetry approximately obtains in various heavy nuclei, for example the particle or hole states in double magic ^{208}Pb, which can be regarded as one–body systems [3].

---

[1] *E-mail:* prp@lanl.gov

CP638, *Mapping the Triangle: International Conference on Nuclear Structure*
edited by A. Aprahamian, J. A. Cizewski, S. Pittel, and N. V. Zamfir
© 2002 American Institute of Physics 0-7354-0093-8/02/$19.00

Here we investigate whether the Dirac Hamiltonian has any further dynamical symmetries of the same type as the spin and pseudospin symmetries. These symmetries satisfy the following properties, which will also be assumed for the posited more general symmetries:

- $\hat{S}_i$ are Hermitean: $\hat{S}_i^\dagger = \hat{S}_i$

- $\hat{S}_i$ satisfies the $SU(2)$ algebra : $[\,\hat{S}_i\,,\,\hat{S}_j\,] = i\sum_k \epsilon_{ijk}\,\hat{S}_k$

- $\hat{S}_i$ transforms like a vector: $[\,J_i\,,\,\hat{S}_j\,] = i\sum_k \epsilon_{ijk}\,\hat{S}_k$

- $[H_0, \hat{S}_i] = 0$

- $[H_{int}, \hat{S}_i] = 0$

where the kinetic part of the Hamiltonian is $H_0 = c\,\vec{\alpha}.\vec{p} + mc^2\beta$ and the potential part of the Hamiltonian is

$$H_{int} = \beta\ (V_S(\vec{r}) + \gamma_\mu V_V^\mu(\vec{r}) + \sigma_{\mu\nu} V_T^{\mu\nu}(\vec{r})) \tag{2}$$

and the total angular momentum operator is $\vec{J} = (\frac{\hbar\vec{\sigma}}{2} + \vec{r}\times\vec{p})\,\mathbf{1}$, with

$$\gamma_\mu \text{ the usual Dirac matrices, i.e.}\gamma_0 \equiv \beta, \vec{\gamma} \equiv \beta\vec{\alpha}, \text{ and } \sigma_{\mu\nu} = \frac{i}{2}\,[\gamma_\mu, \gamma_\nu]\,. \tag{3}$$

The Hamiltonian $H = H_0 + H_{int}$ is the most general Hermitean local parity conserving Dirac Hamiltonian, of which the Hamiltonian H' in Eq. 1 is a special case. Parity conservation is a symmetry of QCD (which is relevant to both quark and nucleon systems). Note that the other Lorentz structures possible for the potential part of the Hamiltonian, i.e. γ_5 and $\gamma_5\gamma_\mu$, do not conserve parity.

We have proved that there are no more general symmetries than the spin and pseudospin symmetries that satisfy the properties mentioned above. We also proved that the spin and pseudospin symmetries can only commute with H if $H = H'$.

If one further restricts $H_{int} = 0$ we found a continuous one–parameter family of new symmetries $\hat{S}_i$ which satisfy the remaining properties. At the two extreme values of the parameter range one recovers the usual $\hat{S}_i$ of spin and pseudospin symmetry.

We also proved that $[\,J_i, H\,] = 0$ if an only if $V_V^i(\vec{r}) = 0$, $i \in \{1, 2, 3\}$, and $V_T^{\mu\nu}(\vec{r}) = 0$, and the vector and scalar potentials only depend on $|\vec{r}|$.

If one relaxes the property that $\hat{S}_i$ transforms like a vector, more symmetries have been found [1].

We are supported by the Department of Energy under contract W-7405-ENG-36.

REFERENCES

1. J.S. Bell and H. Ruegg, *Nucl. Phys.* **B98** (1975) 151.
2. P.R. Page, T. Goldman and J.N. Ginocchio, *Phys. Rev. Lett.* **86** (2001) 204.
3. J.N. Ginocchio, *these proceedings*.

$M1$ transitions in the SU(3) shell model

G. Popa*, J. G. Hirsch†, J. P. Draayer** and C. Bahri**

*Department of Physics, Rochester Institute of Technology, Rochester, NY 14623 USA
†Instituto de Ciencias Nucleares, UNAM, Apartado Postal 70-543 México 04510 DF, México
**Department of Physics and Astronomy, Louisiana State University, Baton Rouge, LA 70803 USA

Abstract. $M1$ transitions in isotopes of Gd and Dy are calculated within the framework of the pseudo-SU(3) model. Basis states are built as linear combinations of direct product SU(3) proton and neutron configurations coupled to total pseudo-spin zero. The system Hamiltonian includes spherical Nilsson single-particle energies, the quadrupole-quadrupole and pairing interactions, as well as four SU(3) symmetry preserving rotor terms. The calculated results compare favorably with the experimental data. The $M1$ transitions give rise to a unique interpretation of the outer multiplicity that enters in a reduction of the direct product of SU(3) irreducible representations. For comparison, some results for light sd-shell and transitional fp-shell nuclei (e.g. ^{20}Ne and ^{44}Ti) are also presented.

The prediction of strongly enhanced $M1$ transitions within the framework of the so called two-rotor model, introduced in 1978 by Lo Iudice and F. Palumbo [1], has attracted considerable attention over the last several years. While the predominantly orbital character of the magnetic dipole excitation has been confirmed experimentally, other features cannot be explained using a simple two-rotor descripton. One such feature is the fragmentation of the $M1$ strength distribution among several levels closely packed and clustered around a few strong transition peaks in the energy region between 2 and 4 MeV in heavy deformed nuclei.

Enhanced low-energy $M1$ transitions in even-even heavy deformed nuclei have now been observed and some of their characteristic features explained using various models. In particular, the pseudo-SU(3) model with a realistic Hamiltonian [2] has been envoked to explore the origin of enhanced low-lying $M1$ transition strengths in rare earth and actinide nuclei. The theory yields excitation energy that are in excellent agreement (less than 7% error) with the corresponding experimental numbers for ^{160,162,164}Dy [3] and ^{156}Gd. Within the pseudo-SU(3) framework, basis states are built as linear combinations of direct product SU(3) proton and neutron configurations coupled to total pseudo-spin zero [4]. The system Hamiltonian includes spherical Nilsson single-particle energies, the quadrupole-quadrupole and pairing interactions, as well as four SU(3) symmetry preserving rotor-like terms [5].

The calculated $M1$ transition strength distributions for the Dy [3] nuclei are in good agreement with the experimental numbers [6]. The ground states for these nuclei are described by even-even irreducible representations (irreps) of SU(3) which, due to the truncation of the space to $\tilde{S} = 0$ states only, couple to the 1^+ states through the orbital channel of the $M1$ operator (Table 1). The total summed $M1$ strength is slightly lower when a realistic SU(3) symmetry breaking rather than a pure SU(3) symmetry preserving Hamiltonian is employed. This is due to interference generated through the mixing

CP638, *Mapping the Triangle: International Conference on Nuclear Structure*
edited by A. Aprahamian, J. A. Cizewski, S. Pittel, and N. V. Zamfir
© 2002 American Institute of Physics 0-7354-0093-8/02/$19.00

of SU(3) irreps. A study of $M1$ transitions in the limit of a pure SU(3) symmetry preserving Hamiltonian yields a unique interpretation of the so-called scissors+twist and twist+scissors modes and their relation to the outer multiplicity that enters in a reduction of the direct product of SU(3) irreps [2, 7]. The total $M1$ strengths are slightly underestimated, especially for ^{164}Dy. This may be due to the fact that the present version of the theory suppresses contributions associated with intruder states. Work to enhance the model to include contributions from active intruder configurations is currently underway.

TABLE 1. B(M1) strengths from experiment [6] and theory. The columns labelled "pure" are pure symmetry limits of the respective theory. For heavy nuclei "mixed" means the pseudo-SU(3) model with a realistic Hamiltonian [2], and for the light nuclei it means the SU(3) model with spin-orbit coupling. For light nuclei the $M1$ strengths are given for the full operator along with the separate orbital and spin parts.

Nucleus	$\sum B(M1)[\mu_N^2]$		
	Experiment	Theory	
		pure	mixed
^{156}Gd	3.40	3.52	1.92
^{160}Dy	2.48	4.24	2.32
^{162}Dy	3.29	4.24	2.29
^{164}Dy	5.63	4.36	3.05

		pure			mixed		
		total	L	S	total	L	S
^{20}Ne	1.99	3.89	1.17	3.59	3.32	1.17	3.90
^{44}Ti		4.45	1.64	3.71	4.92	1.70	5.26

A similar analysis has been carried out for light nuclei, namely ^{20}Ne and ^{44}Ti where full $0\hbar\omega$ shell-model calculations can be performed. In this case theory is the Elliott SU(3) shell model with spin-orbit coupling. The latter results in strong mixing of several basis states with spin $S = 0$ and $S = 1$ [8]. Specifically, the $J = 0^+$ ground states of ^{20}Ne and ^{44}Ti contain both $S = 0$ and $S = 1$ components with the latter giving rise to strong $M1$ transitions through the spin channel. Note that the total $M1$ strength due to spin increases slightly when the spin-orbit interaction is included. It is anticipated that the calculated $M1$ strength will be reduced when a more realistic Hamiltonian is used.

REFERENCES

1. N. Lo Iudice and F. Palumbo, Phys. Rev. Lett. **41**, 1532 (1978).
2. T. Beuschel, J. P. Draayer, D. Rompf , and J. G. Hirsch, Phys. Rev. C **57**, 1233 (1998).
3. J. P. Draayer, G. Popa, and J. G. Hirsch, Acta Phys. Pol. **B 32**, 2697 (2001).
4. J. P. Draayer and K. J. Weeks, Ann. Phys. (N.Y.) **156**, 41 (1894).
5. G. Popa, J. G. Hirsch, and J. P. Draayer, Phys. Rev. C **62**, 064313 (2000).
6. National Nuclear Data Center, (http://bnlnd2.dne.bnl.gov).
7. T. Beuschel, J.G. Hirsch, and J.P. Draayer, Phys. Rev. C **61**, 54307 (2000).
8. C. Vargas, J. G. Hirsch, P.O. Hess, and J. P. Draayer, Phys. Rev. C **58**, 1488 (1998).

Skyrme HFB study of high-spin isomers in neutron-deficient Pb and Po

N. A. Smirnova*, P.-H. Heenen† and G. Neyens*

*IKS, University of Leuven, Celestijnenlaan 200 D, B-3001 Leuven, Belgium
†Service de Physique Nucléaire Théorique, Université Libre de Bruxelles, CP229, B-1050 Bruxelles, Belgium

Abstract.
We investigate the deformations of 11^- and 12^+ isomers in neutron-deficient Pb and Po isotopes within the Hartree-Fock-Bogolyubov approach using Skyrme interaction (Sly4) and density-dependent pairing force. The effects of core polarization due to the breaking of the proton $Z = 82$ magic gap in light Pb isotopes are put into evidence and compared to recent experimental measurements of static quadrupole moments of intruder 11^- isomers.

The region of nuclei around the proton-magic Pb is very rich by a variety of isomeric states. Besides the normal spin isomers formed by neutron multi-quasiparticle states, several intruder spin isomers arise from proton particle-hole excitations across the magic $Z = 82$ shell gap [1]. Among them are 8^+ and 11^- isomers in the even-even Pb isotopes, involving the occupation of the $\pi h_{9/2}$ and $\pi i_{13/2}$ intruder orbitals and corresponding to weakly oblate nuclear shape. The knowledge of the particular configurations ($(\pi(h_{9/2}^2)_{8+}$ and $\pi(h_{9/2}i_{13/2})_{11-}$, respectively) and deformations of these isomers is based on the experimentally measured magnetic and quadrupole moments (e.g., Ref. [2] and [3]).

Up to now, these states have been studied theoretically within the shell-correction approach [5], and the particle-core coupling model [6, 4]. In this contribution, we summarize the results of the first systematic fully self-consistent Hartree-Fock-Bogolyubov (HFB) study of properties of the 11^- isomeric states of even-even $^{186-198}$Pb and $^{188-200}$Po. Our aim is to explore the origin and evolution of the intruder state deformation and from comparative analysis of the results for Pb and Po to extract the effects of core polarization, in particular the influence of the proton core excitations onto the nuclear deformation. Another particular interest to the 11^- states is related to the fact that these intruder states are an important ingredient in the rotational shears bands well-known in both even and odd-A neutron-deficient Pb [7].

We use the Skyrme interaction Sly4 [8] to describe the average mean-field and a zero-range density dependent force [9] in the pairing channel. A smooth cut-off energy of 5 MeV above and below the Fermi level is imposed. An approximate projection of particle number has been done by the Lipkin-Nogami technique. The HFB equations have been solved on a three-dimensional cubic mesh [10] using the two-basis method [11].

The deformation energy curves for $^{186-198}$Pb and $^{188-200}$Po have been calculated with a constraint on the axial mass quadrupole moment [12]. For Pb isotopes, the ground state persists to be spherical, while the second and third minima appear on the both

CP638, *Mapping the Triangle: International Conference on Nuclear Structure*
edited by A. Aprahamian, J. A. Cizewski, S. Pittel, and N. V. Zamfir
© 2002 American Institute of Physics 0-7354-0093-8/02/$19.00

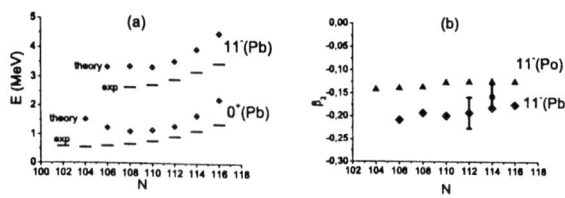

FIGURE 1. (a) Energy systematics of 11^- and 0_2^+ states in $^{198-186}$Pb in comparison with the experiment; (b) Comparative analysis of the deformation of 11^- states in Pb and Po.

oblate and prolate sides and become deeper as the number of neutrons decreases. The deformation energy curves for Po isotopes exhibit a number of coexisting minima at oblate and prolate deformation which compete for the ground state and it is evident that more elaborated methods beyond mean-field should be applied in this case [12].

The isomers 11^- (systematically in Pb and Po) and 12^+ (for ^{196,198}Pb) have been constructed by blocking self consistently two quasiparticles in the specific configurations $(\pi(h_{9/2}i_{13/2})_{11^-}, \nu(i_{13/2})^2_{12^+})$ and their energy has been minimized with respect to the axial quadrupole deformation. The results for Pb isotopes are summarized in Figure 1(a). For comparison, we show also the energy of the 0_2^+ as approximated by the second minimum of the deformation energy curve.

The calculated intrinsic deformation of the 11^- states in Pb and Po is plotted in Figure 1(b), in comparison with the available experimental values extracted from the measured quadrupole moments [3].

For Pb, the calculations slightly overestimate the excitation energy of the high-spin states, but the deformation is well reproduced. The overall trend of increase of oblate deformation of the isomeric states with decreasing neutron number is clearly observed. The larger absolute values of the axial deformation for Pb as compared to those for Po confirm the strong influence of core polarization due to the particle-hole excitations across the $Z = 82$ magic shell gap.

REFERENCES

1. For recent review see Wood, J.L., et al, *Phys. Rep.* **215**, 101 (1992).
2. Chmel, S., *Phys. Rev. Lett.* **79**, 2002 (1997).
3. Vyvey, K., et al., *Phys. Lett.* **B** (2002); Vyvey, K., Ph.D. Thesis, University of Leuven, 2001.
4. Vyvey, K., et al., *Phys. Rev. Lett.* **88**, 102502 (2002)
5. Bentsson, R., and Nazarewicz, W., *Z. Phys.* **A334**, 269-276 (1989); Nazarewicz, W., *Phys. Lett.* **B305**, 195–201 (1993); Clark, R.M., et al, *Nucl. Phys.* **A562**, 121–156 (1993).
6. Oros, A.M., et al., *Nucl. Phys.* **A645**, 107–142 (1999).
7. Clark, R.M., and Macchiaveli, A.O., *Annu. Rev. Nucl. Part. Sci.* **50**, 1 (2000).
8. Chabanat, E., et al, *Phys. Scripta* **T56**, 231 (1995).
9. Terasaki, J., et al, *Nucl. Phys.* **A593**, 1–20 (1995).
10. Bonche, P., et al, *Nucl. Phys.* **A443**, 39–63 (1985).
11. Gall, B., et al, *Z. Phys.* **A348**, 183–197 (1994).
12. Smirnova, N.A., Heenen, P.-H., Neyens, G., (in preparation).

Extended M1 Sum Rule for Mixed-Symmetry States

N. A. Smirnova*, N. Pietralla†, A. Leviatan**, J. N. Ginocchio‡ and C. Fransen§

*Instituut voor Kern- en Stralingsfysica, University of Leuven, Celestijnenlaan 200 D, B-3001 Leuven, Belgium
†Institut für Kernphysik, Universität zu Köln, 50937 Köln, Germany and Wright Nuclear Structure Laboratory, Yale University, New Haven, CT, 06520, U.S.A.
**Racah Institute of Physics, The Hebrew University, Jerusalem 91904, Israel
‡Theoretical Division, Los Alamos National Laboratory, Los Alamos, NM, 87545, U.S.A.
§Department of Physics and Astronomy, University of Kentucky, Lexington, KY, 40506, U.S.A.

Abstract. A generalized $M1$ sum rule for orbital magnetic dipole strength from excited symmetric states to mixed-symmetry states is derived within the interacting boson model of even-even nuclei. The applicability of the sum rule is investigated for the U(5)–SO(6) transition region. By applying the sum rule to the recent extensive data on mixed-symmetry states in ^{94}Mo one obtains valuable structure information in a largely parameter-independent way.

Collective excitation modes of nuclei often appear fragmented due to the coupling to single-particle degrees of freedom. Under such circumstances a sum rule approach constitutes an efficient way to reveal the degree of collectivity of a given state and to link its properties with direct observables.

One particular collective mode which has been studied extensively in recent years is the orbital magnetic dipole mode (for a recent review see Ref. [1]). The experimental systematics of $M1$ strength from the ground state to the $J = 1^+$ scissors state and its deformation dependence agree well with a variety of sum rules [2, 3]. In particular, within the proton-neutron version of the interacting boson model (IBM-2) a sum rule has related this strength to the number of quadrupole d-bosons in the ground state wave function [3]. The $J = 1^+$ scissors state is only one particular example of a class of low-lying isovector collective excitations, or mixed-symmetry states, predicted by the IBM-2 [4]. Within the IBM-2, these states posses a lower symmetry with respect to interchange of proton and neutron bosons (F-spin). The recent extensive data in ^{94}Mo [5] on four mixed-symmetry (*ms*) states with $J = 1^+_{1,ms}, 2^+_{1,ms}, 2^+_{2,ms}, 3^+_{1,ms}$, has paved the way for comparing $M1$ strengths between mixed-symmetry states and different low-lying symmetric states in the same nucleus. This has motivated us to generalize to excited states the above mentioned sum rule.

The total orbital magnetic dipole strength from *any arbitrary* F-spin symmetric state to mixed-symmetry states in even-even nuclei within the sd-IBM-2 framework reads [6]

$$S_J = \sum_{f \neq i} B(M1; i, J \to f, J_f) = \frac{3}{4\pi}(g_\pi - g_\nu)^2 \frac{6N_\pi N_\nu}{N(N-1)} \left[\langle J|\hat{n}_d|J\rangle - \frac{J(J+1)}{6N} \right],$$

CP638, *Mapping the Triangle: International Conference on Nuclear Structure*
edited by A. Aprahamian, J. A. Cizewski, S. Pittel, and N. V. Zamfir
© 2002 American Institute of Physics 0-7354-0093-8/02/$19.00

where J (J_f) is the angular momentum of the initial (final) state and the labels i (f) indicate all quantum numbers that may be needed to specify uniquely the states. As shown, this strength depends on the proton (π) and neutron (v) boson numbers, $N_{\pi,v}$, ($N = N_\pi + N_v$), the boson effective g-factors, $g_{\pi,v}$, and involves the expectation value of the d-boson number operator, $\hat{n}_d = \hat{n}_{d_\pi} + \hat{n}_{d_v}$ in the initial state J.

The main advantage of the sum rule is that it allows one to extract from the data information on the d-boson content of excited states without any assumptions on the effective boson g-factors. This is done by taking ratios of $M1$ strengths from different symmetric states: $R_{J_0}(J) = S_J/S_{J_0}$. In particular,

$$R_{2^+}(0_1^+) = \frac{S_{0_1^+}}{S_{2^+}} = \frac{n_d(0_1^+)}{n_d(2^+) - 1/N} \approx \frac{n_d(0_1^+)}{n_d(2^+)} \quad (\, n_d(2^+) \gg 1/N \,).$$

The d-boson ratio can be reliably extracted directly from the data, provided the actual measured strengths to the mixed-symmetry states are a good approximation to the total strengths S_J in the sum rule. A recent investigation [6] shows that throughout most of the $U(5) - O(6)$ transition region, for moderate boson number, the $J = 0_1^+$ and $J = 2_1^+$ states qualify for a sum rule analysis, since their $M1$ strengths exhausts 100% and about 85% of the respective total strengths. This arises from the fact that the indicated set of mixed-symmetry states span particular irreducible representations of the common $SO(5)$ subgroup. The latter symmetry, which is preserved for γ-soft nuclei such as ^{94}Mo, imposes further constraints on the allowed $M1$ transitions, which in turn enables the observed four mixed-symmetry states to exhaust an appreciable fraction of the sum rules, S_J, for $J = 0_1^+, 2_2^+$.

From the experimental ratio $R_{2_2^+}(0_1^+)_{exp} = (S_{0_1^+})_{exp}/(S_{2_2^+})_{exp} = 0.58^{+11}_{-14}$, we extract a d-boson-content ratio $n_d(0_1^+)/n_d(2_2^+) \approx 0.58$ for ^{94}Mo with $N = 5$. We thus find that the $J = 0_1^+$ ground state of ^{94}Mo contains more than half as many d-bosons as the $J = 2_2^+$ state. This number is considerably higher than that for a spherical vibrator $(n_d(0_1^+)/n_d(2_2^+) = 0$ in the U(5) limit) and is in fact closer to the value expected for a γ-unstable rotor $(n_d(0_1^+)/n_d(2_2^+) = 2/3$ in the SO(6) limit with $N = 5$). The extracted value is independent of any model parameters and suggests the structure of ^{94}Mo being close to the $SO(6)$ dynamic symmetry limit of the IBM-2.

REFERENCES

1. Richter, A., *Prog. Part. Nucl. Phys.*, **34**, 261 (1995);
    Kneissl, U., Pitz, H. H., and Zilges, A., *Prog. Part. Nucl. Phys.* **37**, 349 (1996).
2. Lipparini, E., and Stringari, S., *Phys. Rep.* **175**, 103 (1989).
3. Ginocchio, J. N., *Phys. Lett. B*, **265**, 6 (1991).
4. Iachello, F., and Arima, A., *The Interacting Boson Model*, Cambridge, 1987.
5. Pietralla, N., et. al., *Phys. Rev. Lett.*, **83**, 1303 (1999);
    Pietralla, N., et. al., *Phys. Rev. Lett.*, **84**, 3775 (2000);
    Fransen, C., et. al., *Phys. Lett. B*, **508**, 219 (2001).
6. Smirnova, N. A., et. al., *Phys. Rev. C*, **65**, 024319 (2002).

Investigation of the Gamow-Teller strength near the doubly-magic nucleus ^{100}Sn

A. Stolz*†, T. Faestermann*, R. Schneider*, E. Wefers*, K. Sümmerer**, J. Friese*, H. Geissel**, M. Hellström‡, P. Kienle*, H.-J. Körner*, M. Münch*, G. Münzenberg**, C. Schlegel**, P. Thirolf§ and H. Weick**

*Technische Universität München, 85748 Garching, Germany
†NSCL, Michigan State University, East Lansing, MI 48824
**Gesellschaft für Schwerionenforschung mbH, 64291 Darmstadt, Germany
‡Lund University, 22100 Lund, Sweden
§Ludwig-Maximilians-Universität München, 85748 Garching, Germany

Abstract. Neutron deficient nuclei near ^{100}Sn have been produced by projectile fragmentation of a 1 A·GeV ^{112}Sn beam in a beryllium target. The fragments were separated in the magnetic spectrometer FRS at GSI, Darmstadt. The unambiguously identified ions were stopped in a highly segmented silicon strip detector stack which was designed to measure the halflives and the total energy of emitted particles with very low background and high efficiency. We were able to study ^{102}Sn in decay spectroscopy for the first time and establish a decay scheme for the daughter nuclide ^{102}In. With the measurement of the halflife and the Q-value the Gamow-Teller (GT) strength could be deduced. Additionally the Q-value of ^{98}Cd was measured with high precision.

The precise analysis of β decay properties of the doubly-magic ^{100}Sn and the neighboring nuclei ^{102}Sn and ^{98}Cd with reference to model predictions eventually allows a major contribution to illuminate the question of the missing Gamow-Teller strength [1].

At the fragment separator facility of GSI, Darmstadt, neutron deficient nuclei near ^{100}Sn have been produced by fragmentation of a 1 A·GeV ^{112}Sn beam in a ^{9}Be target, separated in the FRS and unambiguously identified. The ions were stopped in the center of a stack of highly segmented silicon strip detectors placed at the final focal plane. This 4π implantation detector measured the total energy of emitted β^+-particles with high efficiency. For measuring γ-radiation the implantation detector was surrounded by a 6-fold segmented NaI detector covering 2π and a Germanium clover detector in close geometry. We determined the halflives and the positron energies for each implanted nuclide with a maximum likelihood method taking into account three decay generations as well as background events during a fixed correlation time after the implantation.

During a run of 60 hours one ^{100}Sn could be identified resulting in a production cross section of $\sigma = 1.8^{+3.2}_{-1.3}$ pb. The decay data of this event together with 6 events observed in 1994 [3] results in a halflive of $T_{1/2} = 1.0^{+0.54}_{-0.26}$ s and a β endpoint energy of $E_{\beta_0} = 3.8^{+0.7}_{-0.3}$ MeV. Due to the low statistic a meaningful comparison between the experimental Gamow-Teller strength and theoretical values is not yet possible.

With the implantation of 2800 ^{102}Sn we were able to study this nucleus in decay spectroscopy for the first time. Analyzing the correlated beta-decay events with a maximum

CP638, *Mapping the Triangle: International Conference on Nuclear Structure*
edited by A. Aprahamian, J. A. Cizewski, S. Pittel, and N. V. Zamfir
© 2002 American Institute of Physics 0-7354-0093-8/02/$19.00

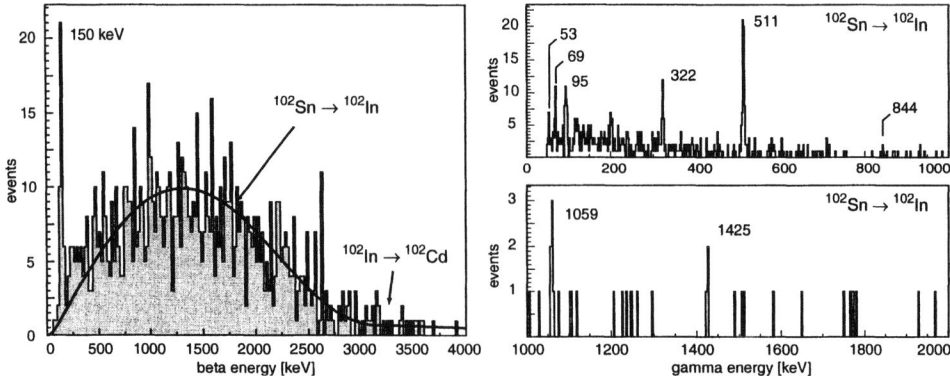

FIGURE 1. Measured beta and gamma energy spectra for the ^{102}Sn $\rightarrow$ ^{102}In decay. Only decay events wich can be assigned to a decay of ^{102}Sn with a probability $p > 0.7$ are shown.

likelihood method the ^{102}Sn halflife was measured as $T_{1/2} = 3.8 \pm 0.2$ s. With the knowledge of the halflive it is possible to calculate the probability that a observed event in a decay chain can be assigned to a decay of the mother isotope. Using only decay events with a probability $p > 0.7$ for a ^{102}Sn decay, 8 gamma transitions could be identified that follow the ^{102}Sn $\rightarrow$ ^{102}In decay (see Fig. 1). In the case of EC decay events it was possible to observe sum lines of conversion electrons of the low energy transitions in the silicon detector. Considering this additional information and guided by theoretical expectations [5] it is possible to construct a decay scheme for the daughter nuclide ^{102}In which is consistent with all experimental observations: the β decay feeds two 1^+ states at 1598 keV and 1964 keV with an intensity ratio 70:30. From these states γ cascades lead down to the (7^+) ground state.

The measured β^+ spectrum was fitted with two components to levels 366 keV apart and with the relative feeding observed in the gamma decay. This determined the endpoint energies to these two levels and also the groundstate decay Q-value as $Q_{EC} = 5760 \pm 90 \pm 50$ keV. The summed Gamow-Teller strength of the two individual transitions observed for the ^{102}Sn decay is $B_{GT} = 4.0 \pm 0.6$. The reference strength value, expected for the $\pi g_{9/2} \rightarrow \nu g_{7/2}$ decay of ^{102}Sn on the basis of the extreme single-particle shell model, is $B_{GT}^{ref} = 17.8$, thus leading to a hinderance factor $h = B_{GT}^{ref}/B_{GT} = 4.45$.

Using the same experimental techniques it was possible to measure the Q-value of ^{98}Cd with high precision ($Q_{EC} = 5430 \pm 40$ keV). Together with the already known decay data [4] the summed Gamow-Teller strength for the four observed transitions results in $B_{GT} = 2.9 \pm 0.2$. This result is in agreement with the published value of $B_{GT} = 3.5^{+0.8}_{-0.7}$ [4].

REFERENCES

1. B.A. Brown et al., Phys. Rev. C50 (1994) 2270
2. H. Geissel et al., NIM B70 (1992) 286
3. R. Schneider et al., Z. Phys. A348 (1994) 241
4. A. Plochocki et al., Z. Phys. A342 (1992) 43
5. B.A. Brown, private communication, 2000

Excited-state g-factor measurements with radioactive ion beams

Andrew E. Stuchbery* and Paul F. Mantica[†]

*NSCL, Michigan State University, East Lansing, Michigan 48824, and
Department of Nuclear Physics, Research School of Physical Sciences and Engineering,
Australian National University, Canberra, ACT 0200, Australia
[†]NSCL, Michigan State University, East Lansing, Michigan 48824, and
Department of Chemistry, Michigan State University, East Lansing, Michigan 48824

Abstract. Techniques that may be applied to measure the g factors of short-lived excited states of fast rare-isotope beams are considered with a view to measuring $g(2_1^+)$ values in Si and S isotopes near the 'island of inversion'.

Measurements of ground-state magnetic moments by β-NMR are well established for light exotic nuclei produced in fragmentation reactions [1]. The time-differential perturbed angular distribution (TDPAD) technique has also been applied to measure the g factors of isomeric states produced by fast fragmentation [2]. For excited states of fragments with lifetimes less than 10 ns, however, there is no established integral perturbed angular distribution/correlation technique. We address this issue by evaluating the prospects for $g(2_1^+)$ measurements in Si and S isotopes near the 'island of inversion' [3, 4]. (Other regions of the nuclear chart will be considered in due course.)

Whereas ground-states and/or isomeric states can be produced in fast-fragmentation reactions, identified in a spectrometer and then delivered to β-NMR or TDPAD apparatus, short-lived excited states cannot. Either in-beam γ-ray spectroscopy on the fragments at the production target prior to the spectrometer or Coulomb excitation of the separated fragments must be considered. An evaluation of the data available suggests that the spin alignment produced in fast-fragmentation reactions is rather small (typically $< 20\%$), whereas nearly full alignment along the beam axis is expected in relativistic Coulomb excitation [5]. Evidently g factors are best measured when the fragments are Coulomb excited. It follows that the experimental set up will be like that used to measure the B(E2) values [4], but with modifications to subject the nucleus to an intense magnetic field and then measure the perturbation of the de-excitation γ-ray angular distribution.

The number of counts required to measure an excited-state g factor can be estimated by taking a standard two-detector transient field IMPAC measurement [6] as the benchmark. It can be shown that the experimental error in the measured precession angle, $\Delta\theta$, is $\sigma_{\Delta\theta} \sim 1/2S\sqrt{N}$, where N is the average number of counts recorded in each detector for each direction of the polarizing field, and S is the logarithmic slope of the angular correlation at the detection angle. Thus if $N = 400$ and $S = 1$, $\sigma_{\Delta\theta} \sim 25$ mrad. It is imperative that the precession angle $\Delta\theta$ be maximized, and at least of the order of a few hundred mrad. For short-lived states ($\tau < 100$ ps) this requires the use of hyperfine fields:

CP638, *Mapping the Triangle: International Conference on Nuclear Structure*
edited by A. Aprahamian, J. A. Cizewski, S. Pittel, and N. V. Zamfir
© 2002 American Institute of Physics 0-7354-0093-8/02/$19.00

either the transient hyperfine field [6] or the fields at the nucleus due to H-like electron configurations [7]. (Static hyperfine fields for impurities with $Z < 20$ in ferromagnetic hosts are too small.) The nuclei of interest must therefore be slowed to the region where a significant fraction of their ions carry an electron, i.e. to about 10 MeV per nucleon.

The maximum transient field strength (i.e. for ion velocities near Zv_0, $v_0 =$ Bohr velocity) has been measured to be ~ 3 kTesla for both Si and S ions in Fe and Gd hosts [8]. It turns out that the transient-field precession is limited by the lifetime of the state and that only when the lifetime exceeds several picoseconds is it possible to achieve precessions of several hundred mrad. The magnitude of the precession also depends on the behavior of the transient field near and above Zv_0, which is not well studied as yet. However the fact that large precessions can be produced for sufficiently long-lived states is encouraging.

The fields at the nucleus of H-like Si and S ions due to the $1s$ electron is 45.8 kTesla for Si and 68.4 kTesla for S. These fields are too large to measure the g factors of states with lifetimes longer than about half a picosecond because the attenuation coefficients have their hard-core values, $G_2 = 0.76$ and $G_4 = 0.2$, essentially independent of the g factor. Shorter-lived states ($\tau < 0.5$ ps), however, should give rise to g-factor dependent attenuation coefficients provided the states of interest do not decay in the target before they can experience the free-ion field. This might be achieved by using a stack of thin targets rather than a single thick target. It has been assumed here that the electronic configuration of the H-like fragment is in its ground-state immediately upon leaving the target. Although this is a reasonable approximation, it should be evaluated more completely because any time-dependent effects in the electronic configuration would reduce the average field and hence increase the sensitivity of the g-factor measurement. If excited fragments can be produced efficiently as Li-like ions, the fields would be even more suitable for g-factor measurements on states with picosecond lifetimes.

We conclude that in favorable cases ($\tau > 5$ ps) transient fields acting on Coulomb-excited relativistic projectiles offer the possibility to measure excited-state g factors in exotic nuclei near the $N = 20$ shell closure. It will be necessary to understand the high-velocity transient field in more detail and, at least initially, to make measurements relative to a known g factor. For shorter-lived states free-ion fields may prove useful. Modeling of the experiments in greater detail is currently underway.

REFERENCES

1. K. Asahi and K. Matsuta, Nucl. Phys. **A693**, 63 (2001).
2. W.-D. Schmidt-Ott *et al.*, Z. Phys. **A350**, 215 (1994);
   G. Neyens *et al.*, Nucl. Phys. **A**, in press.
3. E. Caurier, F. Nowacki and A. Poves, Nucl. Phys. **A 693**, 374 (2001).
4. R.W. Ibbotson *et al.*, Phys. Rev. Lett. **80**, 2081 (1998).
5. H. Scheit, PhD thesis, Michigan State University, unpublished (1998).
6. N. Benczer-Koller, M. Hass and J. Sak, Ann. Rev. Nucl. Part. Sci. 30 (1980) 53.
7. G. Goldring, in *heavy Ion Collisions*, edited by R. Bock (North Holland, Amsterdam, 1982), Vol. 3.
8. K.-H. Speidel *et al.*, Z. Phys. **A 331**, 29 (1988);
   K.-H. Speidel *et al.*, Phys. Lett. B **227**, 16 (1989).

Staggering behavior of isovector 0^+ state energies in even-even and odd-odd nuclei classified according to even representations of $sp(4)$

K. D. Sviratcheva[a], A. I. Georgieva[a,b], and J. P. Draayer[a]

[a]LSU, Baton Rouge, LA, 70803 USA
[b]INRNE, BAS, Sofia 1784, Bulgaria

Abstract. A discrete derivative analysis of the energies of the lowest isovector-paired 0^+ state of even-even and odd-odd nuclei is presented within a classification scheme provided by even representations of a fermion realization of $sp(4)$. The results suggest that the observed energy staggering carries important information about proton-neutron pairing correlations in nuclei.

The observed staggering of energy levels in atomic nuclei requires a theory that goes beyond mean-field considerations [1]. Here we consider even-A nuclei for which the staggering can be explained in terms of isovector pairing correlations, with different pairing strengths for the pn and the pp and nn pairs [2]. The algebraic pairing model we consider is based on a fermion realization of the symplectic $sp(4)$ algebra [3] and includes interactions between pairs with isospin $\tau = 1$ [4, 5, 6]. When considered to be a dynamical symmetry, the $Sp(4) \supset U(1) \otimes SU(2)$ reduction yields a classification of isovector 0^+ states of nuclei in a given j shell with respect to the total number of the valence particles, n, and the third projection of the valence isospin, i. The model Hamiltonian is expressed through the $Sp(4)$ group generators and includes pairing correlations and a symmetry term.

The model can be used to give a detailed explanation of observed staggering in the experimental energies through an investigation of each of the underlying interactions. As shown in Figure 1(a), the $pp + nn$ and pn pairing interactions yield $\triangle n = 2$ staggering patterns that are of opposite phase. Although the total pairing energy has a surprisingly smooth behavior, the symmetry term makes possible the accurate theoretical prediction of the observed regular zig-zag pattern.

The $Sp(4)$ model reproduces the energies of the isovector 0^+ states well. A more detailed investigation is obtained through the discrete derivatives of the total energy

$$Stg_\delta^{(m)}(x) = \frac{Stg_\delta^{(m-1)}(x+\delta) - Stg_\delta^{(m-1)}(x-\delta)}{(2\delta)^m}, \quad Stg_\delta^{(1)}(x) = \begin{cases} \frac{E_0(x+\delta)-E_0(x-\delta)}{2\delta}, m : even \\ \frac{E_0(x+\delta)-E_0(x)}{\delta}, m : odd. \end{cases}$$

where according to the $Sp(4)$ classification the variable x is either n or i. When $m \geq 3$ the staggering is independent of mean-field effects and only provides for a description of higher order terms in x, as well as for discontinuities in the pairing and symmetry terms.

CP638, *Mapping the Triangle: International Conference on Nuclear Structure*
edited by A. Aprahamian, J. A. Cizewski, S. Pittel, and N. V. Zamfir
© 2002 American Institute of Physics 0-7354-0093-8/02/$19.00

Results for even-A nuclei with valence nucleons in the $1f_{7/2}$ ($\Omega = 4$) shell with a core $^{40}_{20}Ca$ are presented. The discrete derivatives for different i-multiplets (Figure 1(b)) (for different isobaric multiplets (Figure 1(c))) show a prominent $\Delta n = 2$ ($\Delta i = 1$) staggering of the experimental energies of the lowest 0^+ isovector states. The theory reproduces this staggering very well. The amplitudes of the $\Delta n = 2$ and $\Delta i = 1$ staggering patterns increase with increasing $2|i| = |Z - N|$. In contrast with the $\Delta n = 2$ pattern, where the staggering amplitude is independent of the total number of particles, the energy difference in an isobaric sequence increases as mid-shell is approached. The analysis shows a more complicated dependence of the energy function on the isospin than on mass number, which is also confirmed by an increase of the $\Delta i = 1$ staggering amplitude in higher-order derivatives. The bigger odd-even energy difference is related to the fact that the pn correlations in odd-odd nuclei are weak compared to the like-particle coupling in even-even neighbors as the number of pairs and $|Z - N|$ increase. The latter dependence is additional evidence that the pn correlations are strongest around the region of $N = Z$ nuclei. A generalization of the theory to multiple-j orbits yields similar staggering patterns for nuclei above the $^{56}_{28}Ni$ core, where the multiplets are larger.

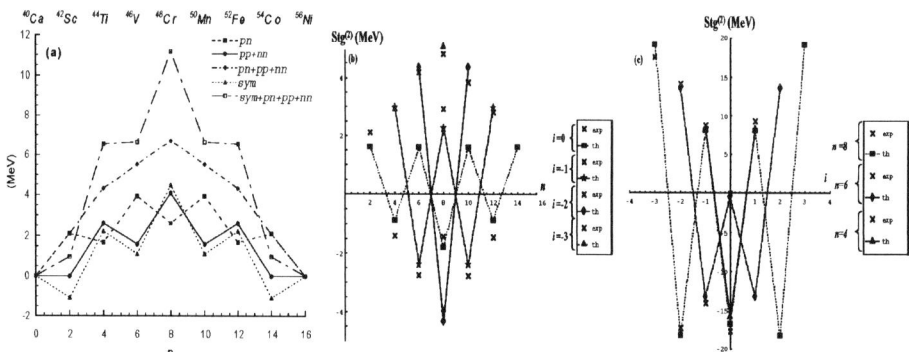

FIGURE 1. (a)Lowest isovector 0^+ state energies for the $N = Z$ nuclei in $1f_{7/2}$; (b) and (c) Second discrete derivatives.

A measure of the difference in energy between even-even and odd-odd nuclei in a given i-multiplet is $\Delta = 2(-1)^{\frac{n}{2}+i}Stg^{(4)}_{\delta=2}(n) \approx 2\Delta_{pp(nn)} - \Delta_{pn}$. The empirical approximation for the like-particle pairing gap is $2\Delta_{pp(nn)} = 24/A^{1/2}$, and for the pn pairing gap it is $\Delta_{pn} = 20/A$ [1]. Within the $Sp(4)$ framework, for the $i = 0$ multiplet we get the average value $\Delta = 2.84MeV$ compared to an experimental approximation that yields $3.05MeV$.

In conclusion, we note that the $Sp(4)$ model, which provides a reliable description of the 0^+ isovector states, has been further tested by considering the observed staggering behavior of the energies of nuclei. The results show that it can be used to describe the mean-field interaction along with the smaller effects driven by pairing correlations, especially in exotic nuclei with $Z \gtrsim N$.

This work was partially supported by the US National Science Foundation through a regular grant (9970769) and a cooperative agreement (9720652) that includes matching from the Louisiana Board of Regents Support Fund. K.D.S. is grateful to the US NSF for the support for young scientists to attend the INSC 2002 on "Mapping the Triangle".

REFERENCES

1. A. Bohr and B. R. Mottelson, *The Nuclear Structure, Benjamin, New York* (1975)
2. K. D. Sviratcheva, J. P. Draayer, and A. I. Georgieva, *nucl-th*/0204070
3. K. D. Sviratcheva *et al.*, *J. Phys. A* **34** (2001) 8365
4. J. N. Ginocchio, *Nucl. Phys.* **74** (1965) 321
5. K. T. Hecht, *Nucl. Phys.* **63** (1965) 177; *Phys. Rev.* **139** (1965) B794; *Nucl. Phys.* **A102** (1967) 11
6. J. Engel, K. Langanke, P. Vogel, *Phys. Lett.* **B389** (1996) 211; J. Dobes, *Phys. Lett.* **B413** (1997) 239

268

g Factor Measurement of the Excited 2^+ State in the *fp* Shell Nucleus ^{44}Ca

M. J. Taylor*, N. Benczer-Koller*, G. Kumbartzki*, T. J. Mertzimekis*,
A. E. Stuchbery[†], K.-H. Speidel** and C. Hutter[‡]

*Department of Physics and Astronomy, Rutgers University, New Brunswick, New Jersey 08903
[†]Department of Nuclear Physics, Australian National University, Canberra, Australia
**Institut für Strahlen und Kernphysik, Universität Bonn, Bonn, Germany
[‡]A. W. Wright Nuclear Structure Laboratory, Yale University, New Haven, Connecticut, 06520

Abstract. The *g* factor of the Coulomb excited 2_1^+ state in ^{44}Ca has been measured using the transient field technique and inverse kinematics. A preliminary value has been established. The technique employed and the implication of the result with regards to *fp* shell model calculations will be discussed.

The measurement of magnetic moments of low lying excited states in nuclei can yield information on the single particle components in the wavefunction as well as their interplay with collective degrees of freedom. Magnetic moments of 2_1^+ states in the *fp* shell nuclei ^{46,48}Ti and ^{50,52,54}Cr have recently been measured [1, 2]. It was shown that the results for the Cr isotopes are in good agreement with full *fp* shell model calculations [3−6] whereas the results for the Ti isotopes are larger than the predictions. To further explore these conclusions and test the nuclear models and large scale shell model calculations a measurement of the *g* factor of the excited 2^+ state in ^{44}Ca has been performed.

The measurement was performed using projectile Coulomb excitation in inverse kinematics in combination with the transient field (TF) technique [7, 8]. An advantage of using inverse kinematics is the high detection efficiency of deexcitation γ rays in coincidence with forward recoiling target nuclei. In addition to the kinematic focusing in the beam direction, the projectile ions have high velocities, a condition which is favourable as the transient field strength generally increases with ion velocity.

In the present work an 85 and subsequently 90 MeV beam of ^{44}Ca ions was provided by the ion source and tandem accelerator at the Wright Nuclear Structure Laboratory at Yale. The ^{44}Ca beam with an average intensity of ~ 0.3 pnA was Coulomb excited by natural carbon. The multilayered target consisted of a 0.174 mg/cm^2 carbon layer deposited onto a 3.808 mg/cm^2 rolled gadolinium foil onto which a 6.28 mg/cm^2 copper backing layer was evaporated. The ^{44}Ca ions, excited in the carbon layer, traverse the ferromagnetic gadolinium layer where they experience the transient field and then stop in the interaction free copper layer. The target was cooled to ~ 50 K and magnetized to saturation by an external field of 0.09 T. The γ rays emitted in coincidence with carbon ions were detected by four 12.7 cm x 12.7 cm NaI(Tl) detectors. The carbon ions were detected in a 100 μm thick Si counter positioned at $0°$ to the beam direction. A Ge

CP638, *Mapping the Triangle: International Conference on Nuclear Structure*
edited by A. Aprahamian, J. A. Cizewski, S. Pittel, and N. V. Zamfir
© 2002 American Institute of Physics 0-7354-0093-8/02/$19.00

detector was also placed at $0°$ behind the Si counter to monitor contaminant lines and target integrity.

The angular correlation of the 2_1^+ state was deduced and a precession measurement was made. A preliminary result for the g factor of the excited 2_1^+ state in ^{44}Ca of $g(2_1^+,^{44}\text{Ca}) = +0.097 \pm 0.048$ has been obtained. The result suggests that the four $f_{7/2}$ neutrons outside of the $N = 20$ closed shell do not contribute significantly and that core polarization may play an important role. This phenomenon has also been observed for Zr isotopes close to the $N = 50$ shell closure. In ^{96}Zr, $g(2_1^+,^{96}\text{Zr}) = +0.23 \pm 0.14$ was measured by this group. This result indicates that the six valence neutrons outside of the $N = 50$ closed shell contribute weakly to the g factor of the excited 2_1^+ state and that excitation on the ^{90}Zr core must be involved to give a positive or close to zero g factor. The $g(2^+)$ for ^{44}Ca has been previously measured by Niv et $al.$ [9], using a surface interaction experiment, $g(2_1^+,^{44}\text{Ca}) = -0.28 \pm 0.11$. The new result does not support a large negative g factor expected for a dominant $(f_{7/2})_\nu$ configuration, $g(f_{7/2})_\nu = -0.546$.

A shell model calculation for ^{44}Ca has been performed by Robinson [10] using an FPD6 interaction considering only the four valence neutrons in the fp shell and a closed ^{40}Ca core. This calculation yielded a value of $g = -0.373$ for the excited 2_1^+ state. Recent calculations by Nowacki [11] assuming 4p−4h and 6p−6h excitation of the ^{40}Ca core yielded g factors $g(4p−4h) < 0$ and $g(6p−6h) = +0.19$. These calculations demonstrate that considerable p−h excitation of the ^{40}Ca core is necessary to understand the wave function of the 2_1^+ state in ^{44}Ca. A similar interpretation was proposed to explain the g factors for ^{46,48}Ti. The neutron effect on the core polarization will be tested with a measurement of $g(2_1^+)$ for ^{42}Ca. Further calculations are under way.

This work was supported in part by the U.S. National Science Foundation.

REFERENCES

1. R. Ernst et $al.$, Phys. Rev. Lett. **84**, 416 (2000)
2. S. Wagner et $al.$, Phys. Rev. C **64**, 034320 (2001)
3. A. Poves and A. Zuker, Phys. Rep. **70**, 235 (1981)
4. E. Caurier et $al.$, Phys. Rev. Lett. **75**, 2466 (1995)
5. G. Martinez-Pinedo et $al.$, Phys. Rev. C **54**, R2150 (1996)
6. A. Pakou et $al.$, Phys. Rev. C **36**, 2088 (1987)
7. K.-H. Speidel et $al.$, Phys. Rev. C **57**, 2181 (1998)
8. N. Benczer-Koller, M. Hass and J. Sak, Annual Rev. Nucl. Part. Sci. **30**, 53 (1980)
9. Y. Niv et $al.$, Phys. Rev. Lett. **43**, 326 (1979)
10. J. Q. Robinson, private communication.
11. F. Nowacki, private communication.

Prediction of Nuclear Masses in the A=80 region of nuclei as a function of P and F-spin

A. Teymurazyan[1], A. Aprahamian[1] and A.Georgieva[2]

[1]University of Notre Dame, Notre Dame, IN 46556, USA
[2]Institute of Nuclear Research, Bulgarian Academy of Sciences, Sofia BG- 1784, Bulgaria

Abstract. Predictions of nuclear masses far from stability are one of the challenges to present day network calculations in simulating various astrophysical scenarios and yielding the appropriate elemental abundances. Various groups have shown that complex nuclear structure properties as well as nuclear masses are correlated with the valence number of neutrons and protons. One such parameter is the promiscuity factor P and a related approach is that of F-spin. Both parameters are used to describe known measurements and used to predict via interpolation methods the masses of nuclei presently unknown in the laboratory. Here we show the exact relationship between P and F-spin in an attempt to strengthen structure based prediction methods for nuclear masses of interest in both the rp-process and the r-process.

Nuclear masses are one of the most important components in nucleosynthesis calculations of elemental abundances for specific stellar scenarios. The rp-process (rapid proton capture) is thought to involve a large number of proton rich nuclei in the A=80 region. Schatz et al. [1] have carried out an extensive comparison of the effects on abundances that result from the use of different mass predictions. One of the approaches used was a semi-empirical model based on the correlation of the nuclear structure component of nuclear masses with the Promiscuity factor P[2,3]. P is equal to NpNn/(Np+Nn) where Np(Nn) are the valence protons(neutrons) counted from the nearest closed shell[4]. The correlation allows the determination of the relationship between P and known nuclear masses. Masses for nuclei far from stability are then given by the relationship already established. The approach is fairly successful but has limitations particularly in nuclei with closed shells. The r-process (rapid neutron capture) is another explosive astrophysical nucleosynthesis scenario that involves nuclei far from stability with extreme neutron to proton ratios and therefore limited available measurements of the essential nuclear properties including masses. One approach that has been used to predict mass differences for the r-process is based on neutron-proton symmetries from IBM-2 in the form of F-spin [5-9]. This approach has been used for the Z=50-82 and N=82-126 region. In this paper, we combine the two approaches (F-spin and P-factor) to show a simple relationship between P and F-spin.

F-spin is also dependent on the number of protons and neutrons but the counting is done slightly differently. F_0 is the third projection of F-spin and it is given by $F_0=1/2(N\pi-N\nu)$ where $N\pi(N\nu)$ are the valence particle bosons counted from the last major closed shell. The total number of bosons $N=N\pi+N\nu$. For example, in the

CP638, *Mapping the Triangle: International Conference on Nuclear Structure*
edited by A. Aprahamian, J. A. Cizewski, S. Pittel, and N. V. Zamfir

nucleus ^{64}Ge, the total number of particles are eight, therefore N=4 and F_0=0. For each N, there are several values of F_0. The same nucleus has Np(Nn) value of 4(4) and a P value of 2. The table below summarizes the relationships between Np(Nn), $N\pi$(Nv), P, and F_0.

The relation between P and F_0 as classification parameters

- $N_t = N_\pi + N_\upsilon$

- $N_\gamma = \dfrac{1}{2}N_t - F_0 = \dfrac{1}{2}N_n$

- $N_\pi = \dfrac{1}{2}N_t + F_0 = \dfrac{1}{2}N_p$

$$P = \frac{1}{2N_t}\left(N_t^2 - 4F_0^2\right)$$

$$F_0^2 = \frac{N_t}{2}\left(\frac{N_t}{2} - P\right)$$

N_t – total number of valance particles

TABLE 1. The resulting relationships of P in terms of F-spin, and F_0 in terms of P.

This simple relationship of F-spin and P puts the promiscuity parameter P on a more firm theoretical footing with respect to masses and allows an entire region of nuclei to be described by one equation. For example, all the nuclei in the A=80 region with N=Z=28-50 can be described by one equation based on measured values and can in turn be used to predict masses of all unknown nuclei of interest. The overall errors between the predicted and measured masses in the A=80 region of nuclei are comparable using the F-spin approach or the P-factor approach. The two approaches combined are quite powerful. The F-spin approach allows one to overcome the challenges associated with closed shell nuclei while P-factor parameterization allows new insights into sub-shell and shell closures using experimental masses.

The support of the National Science Foundation under grant PHY 99-01133 is gratefully acknowledged.

REFERENCES

1. H. Schatz et al., Phys. Rep. 294,167 (1998)
2. P. Haustein, D.S. Brenner, and R.F. Casten, Phys. Rev. C38, 467 (1988)
3. A. Aprahamian, A. Teymurazyan, A. Susalla, and N. Cuka, Hyperfine Interactions 132, 417 (2001); A. Aprahamian, A. Gadala-Maria, N. Cuka, Rev. Mexicana de Fisica, 42, 1 (1996)
4. R. F. Casten, Nuclear Physics, A443, 1 (1985)
5. E. D. Davis et al., Phys. Rev. C 44, 1655 (1991)
6 A. Georgieva, M. Ivanov, P. Raychev and R. Roussev, Int. J. Theor. Phys. 28, 769 (1989); S. Drenska, A. Georgieva, V. Gueorguiev, R. Roussev and P. Raychev, Phys. Rev. C52, 1853(1995).
7 A. Arima, T. Otsuka, F. Iachello and I. Talmi, Phys. Lett, 66B, 205 (1977)205; J. P. Elliott, Rep. Prog. Phys, 48, 171 (1985)
8. H. Harter, P. Von Brentano, A. Gelberg, and R. F. Casten, Phys. Rev.C32, 631, (1985)
9 A. Georgieva, M.Ivanov, P. Raychev and R. Roussev, Int. J. Theor. Phys. 25, 1181 (1985)

Spectroscopy of ^{196}Au, ^{202}Au, ^{194}Ir and ^{193}Os

H.-F. Wirth*, Y. Eisermann*, R. Hertenberger*, G. Graw*, S. Christen†,
O. Möller†, D. Tonev† and J. Jolie†

*Sektion Physik, Ludwig-Maximilians-Universität München, D-85748 Garching, Germany
†Institut für Kernphysik, Universität zu Köln, D-50937 Köln, Germany

Abstract. The reactions ^{198}Hg$(\vec{d}, \alpha)^{196}$Au, ^{196}Pt$(\vec{d}, \alpha)^{194}$Ir and ^{192}Os$(\vec{d},p)^{193}$Os have been studied to test the model of supersymmetry in atomic nulcei. After encouraging results of the $(\vec{d}, \alpha)$ measurements the ^{204}Hg$(\vec{d}, \alpha)$ reaction was performed to gain spectroscopic information on the unknown nucleus ^{202}Au. About 70 levels have been observed in an energy range up to ≈ 2 MeV.

The nuclei ^{196}Au, ^{194}Ir and ^{193}Os are members of two quartets of nuclei, that may be described within the theory of extended supersymmetry [1]. To improve the spectroscopic information on these nuclei we measured angular distributions of the ^{198}Hg$(\vec{d}, \alpha)^{196}$Au ($E_d = 18$ MeV), ^{196}Pt$(\vec{d}, \alpha)^{194}$Ir ($E_d = 18$ MeV) and ^{192}Os$(\vec{d},p)^{193}$Os ($E_d = 22$ MeV) reactions at the tandem accelerator laboratory in Munich. The polarized deuteron beam from our atomic beam source [2] had an intensity of up to $3\,\mu$A on the target at the Q3D magnetic spectrograph, with polarizations of about 65 %. The ejectiles have been detected in the focal plane of the spectrograph by our new position sensitive proportional counter [3] with cathode strip readout, $\Delta E/E_{rest}$ particle identification and an active length of ≈ 1 m. Fig. 1 shows spin up and down spectra for ^{198}Hg$(\vec{d}, \alpha)^{196}$Au and

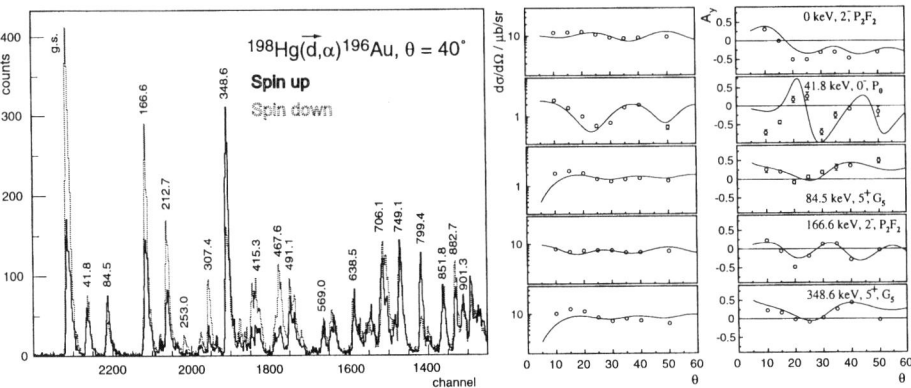

FIGURE 1. Left: Spin up and down spectra of the ^{198}Hg$(\vec{d}, \alpha)^{196}$Au reaction at $\theta_{lab} = 40°$. Many peaks are labeled with their excitation energy in keV. Right: Typical angular distributions of absolute cross section and analyzing power of the $(\vec{d}, \alpha)^{196}$Au measurement. The lines are results of DWBA calculations.

typical angular distributions. The big effect of the polarization is visible when comparing the lines at $E_x = 166.6$ keV and $E_x = 348.6$ keV. The FWHM is 8 keV.

CP638, *Mapping the Triangle: International Conference on Nuclear Structure*
edited by A. Aprahamian, J. A. Cizewski, S. Pittel, and N. V. Zamfir
© 2002 American Institute of Physics 0-7354-0093-8/02/$19.00

Figure 2 shows spectra of ^{196}Pt(d, α)^{194}Ir and ^{204}Hg(d, α)^{202}Au. Both spectra were recorded at the same magnetic setting of the spectrograph. Only the target was changed. Displayed is a part of the ^{202}Au spectrum up to an excitation energy of $\approx$ 1 MeV. The

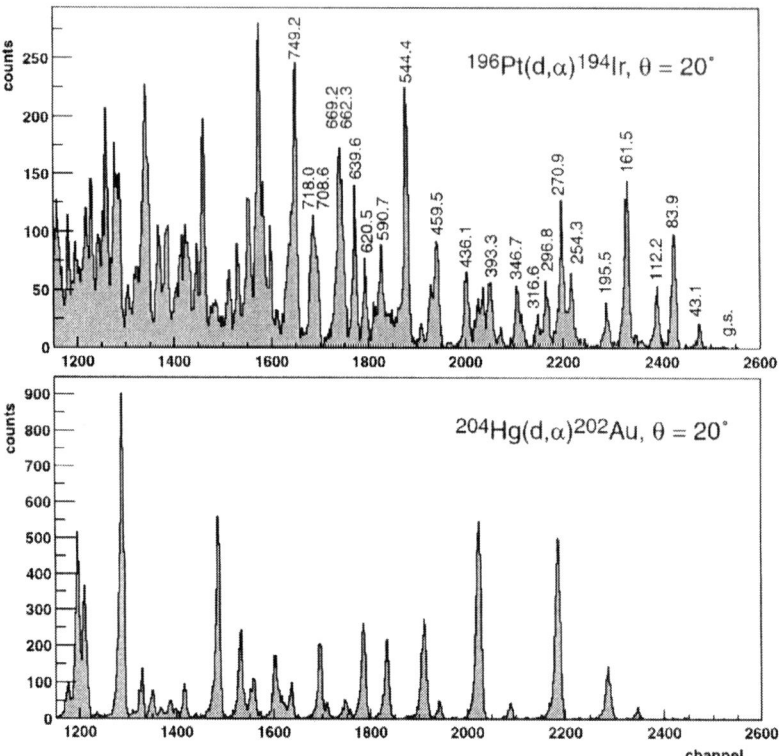

FIGURE 2. Up: Spectrum of the ^{196}Pt(d, α)^{194}Ir reaction at $E_d = 18$ MeV and $\theta_{lab} = 20°$. Some peaks are labeled with their excitation energy in keV. Down: Spectrum of the ^{204}Hg(d, α)^{202}Au reaction at the same magnetic setting of the spectrograph. Only the target was changed. The FWHM is 9 keV.

comparison with the ^{196}Pt(d, α)^{194}Ir spectrum allows an energy calibration. The measurement of angular distributions with polarized beam will give unambigous spin and parity assignments for more than half of the $\approx$ 70 levels that have been observed up to approx. 2 MeV. The first part of the ^{204}Hg($\vec{d}$, α)^{202}Au experiment has been finished one week before this conference.

This work was supported in part by the Deutsche Forschungsgemeinschaft under C4-Gr894/2-3.

REFERENCES

1. J. Gröger *et al.*, Phys. Rev. C **62**, 064304 (2000).
2. R. Hertenberger *et al.*, In *Spin 2000, Osaka 2000, AIP Conf. Proc. 570*, ed. K. Hatanaka, p. 825 (2000).
3. H.-F. Wirth *et al.*, annual report of the *Beschleunigerlaboratorium der Universität und der Technischen Universität München*, p. 71 (2000).

LIST OF PARTICIPANTS

Hanan Amro	Yale University	amro@alf.nbi.dk
Ani Aprahamian	University of Notre Dame	Aprahamian.1@nd.edu
Chairul Bahri	Louisiana State University	bahri@epscor2.phys.lsu.edu
Dipa Bandyopadhyay	University of Kentucky	dipa@vdgpc1.pa.uky.edu
Bruce Barrett	University of Arizona	bbarrett@physics.arizona.edu
Charles Barton	Daresbury Laboratory, UK	C.J.Barton@dl.ac.uk
Jacob Bar-Touv	Ben-Gurion University, Israel	Bartouv@netvision.net.il
Cornelius Beausang	Yale University	cornelius.beausang@yale.edu
John Becker	Lawrence Livermore National Laboratory	jabecker@llnl.gov
Zvi Berant	Nuclear Research Center Negev, Israel	zberant@bgumail.bgu.ac.il
Fred Bertrand	Oak Ridge National Laboratory	feb@ornl.gov
Roelof Bijker	Instituto de Ciencias Nucleares, Mexico	bijker@nuclecu.unam.mx
Carrol Bingham	University of Tennessee	cbingham@utk.edu
Hans Börner	Institut Laue Langevin, France	borner@ill.fr
Plamen Boutachkov	University of Notre Dame	pboutach@nd.edu
Daeg Brenner	Clark University	dbrenner@clarku.edu
Mark Caprio	Yale University	mark.caprio@yale.edu
Richard Casten	Yale University	rick@riviera.physics.yale.edu
Rom Chan	University of Richmond	rc6wh@richmond.edu
Jolie Cizewski	Rutgers University	cizewski@physics.rutgers.edu
Paul Cottle	Florida State University	cottle@nucott.physics.fsu.edu
Patrick Daly	Purdue University	pdaly@purdue.edu
Jorge Dukelsky	Instituto de Estructura de la Materia, Spain	dukelsky@iem.cfmac.csic.es
Mirela Fetea	University of Richmond	mfetea@richmond.edu
Nikolaos Fotiadis	Los Alamos National Laboratory	fotia@lanl.gov
Alejandro Frank	Instituto de Ciencias Nucleares, Mexico	frank@nuclecu.unam.mx
Christoph Fransen	University of Kentucky	fransen@pa.uky.edu
Jose-Enrique Garcia-Ramos	University of Huelva, Spain	jegramos@nucle.us.es
Ana Georgieva	Institute of Nuclear Research, Bulgaria	georgieva@regents.state.la.us
Joe Ginocchio	Los Alamos National Laboratory	gino@lanl.gov
Astrid Gollwitzer	MCI WorldCom, Germany	gollwitzer@uumail.de
Vesselin Gueorguiev	Louisiana State University	vesselin@phys.lsu.edu
Robert Haight	Los Alamos National Laboratory	haight@lanl.gov
Joseph Hamilton	Vanderbilt University	j.h.hamilton@vanderbilt.edu
Gregers Hansen	Michigan State University	hansen@nscl.msu.edu
Daryl Hartley	University of Tennessee	dhartle1@utk.edu
Adam Hecht	Yale University	adam.hecht@yale.edu
Kris Heyde	University of Gent, Belgium	kris.heyde@rug.ac.be
Richard Hoff	Lawrence Livermore National Laboratory	rwhoff@wordnet.att.net
Mark Huyse	University of Leuven, Belgium	mark.huyse@fys.kuleuven.ac.be
Franco Iachello	Yale University	francesco.iachello@yale.edu
Robert Janssens	Argonne National Laboratory	janssens@phy.anl.gov
Micah Johnson	Rutgers University	coolman@physics.rutgers.edu

Noah Johnson	Oak Ridge National Laboratory	noahjohnson28@msn.com
Jan Jolie	University of Cologne, Germany	jolie@ikp.uni-koeln.de
Noemie Koller	Rutgers University	nkoller@physics.rutgers.edu
Gabriele-Elisabeth Körner	NuPECC	sissy.koerner@ph.tum.de
Hans-Joachim Körner	Technische Universität München, Germany	hans-joachim@ph.tum.de
Wojciech Krolas	Oak Ridge National Laboratory	krolas@mail.phy.ornl.gov
Shelly Lesher	University of Kentucky	slesher@pa.uky.edu
Amiram Leviatan	The Hebrew University, Israel	ami@vms.huji.ac.il
Marcus McEllistrem	University of Kentucky	marcus@server1.pa.uky.edu
Deseree Meyer	Yale University	deseree.meyer@yale.edu
Richard Meyer	Clark University	ramruth@adelphia.net
Witek Nazarewicz	University of Tennessee	witek@utkux.utcc.utk.edu
Heather Olliver	Michigan State University	olliver@nscl.msu.edu
Taka Otsuka	University of Tokyo, Japan	otsuka@phys.s.u-tokyo.ac.jp
Philip Page	Los Alamos National Laboratory	prp@lanl.gov
Norbert Pietralla	University of Cologne, Germany	pietrall@ikp.uni-koeln.de
Stuart Pittel	University of Delaware	pittel@bartol.udel.edu
Paddy Regan	Surrey University, UK	paddy@galileo.physics.yale.edu
Jo Ressler	Yale University	jo@galileo.physics.yale.edu
Lee Riedinger	Oak Ridge National Laboratory	riedingerl@ornl.gov
Krzysztof Rykaczewski	Oak Ridge National Laboratory	rykaczew@mail.phy.ornl.gov
Brad Sherrill	Michigan State University	sherrill@nscl.msu.edu
Nadya Smirnova	Belgium	nadya.smirnova@fys.kuleuven.ac.be
Andreas Stolz	Michigan State University	stolz@nscl.msu.edu
Andrew Stuchbery	Australian National University	andrew.stuchbery@anu.edu.au
Kristina Sviratcheva	Louisiana State University	kristina@baton.phys.lsu.edu
Isao Tanihata	RIKEN, Japan	tanihata@postman.riken.go.jp
Michael Taylor	Rutgers University	tmichael@physics.rutgers.edu
Artur Teymurazyan	University of Notre Dame	ateymura@nd.edu
David Thomson	University of Richmond	david1701e@yahoo.com
Piet Van Isacker	GANIL, France	isacker@ganil.fr
Peter von Brentano	University of Cologne, Germany	brentano@ikp.uni-koeln.de
Till von Egidy	Technische Universität München, Germany	egidy@ph.tum.de
Dave Warner	Daresbury Laboratory, UK	d.warner@dl.ac.uk
Cristopher Wesselborg	American Physical Society	chrisw@aps.org
Jeff Allen Winger	Mississippi State University	winger@ph.msstate.edu
Hans-Friedrich Wirth	University of Munich	hans-friedrich.wirth@lmu.de
Samuel Wong	University of Toronto, Canada	swong@physics.utoronto.ca
Brian Wyman	University of Richmond	bw6wd@richmond.edu
Steve Yates	University of Kentucky	yates@uky.edu
Hatim Yousif	Mississippi State University	hatimbareed@hotmail.com
Victor Zamfir	Yale University	victor.zamfir@yale.edu
Nissan Zeldes	The Hebrew University, Israel	zeldes@vms.huji.ac.il
Vladimir Zelevinsky	Michigan State University	zelevinsky@nscl.msu.edu
Jing-ye Zhang	University of Tennessee	jingye@utk.edu

Maier, K. H., 149, 247
Mantica, P. F., 261
McConnell, J. W., 149, 247
McEllistrem, M. T., 231, 239
McNabb, D. P., 237, 245
Mertzimekis, T. J., 269
Meyer, D., 125, 243
Meyer, R. A., 131
Mizusaki, T., 97
Möller, O., 273
Moore, E. F., 119
Mukherjee, S., 43
Mullins, S. M., 43
Münch, M., 259
Münzenberg, G., 259
Mutti, P., 83

N

Navrátil, P., 105
Nazarewicz, W., 29, 149, 247
Newman, H., 243
Newman, R. T., 43
Neyens, G., 255
Nogga, A., 105
Novak, J. R., 51, 233, 243

O

Oganessian, Y. T., 91
Olliver, H., 249
Ormand, W. E., 105
Otsuka, T., 97

P

Padilla, E., 23
Page, P. R., 251
Pearson, C. J., 43
Piechaczek, A., 149, 247
Pietralla, N., 11, 69, 233, 239, 243, 257
Pittel, S., 111
Płoszajczak, M., 29
Podolyák, Z., 43
Popa, G., 253

R

Ramayya, A. V., 91, 149, 247
Rasmussen, J. O., 91
Ressler, J. J., 125, 243
Riedinger, L. L., 117, 119
Riley, M. A., 119
Rykaczewski, K., 149, 247
Rykaczewski, K. P., 149, 247

S

Saleem, K. Abu, 119
Schlegel, C., 259
Schneider, R., 259
Sciacchitano, M., 125, 243
Semmes, P., 117
Seweryniak, D., 119
Shapira, D., 51, 149, 247
Sharpey-Schafer, J. F., 43
Sherrill, B. M., 133
Shestakova, I., 119
Shimizu, N., 97
Sletten, G., 119, 229
Smirnova, N. A., 180, 255, 257
Smit, F. D., 43
Smith, M. B., 245
Smith, M. S., 131
Speidel, K.-H., 269
Stephens, F. S., 245
Stolz, A., 259
Stoyer, M., 91
Stuchbery, A. E., 49, 249, 261, 269
Sümmerer, K., 259
Sviratcheva, K. D., 263

T

Tanihata, I., 141
Tantawy, M. N., 149, 247
Taylor, M. J., 269
Ter-Akopian, G. M., 91
Teymurazyan, A., 271
Thirolf, P., 259
Thomas, J. S., 245
Thompson, J., 229
Tonev, D., 273